REACH-TO-GRASP BEHAVIOR

Reaching for objects in our surroundings is an everyday activity that most humans perform seamlessly a hundred times a day. It is nonetheless a complex behavior that requires the perception of objects' features, action selection, movement planning, multi-joint coordination, force regulation, and the integration of all of these properties during the actions themselves to meet the successful demands of extremely varied task goals. Even though reach-to-grasp behavior has been studied for decades, it has, in recent years, become a particularly growing area of multidisciplinary research because of its crucial role in activities of daily living and broad range of applications to other fields, including physical rehabilitation, prosthetics, and robotics.

This volume brings together novel and exciting research that sheds light into the complex sensory-motor processes involved in the selection and production of reach-to-grasp behaviors. It also offers a unique life-span and multidisciplinary perspective on the development and multiple processes involved in the formation of reach-to-grasp. It covers recent and exciting discoveries from the fields of developmental psychology and learning sciences, neurophysiology and brain sciences, movement sciences, and the dynamic field of developmental robotics, which has become a very active applied field relying on biologically inspired models. This volume is a rich and valuable resource for students and professionals in all of these research fields, as well as cognitive sciences, rehabilitation, and other applied sciences.

Daniela Corbetta is a Professor of Developmental Psychology and the Director of the Infant Perception-Action Laboratory at the University of Tennessee in Knoxville, USA.

Marco Santello is a Professor of Biomedical Engineering and the Director of the Neural Control of Movement Laboratory at Arizona State University in Tempe, USA.

Frontiers of Developmental Science

Series Editors: Martha Ann Bell and Kirby Deater-Deckard

Frontiers of Developmental Science is a series of edited volumes that aims to deliver inclusive developmental perspectives on substantive areas in psychology. Interdisciplinary and life-span oriented in its objectives and coverage, the series underscores the dynamic and exciting status of contemporary developmental science.

Social Cognition
Development Across the Life Span
Jessica A. Sommerville and Jean Decety

Executive Function
Development Across the Life Span
Sandra A. Wiebe and Julia Karbach

Emotion Regulation
A Matter of Time
Pamela M. Cole and Tom Hollenstein

Reach-to-Grasp Behavior
Brain, Behavior, and Modelling Across the Life Span
Edited by Daniela Corbetta and Marco Santello

For more information about this series, please visit: www.routledge.com/Frontiers-of-Developmental-Science/book-series/FRONDEVSCI

REACH-TO-GRASP BEHAVIOR

Brain, Behavior, and Modelling Across the Life Span

Edited by Daniela Corbetta and Marco Santello

Routledge
Taylor & Francis Group
NEW YORK AND LONDON

First published 2019
by Routledge
711 Third Avenue, New York, NY 10017

and by Routledge
2 Park Square, Milton Park, Abingdon, Oxon, OX14 4RN

Routledge is an imprint of the Taylor & Francis Group, an informa business

Library of Congress Cataloging-in-Publication Data
Names: Corbetta, Daniela, editor. | Santello, Marco, editor.
Title: Reach-to-grasp behavior: brain, behavior and modelling across the life span / edited by Daniela Corbetta and Marco Santello.
Description: New York, NY: Routledge, 2018. | Series: Frontiers of developmental science | Includes bibliographical references and index.
Identifiers: LCCN 2018011181 | ISBN 9781138683211 (hbk: alk. paper) | ISBN 9781138683228 (pbk: alk. paper) | ISBN 9780429467875 (ebk)
Subjects: LCSH: Habituation (Neuropsychology) | Developmental psychobiology. | Developmental biology.
Classification: LCC QP374.R43 2018 | DDC 612.8–dc23
LC record available at https://lccn.loc.

ISBN: 978-1-138-68321-1 (hbk)
ISBN: 978-1-138-68322-8 (pbk)
ISBN: 978-0-429-46787-5 (ebk)

Typeset in Bembo
by Wearset Ltd, Boldon, Tyne and Wear

CONTENTS

FRONTIERS IN DEVELOPMENTAL SCIENCE
EXECUTIVE FUNCTION VOLUME
(ALL VOLUMES HAVE A LIFE-SPAN FOCUS)

Theme	*Chapters Focused on That Theme*
Atypical development	Chapter 3 Karl, Sacrey, & Whishaw
Cognition	Chapter 1 Reissland & Austen Chapter 2 Corbetta, Wiener, & Thurman Chapter 3 Karl, Sacrey, & Whishaw Chapter 4 Weigelt Chapter 5 Cisek & Thura Chapter 6 Rouse et. al Chapter 7 Valyear Chapter 8 Smeets & Brenner Chapter 9 Castiello Chapter 10 Rand & Shimansky Chapter 11 Santello Chapter 12 Schöner, Tekülve, & Zibner Chapter 13 Caligiore & Baldassarre Chapter 14 Gaussier & Pitti
Communication/language	Chapter 9 Castiello
Computational modeling	Chapter 1 Reissland & Austen Chapter 2 Corbetta, Wiener, & Thurman Chapter 12 Schöner, Tekülve, & Zibner Chapter 13 Caligiore & Baldassarre Chapter 14 Gaussier & Pitti
Continuity/discontinuity	Chapter 2 Corbetta, Wiener, & Thurman
Cross species	Chapter 3 Karl, Sacrey, & Whishaw Chapter 5 Cisek & Thura Chapter 6 Rouse et. al Chapter 7 Valyear Chapter 11 Santello
Cultural context	Chapter 1 Reissland & Austen
Developmental robotics	Chapter 12 Schöner, Tekülve, & Zibner Chapter 13 Caligiore & Baldassarre Chapter 14 Gaussier & Pitti
Emotion/affect	None
Family/parenting	None
Gene-environment	None
Individual differences	Chapter 8 Smeets & Brenner
Intergenerational transmission	None

Theme	*Chapters Focused on That Theme*
Mechanisms of developmental change	Chapter 1 Reissland & Austen Chapter 2 Corbetta, Wiener, & Thurman Chapter 3 Karl, Sacrey, & Whishaw Chapter 4 Weigelt Chapter 7 Valyear Chapter 12 Schöner, Tekülve, & Zibner Chapter 13 Caligiore & Baldassarre Chapter 14 Gaussier & Pitti
Neuroscience	Chapter 3 Karl, Sacrey, & Whishaw Chapter 5 Cisek & Thura Chapter 6 Rouse et. al Chapter 7 Valyear Chapter 11 Santello Chapter 12 Schöner, Tekülve, & Zibner Chapter 13 Caligiore & Baldassarre Chapter 14 Gaussier & Pitti
Ontogeny	Chapter 1 Reissland & Austen Chapter 2 Corbetta, Wiener, & Thurman Chapter 3 Karl, Sacrey, & Whishaw Chapter 4 Weigelt Chapter 12 Schöner, Tekülve, & Zibner Chapter 13 Caligiore & Baldassarre Chapter 14 Gaussier & Pitti
Plasticity/repair	None
Sensory/motor	Chapter 1 Reissland & Austen Chapter 2 Corbetta, Wiener, & Thurman Chapter 3 Karl, Sacrey, & Whishaw Chapter 4 Weigelt Chapter 5 Cisek & Thura Chapter 6 Rouse et. al Chapter 7 Valyear Chapter 8 Smeets & Brenner Chapter 9 Castiello Chapter 10 Rand & Shimansky Chapter 11 Santello Chapter 12 Schöner, Tekülve, & Zibner Chapter 13 Caligiore & Baldassarre Chapter 14 Gaussier & Pitti
Social	Chapter 1 Reissland & Austen Chapter 9 Castiello Chapter 14 Gaussier & Pitti

CONTRIBUTORS

Joe Austen, *Durham University, United Kingdom*

Gianluca Baldassarre, *Istituto di Scienze e Tecnologie della Cognizione CNR, Italy*

Eli Brenner, *Vrije Universiteit Amsterdam, The Netherlands*

Daniele Caligiore, *Istituto di Scienze e Tecnologie della Cognizione CNR, Italy*

Umberto Castiello, *Università di Padova, Italy*

Paul Cisek, *Université de Montréal, Canada*

Philippe Gaussier, *Université de Cergy-Pontoise, France*

Jenni M. Karl, *Thompson Rivers University, Canada*

Zheng Liu, *University of Rochester, USA*

Kevin A. Mazurek, *University of Rochester, USA*

Alexandre Pitti, *Université de Cergy-Pontoise, France*

Miya K. Rand, *Leibniz-Institut für Arbeitsforschung an der TU Dortmund, Germany*

Nadja Reissland, *Durham University, United Kingdom*

Gil Rivlis, *University of Rochester, USA*

Adam G. Rouse, *University of Rochester, USA*

Lori-Ann R. Sacrey, *University of Alberta, Canada*

Marc H. Schieber, *University of Rochester, USA*

Gregor Schöner, *Ruhr-Universität Bochum, Germany*

Yury P. Shimansky, *Arizona State University, USA*

Jeroen B. J. Smeets, *Vrije Universiteit Amsterdam, The Netherlands*

Jan Tekülve, *Ruhr-Universität Bochum, Germany*

David Thura, *Université de Montréal, Canada*

Sabrina L. Thurman, *Elon University, USA*

Kenneth F. Valyear, *Bangor University, United Kingdom*

Matthias Weigelt, *Universität Paderborn, Germany*

Ian Q. Whishaw, *University of Lethbridge, Canada*

Rebecca F. Wiener, *University of Tennessee, USA*

Stephan Zibner, *Ruhr-Universität Bochum, Germany*

PREFACE

Reach-to-Grasp: Easy but Not Simple

The reach-to-grasp behavior is one of the most essential and broadly used behaviors of daily living. It is one of the earliest goal-directed behaviors to develop in infancy, and remains central throughout life span to ensure continued and effective interactions with our environment. Through reaching and grasping, infants, children, and adults learn about the many manipulable "things" (objects, people, and animals) that are present in our world, they learn how to act and diversify their actions as a function of task requirements and constraints, and ultimately learn how to utilize the critical cognitive skills associated with the selection and decision-making processes involved in the planning and production of goal-directed actions by leveraging memory and online feedback. Interactions between perception and action play an undeniable role in this process, but that is only one piece of a complex behavioral process. Many other questions, such as: how does this behavior come about in early development? how does selection in the planning and execution of the reach-to-grasp behavior play out in the brain? and, how can the control of complex multi-joints in the arm and hand occur? are all increasingly active and hotly debated areas of investigation in these recent years.

The goal of this volume is to bring together novel and exciting research that sheds light on the developmental, neural, and sensory-motor processes involved in the selection, production, and execution of reach-to-grasp behaviors. The last authoritative text on this subject dates back to 1988 (Marc Jeannerod, *The Neural and Behavioural Organization of Goal-directed Movements*). Since then, many advances in technology and brain sciences have significantly moved the field forward and offer a much more comprehensive understanding of the cognitive, perceptual, and motor processes involved in the selection and production of goal-directed movements. This book also aims to provide a unique life-span perspective on this problem (a perspective that was not part of Jeannerod's book), covering very recent and exciting discoveries from the fields of developmental and learning sciences, neurophysiology and brain sciences, motor control, and even from the newly growing field of developmental robotics which has become a very active applied field that leverages biologically inspired models.

Nowadays research is becoming more and more multidisciplinary, and researchers increasingly borrow from other fields to advance knowledge in their respective areas. Here we provide these multiple and complementary perspectives in one volume. Specifically, the book is divided into four areas reflecting distinct, yet complementary discoveries and insights from different research areas. The themes covered across these areas are all connected to the central theme of the book – the reach-to-grasp behavior, and strive to provide a coherent and fully integrated examination of the topic.

Part I of the book consists of four chapters on the origins, acquisition, and planning of the reach-to-grasp behavior in development. Reissland and Austen (Chapter 1) examine the origins of goal-directed behavior in prenatal development. They stress the importance of prenatal touch and somatosensory experience while embryos and fetuses move their arms and legs in the womb. They discuss how prenatal movements are increasingly being directed toward the self and specific parts of the body during the last months of gestation. Importantly, they explain how such prenatal sensory-motor experiences, including exploring the womb environment, are critical contributors of building knowledge of the body, a factor increasingly acknowledged as necessary for the development of goal-directed reaching.

Corbetta, Wiener, and Thurman (Chapter 2) focus on the multiple components that need to be assembled in the first months of life in order for the reach-to-grasp behavior to form. Developing a strong sense of the body via somatosensory experience continues to be a theme in this chapter, as is also discussed in Chapter 3 (Karl, Sacrey, and Whishaw), but many other components are discussed by Corbetta et al., thus providing a comprehensive picture of all the elements that are contributing to learning to reach in infancy. Particularly, learning to reach is seen as the combination of several processes, and not limited to single eye-hand coordination. Other mechanisms and processes in infant learning to reach are also discussed in relation to the emergence of reaching and its subsequent refinement in the first two years of life.

Karl, Sacrey, and Whishaw (Chapter 3) propose a multiple motor channel theory to explain the development of skilled hand movements in infancy. The multiple motor channel theory stipulates that reach, grasp, and withdraw movements are separate movements with distinct neural bases and pathways. Hand movements toward the face and the mouth (see also Reissland and Austen,

Chapter 1) precede goal-directed hand movements toward objects. Grasping movements toward the self also precede grasping behavior directed towards objects. These precursory actions are performed independently and without guidance of vision. They initially rely on somatosensory (proprioceptive and haptic) experience (see also Corbetta et al., Chapter 2) before being integrated in one continuous movement pattern under visual guidance much later in the first year of life.

Weigelt (Chapter 4) examines the development of the reach-to-grasp behavior beyond infancy. Specifically, his chapter addresses the development of anticipatory planning skills in children in the 3 to 12 year-old range in the context of end-state comfort tasks, which requires more sophisticated motor planning processes and the representation of future movement states. Many end-state comfort tasks involve adopting awkward movement postures at movement initiation in order to reach a comfortable state at movement completion. Weigelt proposes a three-stage developmental model revealing how children increasingly integrate and associate future action effects into their planning to attain the end comfort state.

Part II of the book focuses on the neurophysiological bases of action selection, reaching, and grasping. Cisek and Thura (Chapter 5) discuss the neural circuitry responsible for action selection. They review neurophysiological evidence indicating the modulating and regulating roles of the sensorimotor cortex and basal ganglia in action decision-making when multiple alternative action solutions are available. They also provide a unique window into the evolution of decision systems (see also Karl et al., Chapter 3, for evolutionary arguments on reach to grasp). Furthermore, they provide mounting evidence that the neural circuits involved in action selection are the same as the ones involved in the movement execution and are not separate as has been proposed by classic cognitive psychology theories.

Rouse, Mazurek, Liu, Rivlis, and Schieber (Chapter 6) provide compelling behavioral and neurophysiological evidence that the reach and grasp may not be separate but rather overlapping movement processes (but see Karl et al., Chapter 3, for one argument in favor of reach versus grasp independence). They take the reader through several levels of movement analyses (kinematics, EMG, and cortical activity) to argue that reach-to-grasp movements should be studied as one single movement involving the entire upper extremity, rather than being approached as a two-phase movement entailing a reach-then-grasp movement, as has been proposed in the past.

Valyear (Chapter 7) focuses on the grasping action and discusses the neural mechanisms of visual-motor transformation necessary to lead to the control of grasping. He reviews research on the involvement of the dorsal and ventral pathways in grasping performance and examines literature providing insights onto visual-motor transformations related to varied object physical properties. Finally, he stresses the importance of previous experience or recent motor history on movement planning efficiency.

Part III of the book examines the control aspects of the reach-to-grasp movement in human adults. Smeets and Brenner (Chapter 8) address the problem of planning and orienting several digits in space in relation to the object to be grasped. They illustrate this through a broad number of tasks in the context of objects of varying shapes, movement trajectories, the presence of obstacles in the movement path, moving targets, and even in the context of visual illusions. They show that endpoint selection of digit contact for grasp responds to a number of factors that include the movement goal, whether visual guidance is available, and depends on physical constraints such as obstacles or task precision.

Castiello (Chapter 9) focuses on what he defines the "cognitive" substrate of prehension. In his chapter, he shows how many sensory modalities (vision, tactile, proprioception, audition, taste, and olfaction) can affect action selection and execution. He also discusses how social intention can influence reach-to-grasp behavior. Thus, Castiello provides to the reader a novel view of prehensile behaviors in humans where context, intentions, and sensory information all play a role in the selection and execution of goal-directed actions. This chapter also addresses the role of attention, and how this important mechanism is essential for the transformation involving the sensory and motor systems in the planning of actions (see also Valyer, Chapter 7, about sensory-motor transformation).

Rand and Shimansky (Chapter 10) review the mechanisms underlying sensorimotor integration of the reach-to-grasp behavior in relation to the coordination between the transport and aperture phases of the hand during the goal-directed movement. They show that transport and aperture tested in varied conditions always respond to a consistent movement law that is determined by grip aperture, wrist velocity, and initiation of acceleration in a variety of task contexts. They also find that vision plays an important role in modulating grip aperture during the reach, and extend their computational model on sensorimotor integration to reaching tasks using electronic tools such as computer mouse. Lastly, they emphasize the sensitivity of multi-modal sensory integration for reach-to-grasp movements on the relative reliability of each sensory modality.

Santello (Chapter 11) offers a detailed account of the processes involved in dexterous manipulation. He reviews how digit points of contact for grasping are determined by the object's features (a point also made by Smeets and Brenner, Chapter 8), while stressing the complexity of manipulatory behaviors by addressing the additional kinetic control following object contact. This framework proposes that grasp kinematics and kinetics are part of a continuum, where choice of contacts impacts planning and/or online modulation of digit forces. Establishing digit force distribution is critically important to ensure that the object can be manipulated as planned. Such force distribution patterns must take into account the object's properties, such as object weight and center of mass. Determining the proper contact point and force distributions requires sensorimotor integration and is learned via trials and errors (see also Valyer, Chapter 7 on learning in sensorimotor transformation). Importantly, this process is primarily driven by memory when contacts are constrained, or by online feedback of digit position when contact points can be chosen by the subject.

The last section of the book, Part IV, covers three approaches of biologically inspired models and robots in the context of the reach to grasp. This is a fast-growing area of research that cuts across research in human behavior and machine learning. Schöner, Tekülke, and Zibner (Chapter 12) propose a neural process accounting for learning to reach based on the Dynamic Field Theory. The chapter is written as a tutorial and discusses how the model simulates each and every step of the reach-to-grasp movement from scene perception, to movement preparation, movement timing, and control, at the neural level. The model is at an early stage to allow for investigations of learning and developmental mechanisms.

Caligiore and Baldassare (Chapter 13) present an integrated framework for modelling and understanding reaching and grasping in development based on a critical review of computational and robotic models. They discuss how active vision, motor babbling, associative learning processes, learning by trial-and-error, intrinsic motivations, control architectures, and embodiment are all essential elements for understanding the computational mechanisms of learning to reach and cognition. They illustrate how reinforcement learning can be a common process for driving behavioral change over time across these elements.

Lastly, Gaussier and Pitti (Chapter 14) review some neuro-computational mechanisms underlying the multimodal integration processes for reaching and grasping actions. They describe goal-directed actions as emergent properties of visuo-motor coordination and the product of neural fields that are tuning toward a goal. Much of the architectures they describe are inspired from neuroscience and are implemented through dynamic fields with target objects forming attractor basins toward which movement trajectories unfold.

Throughout the chapters and different sections, whether they relate to human development, brain processing, motor control, or to the designing of models or control architectures for reach-to-grasp, common themes clearly recur. These themes include sensorimotor integration between multiple sensory modalities, vision-based extraction of salient features about object properties, memory recall of past hand-object interactions, and how development and learning shapes and refine these processes from infancy to adulthood.

We hope this book will offer a valuable source to researchers in fields as varied as psychology, kinesiology, engineering, neuroscience, cognitive and learning sciences.

Daniela Corbetta
Marco Santello

PART I

Developmental Studies on the Origins and Acquisition of Reach-to-Grasp

1

GOAL DIRECTED BEHAVIOURS

The Development of Pre-Natal Touch Behaviours

Nadja Reissland and Joe Austen

Introduction

Studies indicate that the sensory system becomes functional in a specific and invariant order with tactile sensing abilities to develop first (e.g. Lickliter & Bahrick, 2016). The face becomes sensitive to touch by around 7 weeks' gestational age (Hooker, 1952; Humphrey, 1972) and the hands by around 10 weeks gestational age (Hooker, 1952; Humphrey, 1972). Hence, tactile stimulation is the first stimulation a fetus will be exposed to. Research suggests that touch behaviours are intentional and possibly planned from an early age onward in that neonates and infants move their hands toward their mouths either directly or indirectly by initially making contact with the perioral region of the face and opening their mouths in anticipation of the arrival of their hands (e.g. Blass, Fillion, Rochat, Hoffmeyer, & Metzger, 1989; Butterworth & Hopkins, 1988; Rochat, Blass, & Hoffmeyer, 1988; Takaya, Konishi, Bos, & Einspieler, 2003). Similar claims have been made for fetuses (e.g. Reissland, Francis, Aydin, Mason, & Schaal, 2014).

Prenatally, the functional significance of touch could be that it affords fetuses the first opportunity to learn about the areas and variations of sensitivity of their body as well as define their body limit in relation to the intra-uterine environment (Kravitz, Goldenberg, & Neyhus, 1978) and potentially the extra-uterine environment as suggested by Marx and Nagy (2015). Through touch fetuses might learn a sense of self by differentiating between their own body and others, including the external environment and other bodies. This suggestion is based on studies indicating that somatic sensation allows the animal or person through their touch to generate information necessary to identify or differentiate between objects and physical stimuli that come into contact with their body (e.g. Grunwald, 2017). Khazipov et al. (2004) found that sensory feedback from spontaneous fetal movements in rats played an important role in establishing a mental representation of the body in the somatosensory cortex. Yamada et al. (2016) developed this research further by developing a robot model based on an integrated system of brain-body and environmental interactions which the researchers based on human anatomical and physiological data. They tested their model under human intra-uterine conditions and suggest that intra-uterine stimulation through touch could provide the foundations for cortical learning of body representations which are hypothesized to provide foundations for post-natal visual-somatosensory integration.

After birth, the fetus will seek out the nipple by rooting, namely crawling towards the nipple, which is a behaviour related to their neurobehavioural maturity (Radzyminski, 2005). Hence one potential functional development is related to feeding abilities in general and specifically to their ability to initiate breastfeeding successfully. Others argue that the reason for self-exploration of their bodies leading for example to thumb sucking (Feldman & Brody, 1978) or general self-touch (Rock, Trainor, & Addison, 1999) could have the function of arousal regulation. Weiss (2005) suggests that, if prenatally there is insufficient experience with touch sensations, neurosensory organization could be compromised.

More cognitive explanations have been offered by Butterworth and Hopkins (1988), who propose that self-touch and specifically hand-mouth touch is evidence for goal directed behaviour and the development of intention. This suggestion is supported by research with neonates, which demonstrates that new-born infants perform hand-mouth touch action only when they can taste a drop of sweet solution but not when receiving a drop of water. Casual observations by a number of researchers (e.g. Hata, Kanenishi, & Sasaki, 2010; Hepper, Shahidullah, & White, 1991; Kurjak et al., 2003; de Vries, Visser, & Prechtl, 1982) of fetal ultrasound images indicating fetal ability to introduce a finger, part of the arm or umbilical cord into the mouth, suggest that ability to coordinate movements develops prenatally.

Fetal Sensitivity to Touch and the Effects of Touch on Body Movement

Although historically fetuses were said to live in a sensory deprived environment (Windle, 1940), Bradley and Mistretta (1975) argued that, even though they are to some extent buffered by amniotic fluid, fetuses experience tactile stimulation through their own, as well as maternal, movements. In fact, research indicates that tactile stimulation is essential for healthy development (e.g. Ardiel & Rankin, 2010). For example, Kuniyoshi and Sangawa (2006) developed self-organizing neural robot models of the human body, which were able to learn through movements the sensory and motor structure of normal human infant motor development. In studies examining tactile functions, Mori and Kuniyoshi

(2010) simulated self-organizing behaviours of movements in the human fetus. They argued that tactile stimulation is the dominant sensory experience because tactile stimulation can be elicited through self-stimulation in response to fetal movement. Hence, they suggest that the fetus learns tactile-motor contingencies. Based on this assumption they developed a simulation of fetal motor development induced by tactile sensations (e.g. Kuniyoshi, 2006; Kuniyoshi & Sangawa, 2006). Yamada et al. (2016) found that the intra-uterine environment which is spatially restricted exerts pressure which leads to tactile sensory feedback correlated across body parts. They argue that these physical restrictions may be beneficial and enable the development of a cortical map of the body. The development of a cortical map is essential for normal human development (Dazzan et al., 2006). Without such cortical maps the infant might not be able to direct touch intentionally to their own body or reach other bodies (e.g. hands) or objects. Additionally, the area which the fetus touches, such as the mouth region, corresponds to developmental sensitivity of the skin as fetuses mature. Sensitivity of various areas of the body has been tested in humans and can be seen in Figure 1.1, with the more sensitive areas showing more densely spaced nerve endings.

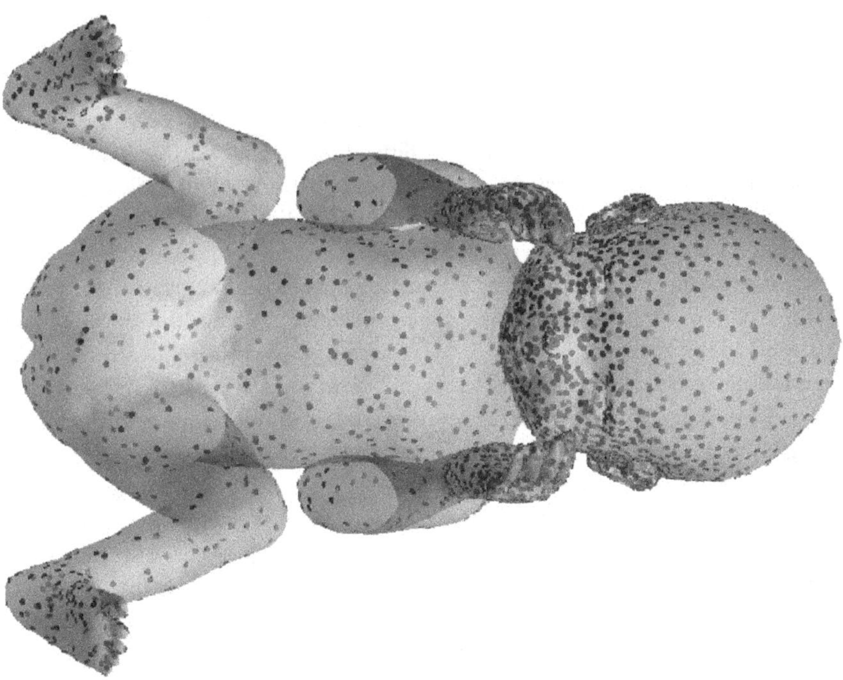

FIGURE 1.1 Tactile points of the robotic simulation of sensitivity of the skin developed by Mori and Kuniyoshi (2010)

Yamada et al. (2016, see Figure 1.1) tested the intra-uterine sensorimotor experiences necessary for learning mechanisms guiding development using a robot model integrating interactions between the brain, body, and environment, which they called 'embodied interactions'. Using this model, they were able to show the mechanistic link between sensorimotor experiences via embodied interactions and the cortical learning of body representation. They demonstrated with their model that intra-uterine movements provide specific organizational patterns relating to the body parts based on both proprioception and tactile sensory feedback which resulted in the learning of body representations. This is the basis then on which we can argue that as fetuses move they will learn about their bodies.

Cross Cultural Differences in Fetal Touch Behaviour and General Movement

Cross cultural studies have concentrated on gait, such as crawling, standing, walking, or sitting (Karasik, Adolph, Tamis-Lemonda, & Bornstein, 2010) and found that infant massage with vigorous movements such as stretching and tossing the infant up in the air, such as used in India and Nepal (Reissland & Burghart, 1987), results in sitting and walking at earlier ages than infants not exposed to such movements (Bril & Sabatier, 1986; Hopkins & Westra, 1988). However, other types of movements are rarely studied cross culturally in prenatal populations.

Reissland et al. (2015) analysed fetal self-touch behaviours comparing 4 low birth weight (<2500 grammes) Japanese fetuses with 3 low birth weight English fetuses and found an indication of differences in touch, with Japanese fetuses seemingly showing more touch compared with English fetuses. The reasons for the observed differences are unclear. However, clues to the differences in behaviour might relate to maternal factors. Lai, Nowlan, Vaidyanathan, Shaw, & Lees (2016) suggest that the key determinant of fetal healthy development relates to adequate perfusion and oxygen transfer which is dependent on a well-functioning placenta which results in a full-term fetus with healthy weight as assessed by Apgar scores. Apgar scores indicate healthy development, through a rating of colour, heart rate, muscle tone, etc. of the new-born at 1 and 5 minutes after birth, with a maximum score of 10 indicating optimal health of the new-born. Hanaoka et al. (2016), testing fetal movements in Japanese and Croatian fetuses using Kurjak's antenatal neurodevelopmental test (KANET; Kurjak et al., 2008), found that although the KANET score was in the normal range in both populations, there was a significant difference in total KANET scores between Japanese (median, 14; range, 10–16) and Croatian fetuses (median, 12; range, 10–15). These authors also recorded maternal factors and infant outcome at birth. There were significant differences in the Japanese and Croatian groups in terms of maternal age, parity, infant birth weight, as well as infant Apgar scores at 1 and 5 min recorded for Japanese and Croatian offspring, which all might be factors contributing to the cultural differences observed. DiPietro et al. (2004) studied fetal general movements using a fetal actocardiograph rather than analysing fetal movements in ultrasound as was the case in the studies by Reissland et al. (2015) and Hanaoka et al. (2016). They recorded fetal general movements with the actocardiograph from 20 to 38 weeks gestation. In their analysis of fetal general movements they found significant differences between fetuses observed in Lima (Peru) and Baltimore, Maryland (USA) with the fetuses from Baltimore moving more compared with fetuses in Lima but with a similar trend of a decrease in general movements from 20–38 weeks gestation. Again, as in previous studies, there were differences in maternal factors in the two groups with women from the US having a heavier pre-pregnancy weight of 68.2 kg compared with 54.8 kg in Lima. The infants' weights also differed, with US infants being slightly heavier at a mean of 3,520 grammes compared with infants from Lima with a mean weight of 3,214 grammes at birth. In sum, the results on cross cultural studies are uncertain and need to be examined in more detail using the same coding system of fetal movements and controlling for maternal factors.

Fetal Action Planning and Goal Directed Action

In a kinematic study Zoia et al. (2007) examined whether fetal hand movements were planned. They performed a kinematic analysis of hand movements directed towards the mouth and eyes by 8 fetuses observed at 14, 18, and 22 weeks of gestation. They found that by 22 weeks, kinematic patterns of fetal movements varied depending on the goal of the action. Specifically the speed with which fetuses either moved their hands toward either the eye or the mouth varied, being slower for movements directed toward the eye compared with movements directed towards the mouth.

One might surmise that fetal facial touch is constrained by gestational age with fetuses at later gestational ages being able to touch fewer areas of the head and face. However, research indicates that fetuses are able to flex their elbows; necessary for touching various parts of the head, face, and mouth. For example, Ververs, Van Gelder-Hasker, De Vries, Hopkins, and Van Geijn (1998) reported that 10 fetuses, which they observed longitudinally, were able to flex their elbows at 24 weeks 93% of the time observed increasing by 3% to 96% of the time observed at 36 weeks. In a study by Reissland et al. (2014) the development of fetal touch behaviours was observed in terms of region of touch, either top of the head, side of the face, lower face or mouth, and timing of touch in relation to mouth opening (see Figure 1.2).

From 24–36 weeks of gestation fetuses in this study touched the upper part of their head less as they developed with rate of touch declining by around 9% of each week of gestational age. The rate of decrease of the side of the face

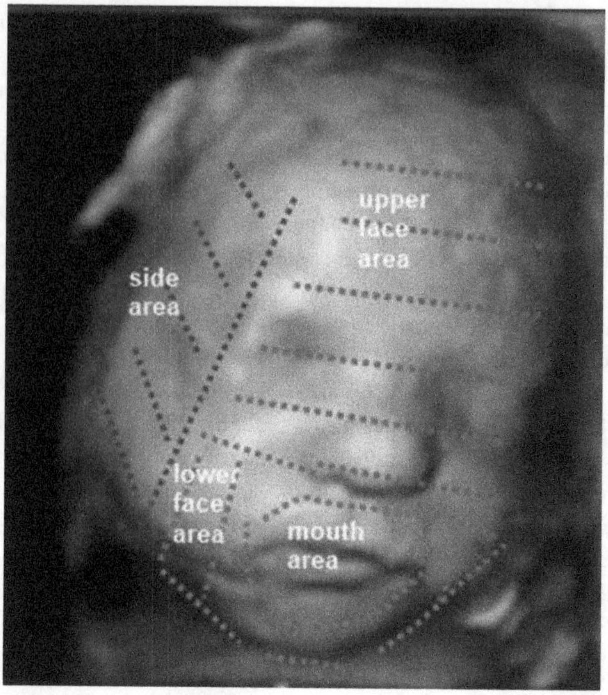

FIGURE 1.2 Regions of touch, upper, side, lower, and mouth areas, coded in the fetal face

(see Figure 1.3 as an example of side of the face touch) was around 10% for each week of gestational age.

In contrast touches of the lower part of the face increased with age by around 4% for each week of gestational age but the most significant increase was touch of the perioral region of the face with the rate of touch increasing by around 7% for each week of gestational age (see Figure 1.4). Hence, it is suggested that fetuses moving their hands and exploring the region of the face is of prime importance for postnatal life, namely the mouth, getting them ready to feed from breast or bottle after birth.

Some researchers report that fetal hand-to-face movements, although possible, appear to occur less frequently than those observed in neonates. According to Butterworth and Hopkins (1988), human neonates showed arm movements toward the face 3.48 times on average per minute observation. On the other hand, fetal arm movements observed in 2D scans occurred 0.2 times on average per minute of observation. Sparling, van Tol, and Chescheit (1999) also counted the frequencies of 'hand to/at/near mouth' movements in 1-day-old neonates in addition to those of the fetuses observed in 2D scans. They reported that although the frequencies of hand-to-face movements decreased in fetuses from 26 weeks to 37 weeks gestation to approximately 0.2 hand-to-face

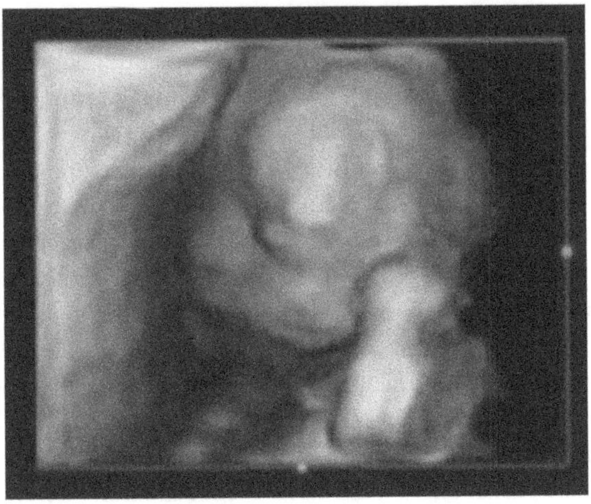

FIGURE 1.3 Fetus touching side of the face

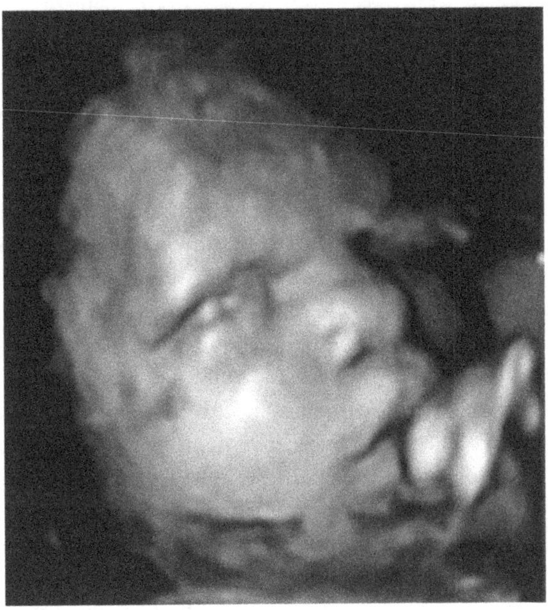

FIGURE 1.4 Fetus touching perioral region of the face

movements per minute observed at 37 weeks, when observed after birth at the age of 1 day, neonatal arm movements increased to 1.2 movements per minute. Nevertheless, prenatal practice of hand-face contacts may be responsible for highly organized inter-sensorimotor patterns observed immediately after birth (Butterworth & Hopkins, 1988). Indeed, prenatal practice starts early in life with reactive touch being observed from the first trimester of fetal life (e.g. Piontelli, 2010). However, the coordination of anticipated touch is neurologically more complex than reactive touch necessitating an increased level of motor control (Zoia et al., 2007). This increased level of motor control implies some action planning (Zoia et al., 2007) and hence arguably cognitive maturation which would be expected to occur later in fetal development.

Regarding specific 'intentional' goal directed behaviours, if neonates already have some knowledge of their own bodies, the well-coordinated movements between their hands and mouths might have developed in the uterus. Indeed, human fetal thumb (or other finger) sucking can be observed as early as 10–15 weeks of gestation (e.g. Hepper et al., 1991; de Vries et al., 1982). Whether such fetal hand-mouth coordination might occur accidentally was investigated by Myowa-Yamakoshi and Takeshita (2006). If human fetuses were capable of coordinating hand and oral activities, then they would observe well-organized or controlled patterns in their hand-mouth coordination. They observed in a cross sectional study of 27 fetuses, divided into a younger group with a mean age of 24.6 weeks (range 19–27 weeks) and an older group with a mean age of 31.6 weeks (range 28–35 weeks), that approximately half of the observed arm movements resulted in hands touching the mouth with no developmental differences between the two groups. Additionally, they report that the results showed that a minority of 30% of the observed fetuses did not show any directional movements of the hand to the mouth. However, in this study it was not clear if fetuses touched the mouth region 'intentionally' or 'accidentally'. This question was investigated in the following longitudinal study from 24–36 weeks gestation (Reissland et al., 2014). In this study fetuses were observed in 4D scans. Results indicated that fetuses increased significantly the proportion of anticipatory mouth movements. These fetuses opened their mouth before touch occurred, which was related to gestational age with an increase of around 8% with each week of gestational age. Additionally, there was a decrease in the proportion of reactive mouth movements, mouth movements which occurred in reaction to a touch. These reactive movements decreased by around 3% for each week of gestational age. Furthermore, fetuses increased the touch of the mouth area compared to the upper and side areas of the face with the rate of touch of the upper face area declining with age by around 9% for each week of gestational age and touch of the side area of the face decreasing with age by around 10% for each week of gestational age. These results indicate that fetuses, as they develop from 24 to 36 weeks gestation, do not only increasingly touch the sensitive part of their face, namely the mouth region and lower part of their face, but also they touch these sensitive areas more frequently compared to the relatively less sensitive areas of their head and face.

Furthermore, fetuses seem not only to anticipate touch opening their mouth when the hand is in front of their face; rather, fetuses actively turn their head, as can be observed in Figure 1.5 of a 32-week-old fetus, opening the mouth in order to suck on the thumb.

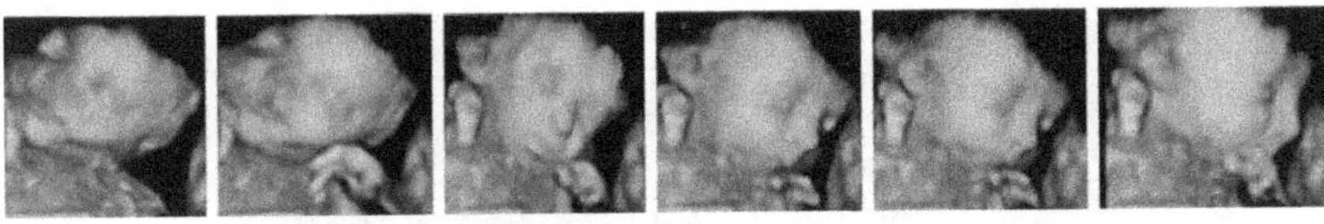

FIGURE 1.5 32-week-old fetus turning the head toward hand and sucking a thumb

According to Castiello et al. (2010), twins provide a unique opportunity to investigate differences in touch behaviour in terms of self-touch, other human touch, and touch of the uterine environment, such as uterine wall or placenta. Castiello et al. (2010) hypothesized that the kinematics for arm actions aimed at a social target which in adults have been found to differ from those aimed at a non-social objects (Becchio, Sartori, & Umberto Castiello, 2010), would differ in twin fetuses. They found that movements toward the uterine wall at 14 and 18 weeks gestation did not differ. However, the proportion of self-directed movements was greater at 14 than 18 weeks gestation but movements directed toward the twin were greater at 18 than 14 weeks gestation. Hence, twin fetuses, as they grow older, seem to direct touch in preference to the other fetus rather than themselves. Kurjak et al. (2013) examining twin pregnancies in 4D in all three trimesters using the KANET scoring system found no statistically significant differences between singleton and twin pregnancies. However, they also reported increasingly complex touch behaviour as the pregnancy progressed. Hata et al. (e.g. 2011, 2012, 2016) published a series of 4D ultrasound studies on inter-twin contact and reactions to touch during the first trimester of pregnancy. They found, comparing mono-chorionic di-amniotic (MD) and di-chorionic di-amniotic (DD) twins at 10–11 weeks gestational age, a significant difference in the total number of all types of contact. Based on these studies Hata et al. (2011, 2012) suggested that the frequency of the type of inter-twin contact at the earlier gestations from 10–13 weeks might be due to early fetal neuromuscular development and differentiation of the neuromuscular system. Arabin, Bos, Rijlaarsdam, Mohnhaupt, and van Eyck (1996) analysed touch reactions of twins from 8 up to 16 weeks gestation using 2D scans. They report increasingly complex interactions from 8–16 weeks gestation with touch of the twin observed at 12 weeks not eliciting any reaction from the twin and complex touch which they called 'embrace' being observed in diachronic twins at 16 weeks gestation. Such embrace can be seen in the figure below of dizygotic twins (25 weeks 2 days old) where the girl fetus wraps her arm around the shoulder of her twin brother (see Figure 1.6).

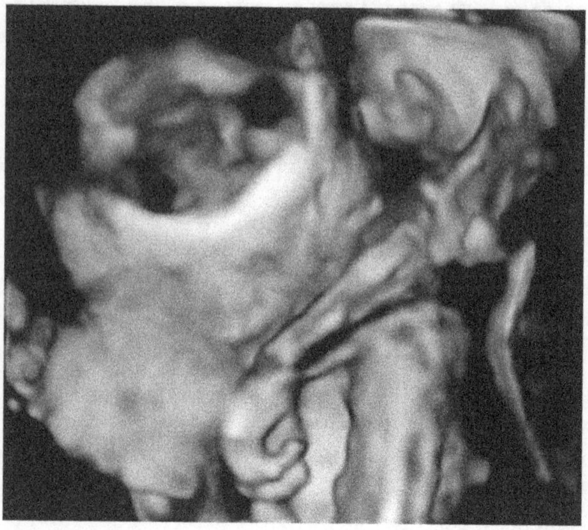

FIGURE 1.6 Twins at 25 weeks 2 days old: sister fetus behind brother fetus with her left arm draped over the left should of her brother

Visually Guided Touch

Some suggest that prenatal hand-to-mouth coordination should occur without visual cues because even if the fetal eyes are actually open, there is little light in the uterus (Butterworth & Hopkins, 1988). In fact, more recent research has shown that there is enough light in the womb permitting the fetus to see his or her hands. Specifically eye blinking which can be observed in utero (see Figure 1.7), is an indication that fetuses might react to different light conditions (e.g. Del Giudice, 2011).

Del Giudice (2011) analysed the amount of light entering the womb. He reports that fetuses will be exposed to ample light for optimal fetal vison of 50 lux intra-uterine illuminance when the mother with an abdominal thickness of 20 mm (measured as the distance between the abdominal surface and the anterior wall of the amniotic sac) stands indoors near a window or outdoors in shadow. Therefore it is likely that during the summer months fetuses can be predicted to develop in conditions allowing for ample visual experience before birth which could result in visually guided touch (see Figure 1.8).

In a 2D scan analysis of 18 fetuses, Petrikovsky, Kaplan, and Holsten (2003) examined fetuses between 33 and 42 weeks' menstrual age and found that fetuses blinked with a mean frequency of 0.1 movements per minute of observation. This increased to 0.25 movements per minute observation when fetuses received vibro-acoustic stimulation Hence touch and sound sensations affected amount of blinking.

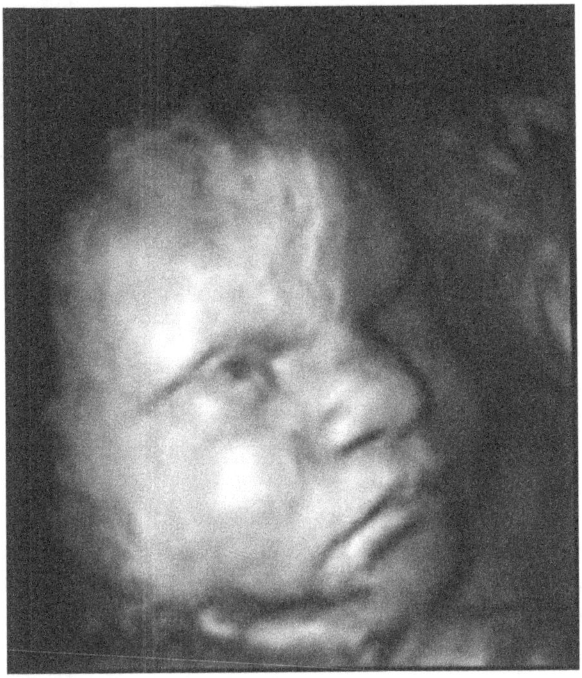

FIGURE 1.7 32-week-old fetus with eyes open

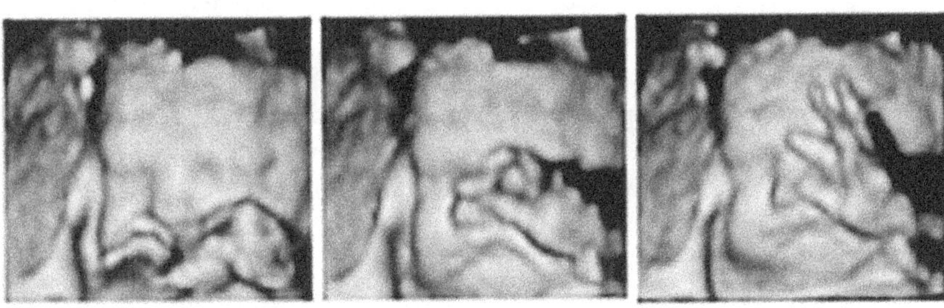

FIGURE 1.8 32-week-old fetus raising the left hand towards the eye during exposure to light

Discussion

This short chapter has provided the reader with an overview of prenatal touch in singleton and twin fetuses and illustrates how far we have come from the early days of fetal research when fetuses were seen to be as sensory deprived. Instead, we can now have a clearer understanding of fetal development and brain function through analysis of prenatal touch and movement.

As the first sense to develop, touch plays a critical role in the early stages of fetal development. The new research examining brain function in utero as well as various types of stimulation, such as vibro-acoustic and different types of sound and light stimulation, indicate that fetuses are prepared for life outside of the womb.

Lickliter (2011) argued that the fetus is prepared for life post-birth in the constrained environment of the uterus. Not only is there the constraint of the womb limiting and regulating the type of stimulation but also the amniotic fluid buffers stimulation in terms of relative amount, type, and timing. These patterns of prenatal controlled stimulation are essential for healthy development and any alteration will have an impact on brain development, as can be seen in premature birth. For example, consequences of an abnormal touch environment provided in the Neonatal Intensive care unit has profound effects on the developing premature brain (e.g. Als et al., 2003).

Castiello et al. (2010) go so far as to suggest that fetuses might plan actions and anticipate the consequences of their touch behaviours and thereby demonstrate 'social actions' from the 14th week of gestation. Although I agree with Castiello et al. (2010) in principle that fetuses show behaviour which can be interpreted as action planning, I would agree with Pezzulo and Castelfranchi (2007), that the observed behaviours do not include cognitive representations.

Hence fetuses behave as if they planned an action, as a precursor to later development which includes the cognitive representation of that planned action.

Prenatal touch serves the function of mapping the self vis-à-vis the other, where 'other' can be either another human being or the environment of the uterus. It is therefore one of the earliest instances of human engagement with the world and, as such, more research must be done in order to understand it better. A next stage in advancing this research would be to include compromised fetuses, such as fetuses restricted in their ability to feel, touch, and move in order to establish whether they will be delayed in their action planning behaviours.

Acknowledgement

We would like to thank Emma Reames for comments on the chapter.

References

Als, H., Gilkerson, L., Duffy, F.H., McAnulty, G.B., Buehler, D.M., Vandenberg, K., Sweet, N., Self, E., Parad, R.B., Ringer, S.A., Butler, S.C., Blickman, J.G., & Jones, K.J. (2003). A three-center, randomized, controlled trial of individualized developmental care for very low birth weight preterm infants: medical, neurodevelopmental, parenting, and caregiving effects. *Journal of Behavioral Pediatrics, 24*, 399–408.

Arabin, B., Bos, R., Rijlaarsdam, R., Mohnhaupt, A., & van Eyck, J. (1996). The onset of inter-human contacts: longitudinal ultrasound observations in early twin pregnancies. *Ultrasound in Obstetrics & Gynecology, 8*, 166–173.

Ardiel, E.L., & Rankin, C.H. (2010). The importance of touch in development. *Journal of Paediatrics & Child Health, 15*, 153–156.

Becchio, C., Sartori, L., & Umberto Castiello, U. (2010). Toward you: the social side of actions. *Current Directions in Psychological Science, 19*, 183–188. doi: 10.1177/0963721410370131.

Blass, E.M., Fillion, T.J., Rochat, P., Hoffmeyer, L.B., & Metzger, M.A. (1989). Sensorimotor and motivational determinations of hand–mouth coordination in 1–3-day-old human infants. *Developmental Psychology, 25*, 963–975.

Bradley, R.M., & Mistretta, C.M. (1975). Fetal sensory receptors. *Physiological Review, 55*, 352–382.

Bril, B., & Sabatier, C. (1986). The cultural context of motor development: postural manipulations in the daily life of Bambara babies (Mali). *International Journal of Behavioural Development, 9*, 439–453.

Butterworth, G., & Hopkins, B. (1988). Hand–mouth coordination in the new-born baby. *British Journal of Developmental Psychology, 6*, 303–314.

Castiello, U., Becchio, C., Zoia, S., Nelini, C., Sartori, L., Blason, L., D'Ottavio, G., Bulgheroni, M., Gallese, V., & Rustichini, A. (2010). Wired to be social: the ontogeny of human interaction. *PLoS ONE, 5*, http://dx.doi.org/10.1371/journal.pone.0013199.

Dazzan, P., Morgan, K.D., Chitnis, X., Suckling, J., Morgan, C., Fearon, P., McGuire, P.K., Jones, P.B., Leff, J., & Murray R.M. (2006). The structural brain correlates of neurological soft signs in healthy individuals. *Cerebral Cortex, 16*, 1225–1231.

Del Giudice, M. (2011). Alone in the dark? Modeling the conditions for visual experience in human foetuses. *Developmental Psychobiology, 53*, 214–219.

DiPietro, J.A., Caulfield, L.E., Costigan, K., Merialdi, M., Nguyen, R., Zavaleta, N., & Gurewitsch, E. (2004). Fetal neurobehavioral development: a tale of two cities. *Developmental Psychology, 40*, 445–456.

Feldman, J.F., & Brody, N. (1978). Non-elicited newborn behaviors in relation to state and prandial condition. *Merrill-Palmer Quarterly, 24*, 79–84.

Grunwald, M. (2017). *Homo hapticus*. Verlag: Droemer, HC.

Hanaoka, U., Hata, T., Kanenishi, K., AboEllail, M.A.M., Uematsu, R., Konishi, Y., Kusaka, T., Noguchi, J., Marumo, G., Vasilj, O., & Kurjak, A. (2016). Does ethnicity have an effect on fetal behavior? A comparison of Asian and Caucasian populations. *Journal of Perinatal Medicine, 44*, 217–221.

Hata, T., Kanenishi, K., Hanaoka, U., Uematsu, R., & Marumo, G. (2016). HDlive in the assessment of twin pregnancy. In N. Reissland & B. Kisilevsky (Eds), *Fetal development: research on brain and behavior, environmental influences, and emerging technologies* (pp. 385–405). Dordrecht: Springer.

Hata, T., Kanenishi, K., & Sasaki, M. (2010). Four-dimensional sonographic assessment of fetal movement in the late first trimester. *International Journal of Gynecology and Obstetrics, 109*, 190–193.

Hata, T., Kanenishi, K., Sasaki, M., & Yanagihara, T. (2011). Fetal reflex movement in twin pregnancies late in the first trimester: 4-D sonographic study. *Ultrasound Med Biol, 37*, 1948–1951.

Hata, T., Sasaki, M., & Yanagihara, T. (2012). Difference in the frequency of types of inter-twin contact at 10–13 weeks' gestation: preliminary four-dimensional sonographic study. *JMatern Fetal Neonatal Med., 25*, 226–230.

Hepper, P.G., Shahidullah, S., & White, R. (1991). Handedness in the human fetuses. *Neuropsychologia, 29*, 1107–1111.

Hooker, D. (1952). *The prenatal origin of behavior.* Lawrence, KS: University of Kansas Press.

Hopkins, B., & Westra, T. (1988). Maternal handling and motor development: an intra-cultural study. *Genetic, Social and General Psychology Monographs, 114*, 379–408.

Humphrey, T. (1972). Central representation of the oral and facial areas of human fetuses. In J.F. Bosma (Ed.), *Third symposium on oral sensation and perception: the mouth of the infant* (pp. 98–136). Springfield, IL: Thomas.

Karasik, L.B., Adolph, K.E., Tamis-Lemonda, C.S., & Bornstein, M.H. (2010). WEIRD walking: cross-cultural research on motor development. *Behavioral and Brain Sciences, 33*, 95–96.

Khazipov, R., Sirota, A., Leinekugel, X., Holmes, G.L., Ben-Ari, Y., & Buzsáki, G. (2004). Early motor activity drives spindle bursts in the developing somatosensory cortex. *Nature, 432*, 758–761.

Kravitz, H., Goldenberg, D., & Neyhus, A. (1978). Tactual exploration by normal infants. *Developmental Medicine and Child Neurology, 20*, 720–726.

Kuniyoshi, Y. (2006). Emergence of embodied behavior: coupled chaos system and fetal motor development. *ALIFE X Workshop Proceedings*, pp. 195–196.

Kuniyoshi, Y., & Sangawa, S. (2006). Early motor development from partially ordered neural-body dynamics: experiments with a cortico-spinal-musculo-skeletal model. *Biological Cybernetics, 95*(6), 589–605.

Kurjak, A., Azumendi, G., Vecek, N., Kupesic, S., Solak, M., Varga, D., & Chervenak, F. (2003). Fetal hand movements and facial expression in normal pregnancy studied by four-dimensional sonography. *Journal of Perinatal Medicine, 31*, 496–508.

Kurjak, A., Miskovic, B., Stanojevic, M., Amiel-Tison, C., Ahmed, B., Azumendi, G., Vasilj, O., Andonotopo, W., Turudic, T., & Salihagic-Kadic, A. (2008). New scoring system for fetal neurobehavior assessed by three- and four-dimensional sonography. *Journal of Perinatal Medicine, 36*, 73–81.

Kurjak, A., Talic, A., Stanojevic, M., Honemeyer, U., Serra, B., Prats, P., & Di Renzo, G.C. (2013). The study of fetal neurobehavior in twins in all three trimesters of pregnancy. *Journal of Maternal Fetal and Neonatal Medicine, 26*, 1186–1195. doi: 10.3109/14767058.2013.773306.

Lai, J., Nowlan, N.C., Vaidyanathan, R., Shaw, C.J. & Lees, C.C. (2016). Fetal movements as a predictor of health. *Acta Obstetrica Gynecologica Scandinavia, 95*, 968–975.

Lickliter, R. (2011). The integrated development of sensory organization. *Clinics in Perinatology, 38*, 591–603.

Lickliter, R., & Bahrick, L.E. (2016). Using an animal model to explore pre-natal origins of social behavior using animal models. In N. Reissland & B. Kisilevsky (Eds), *Fetal development: research on brain and behavior, environmental influences, and emerging technologies* (pp. 3–14). Dordrecht: Springer.

Marx, V., & Nagy, E. (2015). Fetal behavioural responses to maternal voice and touch. *PLoS ONE, 10*(6), e0129118. doi:10.1371/journal.pone.0129118.

Mori, H., & Kuniyoshi, Y. (2010). A human fetus development simulation: self-organization of behaviors through tactile sensations. *IEEE9th International Conference on Development and Learning*, 82–87.

Myowa-Yamakoshi, M., & Takeshita, H. (2006). Do human fetuses anticipate self-oriented actions? A study by four dimensional (4D) ultrasonography. *Infancy, 10*, 289–301.

Petrikovsky, B.M., Kaplan, G., & Holsten, N. (2003). Eyelid movements in normal human fetuses. *Journal of Clinical Ultrasound, 31*, 299–301.

Pezzulo, G., & Castelfranchi, C. (2007). The symbol detachment problem. *Cognitive Processes, 8*, 115–131.

Piontelli, A. (2010). *Development of normal fetal movements: the first 25 weeks of gestation.* Milan, Italy: Springer.

Radzyminski, S. (2005). Neurobehavioral functioning and breastfeeding behavior in the newborn. *Journal of Obstetric, Gynecologic & Neonatal Nursing, 34*, 335–341.

Reissland, N., Austen, J.M., Hanaoka, U., AboEllail, M.A.M., Uematsu, R., & Hata, T. (2015). The potential use of the fetal observable movement system in clinical practice. *Donald School Journal of Ultrasound and Obstetrics and Gynecology, 9*, 426–433.

Reissland, N., & Burghart, R. (1987). The role of massage in South Asia: child health and development. *Social Science and Medicine, 25*, 231–239.

Reissland, N., Francis, B., Aydin, E., Mason, J., & Schaal (2014). The development of anticipation in the fetus: a longitudinal account of fetal mouth movements *Developmental Psychobiology, 56*(5), 955–963. doi:10.1002/dev.21172.

Rochat, P., Blass, E.M., & Hoffmeyer, L.B. (1988). Oropharyngeal control of hand–mouth coordination in newborn infants. *Developmental Psychology, 24*, 459–463.

Rock, A.M.L., Trainor, L.J., & Addison, T.L. (1999). Distinctive messages in infant-directed lullabies and play songs. *Developmental Psychology, 35*, 527–534.

Sparling, J.W., van Tol, J., & Chescheit, N.C. (1999). Fetal and neonatal hand movement. *Physical Therapy, 79*(1), 24–39.

Sparling, J.W., & Wilhelm, I.J. (1993). Quantitative measurement of fetal movement: Fetal-Posture and Movement Assessment (F-PAM). *Physical and Occupational Therapy in Pediatrics, 12*, 97-I 14.

Takaya, R., Konishi, Y., Bos, A.F., & Einspieler, C. (2003). Preterm to early post-term changes in the development of hand–mouth contact and other motor patterns. *Early Human Development, 75*, S193–S202.

Ververs, I.A.P., Van Gelder-Hasker, M.R., De Vries, J.I.P., Hopkins, B., & Van Geijn, H.P. (1998). Prenatal development of arm posture. *Early Human Development, 51*, 61–67.

de Vries, J.I.P., Visser, G.H.A., & Prechtl, H.F.R. (1982). The emergence of fetal behavior: I. Quantitative aspects. *Early Human Development, 7*, 301–322.

Weiss, S.J. (2005). Haptic perception and the psychosocial functioning of preterm, low birth weight infants. *Infant Behavior and Development, 28*, 329–359.

Windle, W.F. (1940). *Physiology of the fetus: origin and extent of function in prenatal life.* Philadelphia, PA: W B Saunders Co.

Yamada, Y., Kanazawa, H., Iwasaki, S., Tsukahara, Y., Iwata, O., Yamada, S., & Kuniyoshi, Y. (2016). An embodied brain model of the human foetus. *Scientific Reports, 6*(27893). doi:10.1038/srep27893.

Zoia, S., Blason, L., D'Ottavio, G., Bulgheroni, M., Pezzetta, E., Scabar, A., & Castiello, U. (2007). Evidence of early development of action planning in the human foetus: a kinematic study. *Experimental Brain Research, 176*, 217–226.

2

LEARNING TO REACH IN INFANCY

Daniela Corbetta, Rebecca F. Wiener, and Sabrina L. Thurman

Reaching for objects in our surroundings is a fundamental behavior that adults perform easily and adaptively hundreds of times a day. Adults flexibly and effectively reach for a wide variety of objects across a large range of situations. In most cases, they reach to achieve a subsequent goal. For example, they reach for a pen for writing, for a glass for drinking, or for a doorknob to open a door. Each of these reaching actions requires different levels of planning, distinct arm, hand, and digit configurations, diverse movement speeds and finger pressure on the objects, as well as distinct levels of dexterity that are quite unique to each task. Despite such wide task variations, adults perform these reaching actions with great ease and seemingly automatically. In fact, reaching is such a routine behavior in adults that we tend to take it for granted. However, the wide goal-directed adaptability, flexibility, and applicability of adult reaching is not an ever-present trait across the life span. If we look back at the developmental origins of this fundamental object-directed behavior, we find that the formation of reaching in infancy is the product of a long process, involving millions of practice trials, and comprising a myriad of repeated perceptions and actions on a variety of objects and situations. Infants begin reaching with roughly undifferentiated arm and hand synergies before being able to develop better-refined actions tailored to specific physical properties of objects. Further, reaching in infancy forms as an action goal by itself, before becoming an intermediate step toward pursuing a subsequent aim. However, as infants interact with objects and experience their own movement capabilities and limitations, they can explore and discover the unique features of objects in conjunction with their own actions, and can learn to refine their movements to apply them to an ever-wider range of situations.

This chapter reviews the early developmental progression of infants' reaching for objects in the first year of life. We discuss research to highlight what we currently know about the processes involved in the formation of this fundamental skill. Specifically, we illustrate how our current understanding of the formation of this behavior is far more complex than thought decades ago. We begin with a review of the behaviors considered to be the precursors of infant reaching, then we present the different processes involved in the emergence of this early goal-directed behavior, and finally, we delineate the main steps relating to the progression and refinement of infant reaching up to the second year of life.

The Behavioral Precursors of Infant Reaching

Automatic Responses of the Arm and Hand

Newborns come to life with a set of reflexes that are considered essential for survival, are fundamental for subsequent development, and are used as critical markers of healthy neurological function (Zafeiriou, 2004). For a long time, these reflexes were thought to be the main precursors of later forming functional skills, but recent research has shown that many other behavioral elements come into play in the formation of purposeful actions, and that is clearly the case for the development of reaching in infancy.

Twitchell (1965) described in great detail the specific automatic responses of the hand and arm of infants that he observed from birth up to 1 year of age. He considered these early responses and their maturation the essential physiological substrate from which goal-directed reaching developed. One of the earliest automatic responses – *the traction response* – present from birth, is observed when the arms of the neonate are being stretched at the shoulders. This stretching triggers a strong synergistic flexion of all the joints of the arms and hands (i.e., a traction) and is thought to be mainly driven by proprioceptive feedback from the stretched flexor muscles. Twitchell (1965) observed that the traction response, extremely strong at birth, progressively declines over the first 4 months of life and vanishes completely by the time infants are 5 months old.

About 1 month after birth, *the grasping reflex* appears. This response becomes fully developed by 3–4 months of age. The grasping reflex is elicited by a light pressure of the palm of the infant's hands, causing the synergistic closure of all the fingers and thumb around the source of stimulation. This palmar grasp reflex results in a strong grip, which further intensifies when pulling on the arm (in conjunction with the traction response) or when trying to remove the object from the newborn's hand. The grasping reflex can be elicited throughout infancy and becomes a voluntary controlled grasping response a few weeks after the emergence of reaching.

Lastly, Twitchell (1965) described a third early automatic response – *the instinctive grasp reaction*. This response corresponds to a reorientation of the hand caused by a light contact or stimulation of any part of the hand. The palm of the hand turns in the direction of the source of stimulation and eventually grasps the stimulus after some exploratory contacts or probing of the source of stimulation. Twitchell noticed that the instinctive grasp reaction is a later response that begins to appear around 3 months of age, it becomes increasingly sophisticated over the remainder of the first year, and is produced mainly in response to tactile-proprioceptive information that is independent from visual control.

Although these early automatic responses clearly delineate distinct steps in the development of grasping, researchers nowadays agree that these responses alone do not suffice to promote the emergence of reaching. Many other behavioral components such as vision, eye-hand coordination, posture, sense of the self, body representation, visual acuity, stereopsis, and visual attention all constitute important precursors to the development of infant reaching.

Eye-Hand Coordination

Early eye-hand coordination is one component of the development of reaching that has been at the center of debates. Based on his earlier observations, Piaget (1936/1952) proposed that at birth, newborns' visual and motor activities were unrelated to one another. Thus, eye-hand coordination was assumed to form progressively during the 3–4 months following birth (White, Castle, & Held, 1964). In the 1970s, however, Bower, Broughton, and Moore (1970) reported having observed an early form of eye-hand coordination in 7-day-old newborns. The response was occurring mainly when an object was presented to the child. It came about in an arm extension with a fully opened hand aimed in the direction of the object fixated visually and it was accompanied by an anticipatory hand shaping as the hand approached the object. This response was also followed by successful interceptions of the object. The authors did not observe similar frequencies of responses when they presented a virtual object (see also Bower, 1974).

This startling finding failed to be replicated by a number of researchers (DiFranco, Muir, & Dodwell, 1978; Dodwell, Muir, & DiFranco, 1976; Rader & Stern, 1982; Trevarthen, Hubley, & Sheeran, 1975). These other researchers observed some newborns' arm/hand extensions and fixations of the target, but no successful hand contact with the target, and no differences in the rates of arm extension between conditions that were either presenting an actual object, a pictured object, a blank card, or no object at all. This led to the conclusion that this early form of eye-hand coordination in newborns was not really functional. It was subsequently referred to as a *pre-reaching response.*

The most updated and detailed observations of this early eye-hand coordination response in neonates has been provided by von Hofsten (1982). Von Hofsten (1982) tested the functionality of this response in two studies; first by documenting its occurrence in the presence or absence of an object in the visual field of the baby, and second by analyzing the occurrence of those responses in relation to visual fixations onto the object. Similarly to prior replication attempts, von Hofsten (1982) found very comparable rates of forward arm-hand extensions whether the object was present or absent. However, when he analyzed visual fixations in the presence of the object, he found that the rate of forward arm-hand extensions occurred significantly more often when the neonates were fixating the target, than when they were looking elsewhere, or had their eyes closed. He also found that when neonates were fixating the object, they were more likely to direct their arm closer to the object, even though they rarely touched and never caught or grabbed the object. Von Hofsten (1982) concluded that there is an early form of eye-hand coordination in newborns. It may not be functional, but may be present in preparation of the development of the reaching skill. Similar observations of this early form of eye-hand coordination were reported by Grenier (1981) who demonstrated that stabilizing the head of the neonate helped relax tensions in the body, allowed an alignment of the posture (particularly the trunk), and promoted the orientation of the visual and motor responses toward the object.

More recently, van der Meer and colleagues (van der Meer, 1997; van der Meer, van der Weel, & Lee, 1995) also suggested the existence of an early form of eye-hand coordination in neonates. In this study, a narrow beam of light was shined to the face of neonates while they were lying on their back in a darkened room with their head turned to one side. The researchers found that the neonates moved their arm into the light and kept it there for a short time so they could look at their hand. They also observed that, when doing so, neonates slowed down their movement to keep their hand in the narrow limelight. Although this task did not involve the presence of an object, the researchers concluded that very early, newborns are capable of purposeful arm movements and that these movements are performed in relation to visual input, suggesting again the existence of an early form of eye-hand coordination.

Posture and Eye-Head Coordination

The presence of automatic responses and early eye-hand coordination in newborns, although important precursors to reaching, again do not suffice by themselves to set a clear path toward the development of object-directed behavior. As exemplified by the work of Grenier (1981), posture plays an important role in the development of reaching. Newborns and infants up to the age of 6 months have no control of their upper body. Because both arms and eyes are anchored in an articulated body, the need to stabilize head and trunk is particularly critical to the formation of reaching and its successful execution. Trunk, head, and eyes must work together to provide a steady base of support and a constant spatial frame of reference for reaching (Bertenthal & von Hofsten, 1998; Spencer, Vereijken, Diedrich, & Thelen, 2000).

At birth and for a few months following birth, newborns lack control of their head. The head is heavy and the muscles of the neck are too weak to support it.

Between 2–3 months of age, however, infants begin to develop enough strength in their neck muscles to keep their head upright and stable (Spencer et al., 2000; Thelen & Spencer, 1998). This newly acquired head control has a tremendous impact on various other behavioral components that are also necessary to develop goal-directed reaching. For instance, from this age, infants not only gain the ability to orient and maintain their head and eyes in a steady direction, but they also improve significantly their ability to track visually moving sources more smoothly (von Hofsten & Rosander, 1997). As head and eye coordination emerges, infants' level of attention when attending and tracking objects also develops considerably (Richards & Holley, 1999). Finally, when provided with trunk support, 3- to 5-month-old infants begin performing their first attempts to reach for targets located close to them. Indeed, as soon as infants start holding their head upright and begin directing their visual attention toward objects located in their vicinity, sometimes quite intensely (Thelen et al., 1993; Trevarthen, 1974; von Hofsten, 1984), they will occasionally attempt to move their arms in the direction of the visually fixated object. With the provision of postural support, infants are freed from fighting against gravity to maintain their posture, thus allowing them to redirect their visual attention to the environment and to display much more organized forms of behaviors than without the added postural support. With trunk support, 3- to 5-month-old infants typically display the first forms of voluntary reaching even though they cannot sit on their own at that age.

Visual Acuity, Stereopsis, and Visual Attention

The 2- to 6-month-old period is also a period marked by significant increases in visual acuity (Banks & Salapatek, 1978; Dobson & Teller, 1978) and stereoacuity (Fox, Aslin, Shea, & Dumais, 1980; Held, Birch, & Gwiazda, 1980; Thorn, Gwiazda, Cruz, Bauer, & Held, 1994). Both visual functions significantly contribute to infants' increased ability to direct and sustain their visual attention toward objects in their surroundings.

At birth, infants' visual acuity is very poor, but by 6–8 months of age, it approaches adults' level (Norcia & Tyler, 1985; Sokol, 1978), although it continues to develop until 8–13 months of age (Chandna, 1991; Norcia & Tyler, 1985). There are several reasons for the poor acuity of newborns. The distance between the pupil and retina is very short, the fovea contains a lower density of cone receptors due to their larger and shorter shape, and the synaptic density and myelination of the cortical pathways in the primary visual cortex are also very low (Arterberry & Kellman, 2016). However, many of these structures in the eye and brain develop rapidly in the first 6 months and continue to develop until the first 3–4 years of life (see Braddick & Atkinson, 2011; Chandna, 1991 for reviews). For instance, synaptic density in the visual cortex rises rapidly and reaches its highest peak at around 6 months of age (Huttenlocher, 1994), myelination of the optic nerves and subcortical pathways are achieved between 3 and 24 months, however it is not until 45 months that the cones and macula have fully reached maturity (Chandna, 1991).

These early rapid changes in the visual system and maturation of brain structures translate in increased alertness in the infants, which is noticeable around 2–3 months of age, and reveal much rapid improvement in infants' ability to track targets and shift their visual attention (Colombo, 2001). Also, between 2 and 6 months, infants display substantial changes in object attention and their ability to visually explore, detect, and recognize several features of objects (Colombo, 2001; Gibson, 1988).

Finally, stereopsis – the ability of both eyes to align and foveate on the same point in space thereby creating sensory binocularity and depth perception – also forms around 3 months of age (Thorn et al., 1994). The emergence of binocularity has been shown as important for the emergence of reaching in infancy (von Hofsten, 1977) and is considered a necessary condition for the development of goal-directed movements, although infants are capable of inferring depth cues in several other ways prior to the onset of reaching (Bushnell & Boudreau, 1993).

Body Representation and the Sense of Self

Most recently, researchers have begun acknowledging that in order to reach, infants must also develop a sense of their body and limb movements in space to interact with their environment. Very early on, and even in the womb, fetuses touch their bodies, make contact with their face, and are able to find their mouth to suck their thumb (de Vries, Visser, & Prechtl, 1982; Hepper, Wells, & Lynch, 2005; Reissland & Austen, in this volume). Frequent contacts with the upper body and mouth continue after birth (Lobo & Galloway, 2013; Rochat & Sanders, 1991; Thomas, Karl, & Whishaw, 2015) and hand-to-mouth movements remain quite prevalent after birth (Rochat, 1993).

Studies that have investigated the organization of spontaneous movements in newborns found that these movements are actually quite structured from birth (von Hofsten & Rönnqvist, 1993). Developmental change of these movement patterns, from spontaneous to goal-directed movements, follows a U-shaped trend with a decrease or suppression in movement activity around 2–3 months of age, especially when a toy is presented in the visual field of the infant (McDonnell, Corkum, & Wilson, 1989). A similar U-shaped developmental trend was observed by Bhat and Galloway (2006), even though in a prior study (Bhat, Heathcock, & Galloway, 2005), these authors reported an increase in number of movements when a toy was present in a group of non-reaching infants compared to pre- or near-reaching infants.

Regardless of the developmental trend observed, these data as a whole suggest that prior to the emergence of goal-directed reaching, infants gain a sense of their movement activity and learn about their body and limbs via sensory-motor experience and self-touch during extended self-generated, non-task-oriented movement activity. As argued in prior work, these spontaneously generated movements form the foundation of later purposeful movements (Corbetta & Thelen, 1996, 1999; Hoffmann, 2017; Thomas et al., 2015) and evidence of self-perception in early infancy have clearly been reported (Rochat, 1995, 1998).

Assembling Those Behavioral Components into a Reaching Response

Although the developmental field increasingly acknowledges the contribution of all the above components in the formation of infant reaching, little is known about how these components actually assemble over time to form goal-directed reaching. Here, we speculate that this process may entail a progressive assembly and the integration-by-interaction of a few components at a time. Some longitudinal studies (Bhat & Galloway, 2006; McDonnell et al., 1989) have shown that the developmental path from the earliest spontaneous movement patterns in newborns to the emergence of the first purposeful reaching attempts around 3–5 months of age progresses non-linearly. Von Hofsten (1984) also found a non-linear progression in the development of arm, hand, and eye coordination in infants aged 1–19 weeks. He identified three phases of development. Initially, infants produced synergistic forward, open-hand, and arm extensions. These responses occurred regardless of whether the object was present or not, or whether the object was fixated or not. Around 7 weeks of age, infants' forward arm extensions were no longer accompanied by an opened hand; they occurred with a fisted hand. Also, during that period, fixations to the object continued to not make a difference in the frequency or form of the response. Finally, around 13 weeks of age, arm extensions were accompanied anew with an opening of the hand and occurred significantly more frequently when the object was present and fixated.

Such non-linearities in development have been shown to occur when other behavioral sub-systems form and are being integrated in the existing action repertoire of the child. Specifically, it has been argued that interactions between concomitant developing sub-systems create transitions and new forms of behaviors (Thelen, 1986, 1992, 1993; Thelen & Smith, 2006). Evidence of such interactions leading to non-linear behavioral changes have been well documented in the case of reaching and self-produced locomotion in later infancy (Corbetta & Bojczyk, 2002; Corbetta, Friedman, & Bell, 2014; Corbetta & Thelen, 2002). In this example, the assembly and integration of emerging locomotor skills in the infant sensorimotor repertoire at around 12 months of age were associated with a temporary return to two-handed reaching in infants, until both reaching and locomotion were fully integrated with each other. Such change in reaching with the emergence of locomotion was also accompanied by a cortical reorganization. Following a similar scenario, it is possible that the acquisition of head control and growing acuity occurring around 2 months of age interact with the pre-reaching pattern of arm-hand extension present from birth, possibly causing a temporary reorganization of the arm-hand synergistic response. The change in arm extension and fisted hand noted by von Hofsten (1984) around 7 weeks old also coincides approximately with the time when infants begin to display significant qualitative changes in visual attention. Once attention, head, eye, and hand are newly integrated with each other, the reorganization of the pre-reaching movement, now directed toward a visually attended target, can begin to become more purposeful. Other emerging components such as stereopsis and the increasing sense of the body, aid in further preparing infants to direct their actions toward visually fixated and reachable objects.

The Emergence of Reaching

Our current understanding of how infants succeed in performing their first goal-directed reaches has also changed dramatically in the last decades. For the longest time, the emergence of reaching has been considered the direct product of infants' ability to coordinate their arm movements with their vision (Piaget 1936/1952; White et al., 1964; see also Corbetta, Wiener, Thurman, & McMahon, 2018, for a review). However, as in the prior section on the behavioral precursors of infant reaching, researchers nowadays acknowledge that many factors come into play in the process of learning to reach in infancy.

The First Signs of Reaching Readiness

In our lab, we have studied the emergence of reaching for quite a few years now, and there are a few behavioral signs that are strongly indicative of when babies are coming close to performing their first reaching attempts. One sign that has already been well documented in the literature, is the infant increased, somewhat intense and sustained looking at the object target when it is within reach (Thelen et al., 1993; Trevarthen, 1974; von Hofsten, 1984; Williams, Corbetta, & Guan, 2015). Another sign is the increased arm movement activity of the infants in response to the object placed in or approaching their reaching space (Bhat & Galloway, 2006; Thelen et al., 1993; Williams & Corbetta, 2016; Williams, Corbetta, & Guan, 2015). Bhat and Galloway (2006), in particular, have shown that arm movement speed progressively increased 4 to 6 weeks *prior* to reach onset, especially when infants were presented with a toy in their visual field compared to a condition without a toy. Then, typically, movement speed decreases again a few weeks after reach onset, when infants begin to gain control of their arm (Thelen et al., 1993; Thelen, Corbetta, & Spencer, 1996; Williams & Corbetta, 2016; Williams, Corbetta, & Guan, 2015).

Here is one example of how infants may behave when they are really close to making their first contact with a seen target. In our lab, we investigated the reaching emergence in infants aged 2 months and 21 days who were approaching reach onset (Williams & Corbetta, 2016). Infants were followed over a 16-day intervention. Figure 2.1 illustrates a snapshot sequence of one infant in that study during his first-day visit in our lab over a 1-minute trial. Before the beginning of the trial (image 1), the object is brought into the infant's reaching space by an experimenter. The infant tracks visually the object being placed on the miniature stand. The trial recording starts when the object is set on the miniature stand. At second 4 into the trial, the infant continues to stare at the object on the stand and this frame is also the last before the infant begins moving

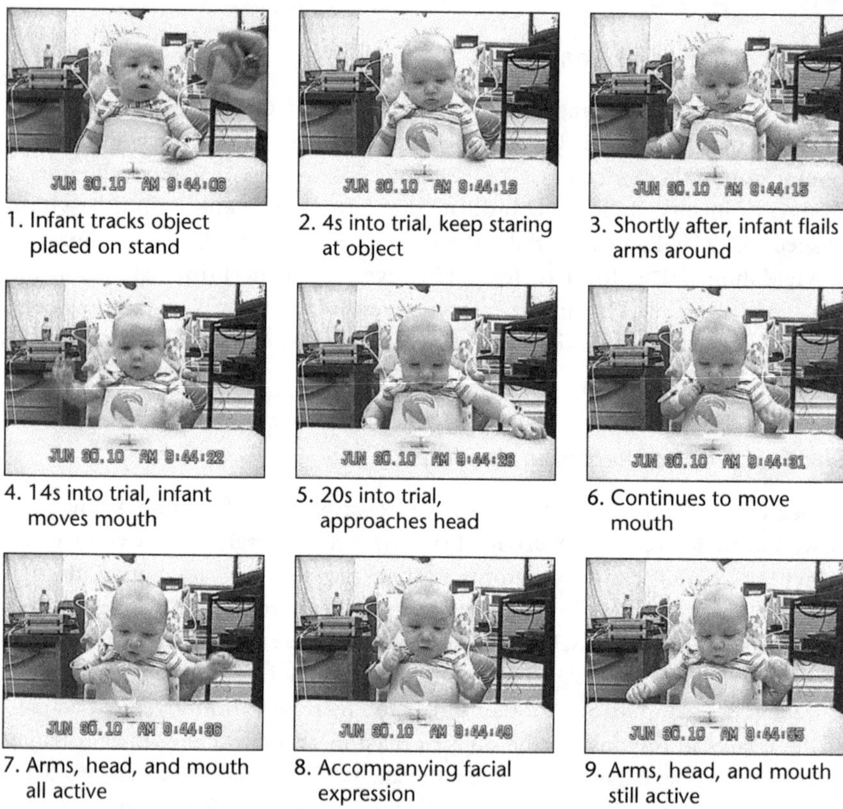

1. Infant tracks object placed on stand
2. 4s into trial, keep staring at object
3. Shortly after, infant flails arms around
4. 14s into trial, infant moves mouth
5. 20s into trial, approaches head
6. Continues to move mouth
7. Arms, head, and mouth all active
8. Accompanying facial expression
9. Arms, head, and mouth still active

FIGURE 2.1 Snapshot sequence from a video of a 2 months 21 days old infant who does not know how to reach for objects. Although actively moving his arms and staying focused during an entire 1-minute trial, this infant did not succeed in contacting the target

his arms actively. During the remainder of the 1-minute trial, this infant never took his eyes off the toy and his arms flailed around without ever intercepting the toy, as illustrated by snapshots 3–9. At 14 seconds in the trial (image 4), the infant's mouth becomes also active, and at second 20 (image 5), the infant leans forward bringing his head closer to the toy as if to grab the object by the mouth. Mouth involvement in such early reaching attempts are also frequent responses from these young babies. Thus, clearly, intense looking at the target and increased arm, head, and mouth activity are strong indicators that an infant is attempting to reach for a seen target.

First Object Contact, Eye-Hand Coordination, and Proprioception

How infants manage to make their first hand-target contact is also somewhat unclear. For several decades, and mainly under the influence of Jean Piaget's observations of his own children (Piaget, 1936/1952), it was assumed that infants needed to see both hand and target in the same visual field in order to visually monitor the position of their hand in relation to the position of the desired object. This led to the assumption that infants needed to glance back and forth between hand and object in order to steer their hand progressively closer to the target (see also White et al., 1964). This assumption was later reinforced by studies that recorded the movement kinematics of infant reaching (e.g., McDonnell, 1975, 1979). These studies found that infants' initial movement trajectories were very circuitous, and contained many corrections and changes in direction before contacting the target. Such indirect trajectories were considered evidence that infants were visually guiding their hand to the target (Bushnell, 1985).

Using the same reasoning, another set of studies utilized mirrors and displacement prisms to perturb and assess the visuomotor coordination capabilities of infant's early reaching. They found that only the reaches of older infants were perturbed by the mirrors and prisms (Lasky, 1977; McDonnell, 1975). Young infants were not affected by the mirror/prism displacement. This finding was interpreted as evidence that infants initially looked at both their hand and target alternately to guide their hand. Since both object and hand were perceived through the same displaced visual field, disruption in perceptual-motor coordination was considered reduced in that earlier age group.

In subsequent research, however, the critical role of visual guidance in the development of infant reaching began to be questioned. Von Hofsten (1979), who used a video camera facing the child, reported not seeing any visual glances between hand and object during reaching. Later, studies from Clifton and colleagues (Clifton, Muir, Ashmead, & Clarkson, 1993; Perris & Clifton, 1988) confirmed that visual guidance of the hand was not occurring in infants' early reaching. These researchers had young infants aiming for glowing or sounding objects in the dark. Thus, infants had visual or auditory information about the location of the object, but they did not have visual information about the position of their hands. Despite this lack of vision of their hand, they were able to contact the target successfully. Furthermore, in a longitudinal study in which infants were followed weekly a month prior to the onset of reaching, the researchers found that infants who had toy exposure in the dark began to reach for objects at approximately the same time as infants who had toy exposure in the light (Clifton et al., 1993). This further confirmed that infants did not need to see their hand for learning to reach.

Finally, in the same vein, another set of studies that took a closer look at the reaching trajectories of young infants revealed that the circuitous hand trajectories typical of early reaching were not the result of visual guidance of the hand, but rather the product of motion-dependent forces that infants could not control very well initially. These uncontrolled motion-dependent forces, rising from fast movements of the arm, created distortions in trajectory formation that dragged the arm away from its intended trajectory (Konczak, Borutta, Topka, & Dichgans, 1995; Thelen et al., 1993, 1996). Thus, from these studies it appeared that, when vision of the hand was available, infants did not use vision to guide their hand to the target, they maintained their visual focus on the target while correcting their movement. Thus, they seemed to rely more on proprioception than vision to steer their arm toward the target (Clifton, Rochat, Robin, & Berthier, 1994; Thelen et al., 1993; von Hofsten, 1979; although see Pogetti, Souza, Tudella, & Teixeira [2013] on the possible role of peripheral vision of the arm during reaching).

The interpretation that proprioception may play more of a predominant role in early reaching has found additional support in a recent longitudinal study (Corbetta, Thurman, Wiener, Guan, & Williams, 2014). Our lab followed 3 young infants over 11 weeks over their transition to reaching (5 weeks prior, the week of reach onset, and 5 weeks after). We used eye-tracking in addition to tracking movements to identify where infants looked on the objects 5 seconds prior to reaching. We found that during the weeks preceding reach onset, infants distributed their looking patterns across the different areas of the object. However, when infants began making their first contact with the target with their hand, the visual patterns progressively shifted toward the head of an object shaped like a drumstick in the remaining weeks of the study. If infants had used vision to guide their reaching hand, we would have anticipated seeing vision leading the shift toward the head of the drumstick over the hand in the subsequent weeks. However, this is not what occurred. We found that from the week of reach onset, infants were steadily contacting the head of the drumstick significantly more than any other area of the object, and vision only progressively spatially aligned to that contacted object area in the following weeks. Thus, it appeared as if the reaching movement "guided" vision, not the reverse. This finding is consistent with the idea that infants may develop a strong sense of their body and limb movements in the months prior to the development of reaching. Once infants contact the object a first time, their prior body movement and proprioceptive experience should allow them to reproduce the movement fairly reasonably without visual guidance of the arm (see also Corbetta et al., 2018, for a more thorough review).

First Object Contact, Haptic Perception, and Movement Formation

Haptic perception occurring in response to the hand contacting the target is also increasingly capturing the attention of researchers as a critical factor in the emergence and subsequent refinement of infant reaching. The very first hand contact that infants make with a target while they are looking at it may well happen by chance as they are actively flailing their arms around trying to attain it, but once that contact happens, the chances of reproducing this event increases rather rapidly (von Hofsten, 1979; Williams & Corbetta, 2016; Williams, Corbetta, & Guan, 2015). One current explanation for this rapid increase in goal-directed responses following the very first contact is that the hand-object contact event, even if occurring by chance initially, may work as a movement reinforcer inciting the infant to attempt to reproduce that action that led to a successful target interception. The contact event may also help aligning the visual and proprioceptive frames of reference and help build the boundaries of the body. Reinforcement occurs also at the reaching movement level (not just the contact) and acts as a movement selection process. In this scenario, more effective, or better-performed reaching actions are being reinforced and are more likely to be attempted in future reaches compared to less successful actions. Such a scenario has been addressed at the theoretical level by a few researchers (Berthier, Rosenstein, & Barto, 2005; Schlesinger, Parisi, & Langer, 2000; Sporns & Edelman, 1993; Thelen, 1995; Thelen & Corbetta, 1994; Williams, Corbetta, & Cobb, 2015). Developmental data agree with this scenario by showing extreme variability in the trajectories of infants' early reaching movements, but after a few weeks and months of practice, reaching trajectories become more stable, more direct, and more alike (Thelen et al., 1993, 1996).

The more specific evidence that haptic information can serve as a drive to help develop reaching and eventually action selection has also been shown through a series of infant studies (Williams & Corbetta, 2016; Williams, Corbetta, & Guan, 2015). In our lab, Williams, Corbetta, and Guan (2015) reused a paradigm, known as the "sticky mittens" paradigm, which was developed by Needham and colleagues to provide grasping simulation to very young infants otherwise unable to grasp objects (Needham, Barrett, & Peterman, 2002). The mittens worn by the infants during the reaching task were covered with one side of the Velcro and the other side of the Velcro was attached to the toys. When infants moved their arms and their hand contacted the object, even if happening by chance, the toy would stick to the mittens simulating object pick-up.

We wanted to disambiguate the relative influence of such grasping simulation versus reaching experience on infants' early reaching progression as measured by the number of hand-target contact performed. Those two dimensions of the intervention were confounded in the Needham studies. Infants, not quite 3 months of age, who did not have any prior reaching experience participated in a microgenetic study where infants were given 10 reaching trials for a single object at daily intervals for 16 consecutive days. We had 3 groups: a "sticky mittens" group (wearing mittens with Velcro), which received daily reaching experience and grasping simulation, a "non-sticky mittens" group (wearing mittens without Velcro to control for the mittens effect), which received daily reaching experience but no grasping simulation, and a control group, which received neither daily reaching experience, nor grasping simulation. This control group did not wear mittens and was tested on day 1 and day 16 only.

Our study, contrary to Needham et al.'s reports (Libertus & Needham, 2010, 2011; Needham et al., 2002), revealed that the "sticky mittens" group did not increase object contact beyond that of the control group on day 16. The "non-sticky mittens" group, on the contrary, which did not receive grasping simulation, turned out to be the only group that actually increased hand-object contact significantly above the performance of the group control on day 16. Figure 2.2 shows how the "non-sticky" (gray symbol, left panel) and "sticky mittens" groups (gray symbol, right panel) performed on day 16 compared to the control group (white symbols).

One important difference with Needham et al.'s (2002) study was that our mittens did not entirely conceal the infants' hands as those used by the Needham group. Our mittens had the thumb and tip of the fingers open. Thus, even in the "sticky mittens" group, where the objects would stick to the mittens at contact as in the Needham studies, infants could potentially contact the object with bare fingers and receive haptic feedback, depending on the movement performed. Between-subject analyses in the "sticky mittens" group identified those infants who made frequent bare fingers contacts with the object and those who did not over the 16-day intervention. Results revealed that bare finger contacts with the object were actually driving the reaching progression in a few infants more than the grasping simulation provided by the sticky mittens. The few highly performing infants in that group made frequent bare finger contacts with the target object and were the only ones to display a significant reaching progression over time. The remaining majority of the infants who touched the object mainly with the sticky part of the mittens thereby receiving the grasping simulation but seldom bare fingers contacts, did not progress significantly over time. These findings underscored the importance of haptic perception in learning to reach (Corbetta, Williams, & Haynes, 2016).

In a subsequent study, we further examined the reinforcing value of the hand contacting the object with bare fingers in 2- to 3-month-old infants ready to begin to reach (Williams & Corbetta, 2016). We performed another 16-day intervention, without mittens this time; however, we varied the impact of the hand-object contacts using a contingent reinforcement paradigm. In one group – the contingent group – the object was mounted on a spring and it would move and produce a bell sound only as a result of being contacted by the hand. In a second group – the continuous group – the object would also move and sound, but the object motion in this condition was continuous and independent from the infant action. In this continuous condition, the object was mounted on a small motor that was moving the object back and forth in a regular motion. Finally, a third group receiving neither daily exposure to the reaching task, and neither continuous, nor action-contingent object motion served as control group.

The results from these studies are presented in Figure 2.2 for object contacts that were made while the infants were looking at the target. The reaching progression of the contingent group (black symbols, left panel) were revealed to be the steepest (Williams & Corbetta, 2016). When comparing performance across the "sticky mittens" study (Williams, Corbetta, & Guan, 2015) and this study on contingent reinforcement, Figure 2.2 shows that the contingent group even outperformed the "non-sticky mittens" group on day 16. The continuous group, on the other hand, did not differ fundamentally from the control and "sticky mittens" groups' performance on day 16 (right panel). Thus, reinforcing the haptic event at contact had a significant impact on infants learning to reach.

Studies modelling the process of learning to reach in infants, that weighed the role of vision, proprioception, and haptic feedback via simulations, have supported the conclusion that haptic feedback from touching the target may

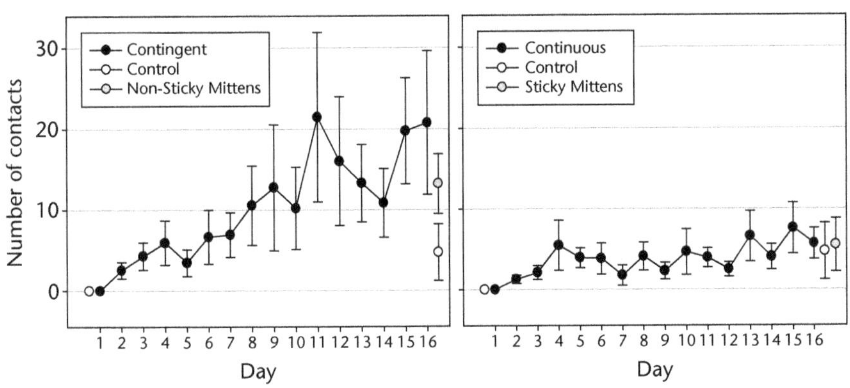

FIGURE 2.2 Frequency of hand-object contacts occurring while the object was visually attended by infants by day and by study group. Infants in the contingent group, who received contingent reinforcement when they contacted the target with their hand, displayed the fastest reaching progression (left panel)

Source: Adapted from Williams and Corbetta (2016) and Williams, Corbetta, and Guan (2015).

help coordinating vision and proprioception in infants (Schlesinger & Parisi, 2001). Thus, contacting the object and receiving haptic feedback does not only reinforce the response, but also contributes to the formation of the movement. Furthermore, this simulation work also revealed that the model preferred haptic information to visual information in learning to reach, a finding that is compatible with our studies with infants, indicating that vision may be mapped onto proprioception at reach onset and not the reverse (Corbetta et al., 2014).

Refining Reaching Over the First Two Years of Life

Many changes in infants' reaching skills occur in the weeks and months that follow reach onset. Progressions occur at all levels: in arm trajectory, reaching anticipation, grasp refinement, and infants increasingly learn to map their actions as a function of the goal and objects' physical properties. We focus on a few of these progressions in this last section.

Exploration, Selection, and Trajectory Formation

Developmental changes in infant reaching have classically been described as a two-phase process. First, from 3–5 to 7–8 months of age, infants mainly refine the transport phase of their goal-directed movement, and then, at 7–8 months of age, they begin to modulate their reaching and grasping response as a function of objects' physical characteristics (Lockman, Ashmead, & Bushnell, 1984; von Hofsten & Fazel-Zandy, 1984). While researchers have agreed on the existence of these two developmental phases, process accounts of how infants progress through these phases have changed drastically in the last decades. As mentioned earlier, the emergence and progression of infant reaching was initially assumed to be largely under the control of vision (Piaget, 1936/1952; McDonnell, 1975; White et al., 1964). When infants transitioned to phase two, it was assumed that visual attention was released from its guiding role of the hand and newly used to pick up specific information about the target object in order to adjust and shape the reach and grasp accordingly (Bushnell, 1985). In this early work, proprioceptive and haptic experience were only seen as subordinate to the control of vision. However, since the discovery that young infants can reach in the dark (Clifton et al., 1993) and that the distortions in hand trajectories are the product of poor speed calibration of the movement to the goal (Thelen et al., 1993, 1996), explanations of the process involved in the first developmental phase of reaching have changed.

Nowadays, the first phase is considered a period of motor exploration aimed at discovering how to calibrate the speed of the movement to the task at hand. Indeed, infants at first often swipe at the object. To counteract the disruptive motion-dependent forces that are generated by such fast swiping movements, infants need to learn how to map the speed of their movement to the goal. This process takes place by trial and error, by exploring different movement speeds in different contexts, and by mapping the resulting proprioceptive-motor experiences to respective outcomes. A similar process of selection by reinforcement, as we discussed in the previous section, is believed to take place during this developmental period. Movement speeds leading to good results are reinforced, while other movements leading to poor outcomes are progressively eliminated (Thelen & Corbetta, 1994; Thelen et al., 1996). Thus, for the last few decades, learning to reach is not anymore considered a top-down process where vision guides the arm, but is rather seen as a deeply embodied dynamic process in which brain, mind (the act of planning), and body work in concert to produce reaching behaviors that become increasingly tuned to their goals (Corbetta, 2009; Thelen, 2000). In this view, movement variability, movement exploration, and reinforcement learning are all considered fundamental to the process of movement selection and learning to reach (Berthier et al., 2005; Sporns & Edelman, 1993; Thelen & Corbetta, 1994; Williams, Corbetta, & Cobb, 2015). Furthermore, other studies have begun to suggest that infants' reliance on proprioceptive and haptic perception may continue to precede the reliance on vision for reaching and grasping even later in the first year (Berthier & Carrico, 2010; Carrico & Berthier, 2008; Corbetta & Snapp-Childs, 2009; Corbetta, Thelen, & Johnson, 2000).

Predictive Reaching, Perceived Reachability, and Movement Refinement

Within a few weeks of beginning to reach for stationary objects, infants develop the ability to reach for moving objects. They start intercepting slow-moving targets first, but progressively, they increasingly succeed intercepting targets that are moving faster (von Hofsten, 1979; von Hofsten & Lindhagen, 1979). By 6 months, infants can even extrapolate the trajectory of objects based on inertia properties of the object motion, particularly if the object follows a linear path (von Hofsten, Vishton, Spelke, Feng, & Rosander, 1998).

Infants also quickly develop the ability to perceive if an object is within reaching distance or not. Studies from Rochat and colleagues (Rochat & Goubet, 1995; Rochat, Goubet, & Senders, 1999) have shown that 5- and 6-month-old infants will not attempt to reach for an object if it is out of reaching space. The acquisition of the sitting posture, which occurs around 6 months of age, on average, will further modulate how far infants will extend their arms out to reach for a distant object. Indeed, when infants are able to sit and have acquired control of their trunk, they can lean forward to reach for objects further out.

Refinement of the hand and arm coordination also begins to take place shortly after reach onset and continues to develop through the second year of life. First, infants' basic grasping abilities form (von Hofsten & Lindhagen,

1979). At reach onset, infants only contact the target with a part of their hand, but within 2–3 weeks, infants successfully grasp targets which fit the size of their hands using a power grip, or synergistic palmar grasp. Grasping very gradually develops into a more refined precision grip or thumb-index pincer grasp toward the end of the first year and through the second year (Butterworth, Verweij, & Hopkins, 1997; Halverson, 1931).

Other important reaching refinements occur around the age of 7–8 months when infants enter their second reaching developmental phase. By that time, infants begin to pick up on the physical properties of objects and begin to integrate them into their movement plan. For example, at that age, infants begin pre-shaping or anticipating the orientation of their hand during the transport phase of the reach to match it to the object orientation or shape (Lockman et al., 1984; Piéraut-Le Bonniec, 1985; von Hofsten & Fazel-Zandy, 1984; Witherington, 2005). They begin using one or two hands to adapt their reach and grasp to the size of objects (Corbetta & Snapp-Childs, 2009; Corbetta *et al.*, 2000; Fagard & Jacquet, 1996). By 10 months, they begin anticipating two concurrent dimensions of objects, although they may do this sequentially by adjusting orientation first and size next (Ransburg, Reiser, Munzert, Jovanovic, & Schwarzer, 2017; Schum, Jovanovic, & Schwarzer, 2011).

By the end of the first year and through the second year of life, more sophisticated reaching skills begin to appear. For example, this is the age range when reaching is not anymore performed as a goal per se, but is beginning to be used as a mean towards accomplishing subsequent goals. For example, infants will reach for cloths or lids to retrieve hidden objects (Bojczyk & Corbetta, 2004; Bruner, 1970; Smith, Thelen, Titzer, & McLin, 1999), they will begin using tools such as a spoon for eating (Connolly & Dalgleish, 1989; McCarty, Clifton, & Collard, 2001), or a hammer for banging (Kahrs, Jung, & Lockman, 2013). By the end of the first year of life infants also begin modulating their reaching phase as a function of the task they are planning to accomplish next. For example, 10- and 14-month-old infants will reach faster for an object if they plan to throw it in a wider container as opposed to inserting it or fitting it into a smaller tube (Claxton, Keen, & McCarty, 2003; Gottwald et al., 2017).

Conclusions

The development of reaching in infancy is a long and complex process that requires the progressive assembly and coordination of many behavioral components throughout the first year of life before infants can use this skill as a means to accomplish subsequent and more sophisticated goals. Even by the end of the second year of life, infants are far from performing reaching behaviors as skillfully as adults (Konczak et al., 1995). In this chapter, we did not expand on the motor accomplishments occurring beyond the second year of life, but clearly, the development of reaching continues to progress through early childhood.

This chapter also provided a review of the processes that are assumed to be involved in the formation of this goal-directed behavior from the first days of life. Vision is not anymore considered as the primary behavioral element allowing and guiding the development of reaching in infancy. It is now seen as one element among several others that contribute to the development of reaching. Particularly, in recent years, there has been a growing emphasis on the role of proprioception and haptic sensing as possibly playing initially a greater critical role in the formation and refinement of early reaching movements. That is not to say that vision does not play a role. Vision is clearly needed for eye-hand coordination, but evidence strongly suggests that vision is not used for guiding the movement. Vision may play a more important role in providing information about object location, distance, and physical characteristics. Thus, eye-hand coordination may be casted as the product of mapping visual information from the world with growing experience in motor coordination. The emergence of reaching in 3- to 5-month-old infants can be considered, after all, as the first behavior in ontogeny in which vision and action converge onto a common aiming goal (see Corbetta et al., 2018). Calibrating proprioceptive and haptic experience (related to the infant's peripersonal space), with visual experience of a world (extending beyond the peripersonal space of the child), may be one challenge to overcome in learning to reach. Clearly, more studies still need to be performed to fully understand how vision and action come together in early infancy.

References

Arterberry, M. E., & Kellman, P. J. (2016). *Development of perception in infancy: The cradle of knowledge revisited.* New York: Oxford University Press.

Banks, M. S., & Salapatek, P. (1978). Acuity and contrast sensitivity in 1-, 2-, and 3-month-old human infants. *Investigative Ophthalmology & Visual Science, 17*(4), 361–365.

Bertenthal, B., & von Hofsten, C. (1998). Eye, head and trunk control: The foundation for manual development. *Neuroscience & Biobehavioral Reviews, 22*(4), 515–520. doi:https://doi.org/10.1016/S0149-7634(97)00038-9.

Berthier, N. E., & Carrico, R. L. (2010). Visual information and object size in infant reaching. *Infant Behavior and Development, 33*(4), 555–566. doi:10.1016/j.infbeh.2010.07.007.

Berthier, N. E., Rosenstein, M. T., & Barto, A. G. (2005). Approximate optimal control as a model for motor learning. *Psychological Review, 112*(2), 329–346.

Bhat, A. N., & Galloway, J. C. (2006). Toy-oriented changes during early arm movements: Hand kinematics. *Infant Behavior and Development, 29*(3), 358–372. doi:10.1016/j.infbeh.2006.01.005.

Bhat, A., Heathcock, J., & Galloway, J. C. (2005). Toy-oriented changes in hand and joint kinematics during the emergence of purposeful reaching. *Infant Behavior and Development, 28*(4), 445–465. doi:10.1016/j.infbeh.2005.03.001.

Bojczyk, K. E., & Corbetta, D. (2004). Object retrieval in the 1st year of life: Learning effects of task exposure and box transparency. *Developmental Psychology, 40*(1), 54–66. doi:http://dx.doi.org/10.1037/0012-1649.40.1.54.

Bower, T. G. R. (1974). *Development in infancy*. San Francisco: Atkinson, Freeman & Thomson.

Bower, T. G. R., Broughton, J., & Moore, M. K. (1970). Demonstration of intention in reaching behavior of the neonate. *Nature, 228*(5272), 679–681.

Braddick, O., & Atkinson, J. (2011). Development of human visual function. *Vision Research, 51*(13), 1588–1609. doi:https://doi.org/10.1016/j.visres.2011.02.018.

Bruner, J. S. (1970). The growth and structure of skill. In K. Connolly (Ed.), *Mechanisms of motor skill development* (pp. 63–94). New York: Academic Press.

Bushnell, E. W. (1985). The decline of visually guided reaching during infancy. *Infant Behavior and Development, 8*(2), 139–155.

Bushnell, E. W., & Boudreau, J. P. (1993). Motor development and the mind: The potential role of motor abilities as a determinant of aspects of perceptual development. *Child Development, 64*(4), 1005–1021.

Butterworth, G., Verweij, E., & Hopkins, B. (1997). The development of prehension in infants: Halverson revisited. *British Journal of Developmental Psychology, 15*(2), 223–236. doi:10.1111/j.2044-835X.1997.tb00736.x.

Carrico, R. L., & Berthier, N. E. (2008). Vision and precision reaching in 15-month-old infants. *Infant Behavior and Development, 31*(1), 62–70. doi:10.1016/j.infbeh.2007.07.005.

Chandna, A. (1991). Natural history of the development of visual acuity in infants. *Eye, 5*, 20–26. doi:10.1038/eye.1991.4.

Claxton, L. J., Keen, R., & McCarty, M. E. (2003). Evidence of motor planning in infant reaching behavior. *Psychological Science, 14*(4), 354–356. doi:psci_24421 [pii].

Clifton, R. K., Muir, D. W., Ashmead, D. H., & Clarkson, M. G. (1993). Is visually guided reaching in early infancy a myth? *Child Development, 64*, 1099–1110.

Clifton, R. K., Rochat, P., Robin, D. J., & Berthier, N. E. (1994). Multimodal perception in the control of infant reaching. *Journal of Experimental Psychology: Human Perception and Performance, 20*(4), 876–886.

Colombo, J. (2001). The development of visual attention in infancy. *Annual Review of Psychology, 52*, 337–367.

Connolly, K., & Dalgleish, M. (1989). The emergence of a tool-using skill in infancy. *Developmental Psychology, 25*(6), 894–912.

Corbetta, D. (2009). Brain, body, and mind: Lessons from infant motor development. In J. P. Spencer, M. Thomas, & J. McClelland (Eds.), *Toward a unified theory of development: Connectionism and dynamic systems theory re-considered* (pp. 51–66). New York: Oxford University Press.

Corbetta, D., & Bojczyk, K. E. (2002). Infants return to two-handed reaching when they are learning to walk. *Journal of Motor Behavior, 34*(1), 83–95. doi:http://dx.doi.org/10.1080/00222890209601933.

Corbetta, D., Friedman, D. R., & Bell, M. A. (2014). Brain reorganization as a function of walking experience in 12 month-old infants: Implications for the development of manual laterality. *Frontiers in Psychology, 5*. doi:10.3389/fpsyg.2014.00245.

Corbetta, D., & Snapp-Childs, W. (2009). Seeing and touching: The role of sensory-motor experience on the development of infant reaching. *Infant Behavior and Development, 32*(1), 44–58. doi:http://dx.doi.org/10.1016/j.infbeh.2008.10.004.

Corbetta, D., & Thelen, E. (1996). The developmental origins of bimanual coordination: A dynamic perspective. *Journal of Experimental Psychology: Human Perception and Performance, 22*(2), 502–522. doi:http://dx.doi.org/10.1037/0096-1523.22.2.502.

Corbetta, D., & Thelen, E. (1999). Lateral biases and fluctuations in infants' spontaneous arm movements and reaching. *Developmental Psychobiology, 34*(4), 237–255.

Corbetta, D., Thelen, E., & Johnson, K. (2000). Motor constraints on the development of perception-action matching in infant reaching. *Infant Behavior and Development, 23*(3–4), 351–374.

Corbetta, D., & Thelen, E. (2002). Behavioral fluctuations and the development of manual asymmetries in infancy: Contribution of the dynamic systems approach. In S. J. Segalowitz & I. Rapin (Eds.), *Handbook of neuropsychology (vol. 8): Child neuropsychology, part I* (pp. 309–328). Amsterdam: Elsevier Science Publishing Co.

Corbetta, D., Thurman, S. L., Wiener, R. F., Guan, Y., & Williams, J. L. (2014). Mapping the feel of the arm with the sight of the object: On the embodied origins of infant reaching. *Frontiers in Psychology, 5*, 18.

Corbetta, D., Wiener, R. F., Thurman, S. L., & McMahon, E. (2018). The embodied origins of infant reaching: Implications for the emergence of eye-hand coordination. *Kinesiology Review*, 7, 10–17. https://doi.org/10.1123/kr.2017-0052.

Corbetta, D., Williams, J. L., & Haynes, J. M. (2016). Bare fingers, but no obvious influence of "prickly" Velcro! In the absence of parents' encouragement, it is not clear that "sticky mittens" provide an advantage to the process of learning to reach. *Infant Behavior and Development, 42*, 168–178. doi:http://dx.doi.org/10.1016/j.infbeh.2015.05.001.

DiFranco, D., Muir, D. W., & Dodwell, D. (1978). Reaching in very young infants. *Perception*, 7, 385–392.

Dobson, V., & Teller, D. Y. (1978). Visual acuity in human infants: A review and comparison of behavioral and electrophysiological studies. *Vision Research, 18*(11), 1469–1483. doi:https://doi.org/10.1016/0042-6989(78)90001-9.

Dodwell, P. C., Muir, D., & DiFranco, D. (1976). Responses of infants to visually presented objects. *Science, 194*, 209–211.

Fagard, J., & Jacquet, A. Y. (1996). Changes in reaching and grasping objects of different sizes between 7 and 13 months of age. *British Journal of Developmental Psychology, 14*, 65–78.

Fox, R., Aslin, R. N., Shea, S. L., & Dumais, S. T. (1980). Stereopsis in human infants. *Science, 207*(4428), 323–324.

Gibson, E. J. (1988). Exploratory behavior in the development of perceiving, acting and the acquiring knowledge. In M. R. Rosenzweig & L. W. Porter (Eds.), *Annual review of psychology*. Palo Alto, CA: Annual Review, Inc.

Gottwald, J. M., De Bortoli Vizioli, A., Lindskog, M., Nyström, P., Ekberg, T. L., von Hofsten, C., & Gredebäck, G. (2017). Infants prospectively control reaching based on the difficulty of future actions: To what extent can infants' multiple-step actions be explained by Fitts' law? *Developmental Psychology, 53*(1), 4–12.

Grenier, A. (1981). La "motricité libérée" par fixation manuelle de la nuque au cours des premières semaines de la vie. *Archives Françaises de Pédiatrie, 38*, 557–562.

Halverson, H. M. (1931). An experimental study of prehension in infants by means of systematic cinema records. *Genetic Psychology Monographs, 10*, 107–283.

Held, R., Birch, E., & Gwiazda, J. (1980). Stereoacuity of human infants. *Proceedings of the National Academy of Sciences of the United States of America*, 77(9), 5572–5574.

Hepper, P. G., Wells, D. L., & Lynch, C. (2005). Prenatal thumb sucking is related to postnatal handedness. *Neuropsychologia, 43*(3), 313–315. doi:S0028-3932(04)00242-8 [pii] 10.1016/j.neuropsychologia.2004.08.009.

Hoffmann, M. (2017). *The role of self-touch experience in the formation of the self, in The Development of the Self Workshop*. Paper presented at the IEEE ICDL-EpiRob 2017, Lisbon, Portugal.

von Hofsten, C. (1977). Binocular convergence as a determinant of reaching behavior in infancy. *Perception, 6*(2), 139–144. doi:http://dx.doi.org/10.1068/p060139.

von Hofsten, C. (1979). Development of visually guided reaching: The approach phase. *Journal of Human Movement Studies, 5*, 160–178.

von Hofsten, C. (1982). Eye–hand coordination in the newborn. *Developmental Psychology, 18*(3), 450–461. doi:http://dx.doi.org/10.1037/0012-1649.18.3.450.

von Hofsten, C. (1984). Developmental changes in the organization of prereaching movements. *Developmental Psychology, 20*(3), 378–388. doi:http://dx.doi.org/10.1037/0012-1649.20.3.378.

von Hofsten, C., & Fazel-Zandy, S. (1984). Development of visually guided hand orientation in reaching. *Journal of Experimental Child Psychology, 38*(2), 208–219. doi:http://dx.doi.org/10.1016/0022-0965(84)90122-X.

von Hofsten, C., & Lindhagen, K. (1979). Observations on the development of reaching for moving objects. *Journal of Experimental Child Psychology, 28*(1), 158–173. doi:http://dx.doi.org/10.1016/0022-0965(79)90109-7.

von Hofsten, C., & Rönnqvist, L. (1993). The structuring of neonatal arm movements. *Child Development, 64*(4), 1046–1057. doi:http://dx.doi.org/10.2307/1131326.

von Hofsten, C., & Rosander, K. (1997). Development of smooth pursuit tracking in young infants. *Vision Research, 37*(13), 1799–1810. doi:http://dx.doi.org/10.1016/S0042-6989(96)00332-X.

von Hofsten, C., Vishton, P., Spelke, E. S., Feng, Q., & Rosander, K. (1998). Predictive action in infancy: Tracking and reaching for moving objects. *Cognition, 67*(3), 255–285. doi:http://dx.doi.org/10.1016/S0010-0277(98)00029-8.

Huttenlocher, P. H. (1994). Synaptogenesis in human cerebral cortex. In G. Dawson & K. W. Fisher (Eds.), *Human behavior and the developing brain* (pp. 137–152). New York: The Guildford Press.

Kahrs, B. A., Jung, W. P., & Lockman, J. J. (2013). Motor origins of tool use. *Child Development, 84*(3), 810–816. doi:10.1111/cdev.12000.

Konczak, J., Borutta, M., Topka, H., & Dichgans, J. (1995). The development of goal-directed reaching in infants: Hand trajectory formation and joint torque control. *Experimental Brain Research, 106*, 156–168.

Lasky, R. E. (1977). The effect of visual feedback of the hand on the reaching and retrieval behavior of young infants. *Child Development, 48*, 112–117.

Libertus, K., & Needham, A. (2010). Teach to reach: The effects of active vs. passive reaching experiences on action and perception. *Vision Research, 50*(24), 2750–2757. doi:S0042-6989(10)00435-9 [pii] 10.1016/j.visres.2010.09.001.

Libertus, K., & Needham, A. (2011). Reaching experience increases face preference in 3-month-old infants. *Developmental Science, 14*(6), 1355–1364. doi:10.1111/j.1467-7687.2011.01084.x.

Lobo, M. A., & Galloway, J. C. (2013). The onset of reaching significantly impacts how infants explore both objects and their bodies. *Infant Behavior and Development, 36*(1), 14–24. doi:10.1016/j.infbeh.2012.09.003.

Lockman, J. J., Ashmead, D. H., & Bushnell, E. W. (1984). The development of anticipatory hand orientation during infancy. *Journal of Experimental Child Psychology, 37*(1), 176–186. doi:0022-0965(84)90065-1 [pii].

McCarty, M. E., Clifton, R. K., & Collard, R. R. (2001). The beginnings of tool use by infants and toddlers. *Infancy, 2*(2), 233–256.

McDonnell, P. M. (1975). The development of visually guided reaching. *Perception and Psychophysics, 18*(3), 181–185.

McDonnell, P. M. (1979). Patterns of eye-hand coordination in the first year of life. *Canadian Journal of Psychology, 33*, 253–267.

McDonnell, P. M., Corkum, V. L., & Wilson, D. L. (1989). Patterns of movement in the first 6 months of life: New directions. *Canadian Journal of Psychology, 43*(2), 320–339.

van der Meer, A. L. (1997). Keeping the arm in the limelight: Advanced visual control of arm movements in neonates. *European Journal of Paediatric Neurology, 4*, 103–108.

van der Meer, A. L. H., van der Weel, F. R., & Lee, D. N. (1995). The functional significance of arm movements in neonates. *Science, 267*(5198), 693–695.

Needham, A., Barrett, T., & Peterman, K. (2002). A pick me up for infants' exploratory skills: Early simulated experiences reaching for objects using "sticky" mittens enhances young infants' object exploration skills. *Infant Behavior and Development, 25*(3), 279–295.

Norcia, A. M., & Tyler, C. W. (1985). Spatial frequency sweep VEP: Visual acuity during the first year of life. *Vision Research, 25*(10), 1399–1408. doi:https://doi.org/10.1016/0042-6989(85)90217-2.

Perris, E., & Clifton, R. (1988). Reaching in the dark toward sound as a measure of auditory localization in infants. *Infant Behavior and Development, 11*, 473–491.

Piaget, J. (1936/1952). *The origins of intelligence in children [La naissance de l'intelligence chez l'enfant]* (M. Cook, Trans.). New York: Basic Books.

Piéraut-Le Bonniec, G. (1985). Hand-eye coordination in infants' construction of convexity and concavity. *British Journal of Developmental Psychology, 3*, 273–280.

Pogetti, L., Souza, R., Tudella, E., & Teixeira, L. (2013). Early infant's use of visual feedback in voluntary reaching for a spatial target. *Frontiers in Psychology, 4*(520). doi:10.3389/fpsyg.2013.00520.

Rader, N., & Stern, J. D. (1982). Visually elicited reaching in neonates. *Child Development, 53*, 1004–1007.

Ransburg, N., Reiser, M., Munzert, J., Jovanovic, B., & Schwarzer, G. (2017). Concurrent anticipation of two object dimensions during grasping in 10-month-old infants: A quantitative analysis. *Infant Behavior and Development, 48*, 164–174. doi:https://doi.org/10.1016/j.infbeh.2017.04.003.

Richards, J. E., & Holley, F. B. (1999). Infant attention and the development of smooth pursuit tracking. *Developmental Psychology, 35*(3), 856–867.

Rochat, P. (1993). Hand-mouth coordination in the newborn: Morphology, determinants, and early development of a basic act. In G. J. P. Savelsbergh (Ed.), *Advances in psychology* (Vol. 97, pp. 265–288). Amsterdam: North-Holland.

Rochat, P. (1995). *The self in infancy: Theory and research*. Amsterdam: North-Holland/Elsevier Science Publishers.

Rochat, P. (1998). Self-perception and action in infancy. *Experimental Brain Research, 123*(1–2), 102–109.

Rochat, P., & Goubet, N. (1995). Development of sitting and reaching in 5- to 6-month-old infants. *Infant Behavior and Development, 18*(1), 53–68.

Rochat, P., Goubet, N., & Senders, S. J. (1999). To reach or not to reach? Perception of body effectivities by young infants. *Infant and Child Development, 8*(3), 129–148.

Rochat, P., & Sanders, S. J. (1991). Active touch in infancy: Action systems in development. In M. J. Weiss & P. R. Zelazo (Eds.), *Infant attention: Biological constraints and the influence of experience* (pp. 412–442). Norwood, NJ: Ablex.

Schlesinger, M., & Parisi, D. (2001). Multimodal control of reaching: Simulating the role of tactile feedback. *IEEE Transactions on Evolutionary Computation, 5*(2), 122–128. doi:10.1109/4235.918433.

Schlesinger, M., Parisi, D., & Langer, J. (2000). Learning to reach by constraining the movement search space. *Developmental Science, 3*(1), 67–80. doi:10.1111/1467-7687.00101.

Schum, N., Jovanovic, B., & Schwarzer, G. (2011). Ten- and twelve-month-olds' visual anticipation of orientation and size during grasping. *Journal of Experimental Child Psychology, 109*(2), 218–231. doi:S0022-0965(11)00018-X [pii] 10.1016/j.jecp. 2011.01.007.

Smith, L. B., Thelen, E., Titzer, R., & McLin, D. (1999). Knowing in the context of acting: The task dynamics of the A-not-B error. *Psychological Review, 106*(2), 235–260. doi:http://dx.doi.org/10.1037/0033-295X.106.2.235.

Sokol, S. (1978). Measurement of infant visual acuity from pattern reversal evoked potentials. *Vision Research, 18*(1), 33–39. doi:https://doi.org/10.1016/0042-6989(78)90074-3.

Spencer, J. P., Vereijken, B., Diedrich, F. J., & Thelen, E. (2000). Posture and the emergence of manual skills. *Developmental Science, 3*(2), 216–233.

Sporns, O., & Edelman, G. M. (1993). Solving Bernstein's problem: A proposal for the development of coordinated movement by selection. *Child Development, 64*(4), 960–981.

Thelen, E. (1986). Development of coordinated movement: Implications for early human development. In M. G. Wade & H. T. A. Whiting (Eds.), *Motor development in children: Aspects of coordination and control* (pp. 107–124). Boston: Martinus Nijhoff Publishers.

Thelen, E. (1992). Development as a dynamic system. *Current Directions in Psychological Science, 1*(6), 189–193. doi:http://dx.doi.org/10.1111/1467-8721.ep10770402.

Thelen, E. (1993). Timing and developmental dynamics in the acquisition of early motor skills. In G. Turkewitz & D. A. Devenny (Eds.), *Developmental time and timing* (pp. 85–104). Hillsdale, NJ: Lawrence Erlbaum.

Thelen, E. (1995). Motor development: A new synthesis. *American Psychologist, 50*(2), 79–95.

Thelen, E. (2000). Grounded in the world: Developmental origins of the embodied mind. *Infancy, 1*(1), 3–28. doi:http://dx.doi.org/10.1207/S15327078IN0101_02.

Thelen, E., & Corbetta, D. (1994). Exploration and selection in the early acquisition of skill. In O. Sporns & G. Tononi (Eds.), *Selectionism and the brain* (pp. 75–102). San Diego, CA: Academic Press.

Thelen, E., Corbetta, D., Kamm, K., Spencer, J. P., Schneider, K., & Zernicke, R. F. (1993). The transition to reaching: Mapping intention and intrinsic dynamics. *Child Development, 64*(4), 1058–1098. doi:http://dx.doi.org/10.2307/1131327.

Thelen, E., Corbetta, D., & Spencer, J. P. (1996). Development of reaching during the first year: Role of movement speed. *Journal of Experimental Psychology: Human Perception and Performance, 22*(5), 1059–1076. doi:http://dx.doi.org/10.1037/0096-1523.22.5.1059.

Thelen, E., & Smith, L. B. (2006). Dynamic systems theories. In R. M. Lerner & W. Damon (Eds.), *Handbook of child psychology: Theories* (6th ed., pp. 258–312). Hoboken, NJ: John Wiley & Sons Inc.

Thelen, E., & Spencer, J. P. (1998). Postural control during reaching in young infants: A dynamic systems approach. *Neuroscience and Biobehavioral Reviews, 22*(4), 507–514. doi:S0149763497000377 [pii].

Thomas, B. L., Karl, J. M., & Whishaw, I. Q. (2015). Independent development of the reach and the grasp in spontaneous self-touching by human infants in the first 6 months. *Frontiers in Psychology, 5*(1526). doi:10.3389/fpsyg.2014.01526.

Thorn, F., Gwiazda, J., Cruz, A. A. V., Bauer, J. A., & Held, R. (1994). The development of eye alignment, convergence, and sensory binocularity in young infants. *Investigative Ophtalmology & Visual Science, 35*(2), 544–553.

Trevarthen, C. (1974). The psychobiology of speech development. *Neurosciences Research Programme Bulletin, 12*, 570–585.

Trevarthen, C., Hubley, P., & Sheeran, L. (1975). Les activités innées du nourrisson. *La Recherche, 6*(56), 447–458.

Twitchell, T. E. (1965). The automatic grasping responses in infants. *Neuropsychologia, 3*, 247–259.

de Vries, J. I. P., Visser, G. H. A., & Prechtl, H. F. R. (1982). The emergence of fetal behaviour. I. Qualitative aspects. *Early Human Development,* 7(4), 301–322. doi:https://doi.org/10.1016/0378-3782(82)90033-0.

White, B. L., Castle, P., & Held, R. (1964). Observations on the development of visually directed reaching. *Child Development, 35*, 349–364.

Williams, J. L., & Corbetta, D. (2016). Assessing the impact of movement consequences on the development of early reaching in infancy. *Frontiers in Psychology,* 7, 587.

Williams, J. L., Corbetta, D., & Cobb, L. (2015). How perception, action, functional value, and context can shape the development of infant reaching. *Movement and Sport Sciences/Science et Motricité, 89*, 5–15.

Williams, J. L., Corbetta, D., & Guan, Y. (2015). Learning to reach with "sticky" or "non-sticky" mittens: A tale of developmental trajectories. *Infant Behavior and Development, 38*, 82–96. doi:10.1016/j.infbeh.2015.01.001.

Witherington, D. C. (2005). The development of prospective grasping control between 5 and 7 months: A longitudinal study. *Infancy,* 7(2), 143–161.

Zafeiriou, D. I. (2004). Primitive reflexes and postural reactions in the neurodevelopmental examination. *Pediatric Neurology, 31*(1), 1–8.

3

MULTIPLE MOTOR CHANNEL THEORY AND THE DEVELOPMENT OF SKILLED HAND MOVEMENTS IN HUMAN INFANTS

Jenni M. Karl, Lori-Ann R. Sacrey, and Ian Q. Whishaw

Introduction

Our hand movements enable us to feed, groom, make and use tools, create art, and communicate through gestural and written language. We share versions of skilled hand movements with other species of animals, including non-human primates, as well as many species of rodents, carnivores, marsupials, and amphibians. Investigation into the neural basis of skilled hand movements reveals that they depend on a network of brain areas. Just as comparative studies and anatomical studies provide insights into the neurobehavioural control of the human hand, so too do developmental studies. Hand movements are amongst the first functional movements to emerge during human development and their developmental trajectory is lengthy and variable. Further, alterations in the developmental trajectory of human hand movements can predict future sensorimotor, cognitive, and developmental disorders. Currently, there is a need to organize the diverse array of information derived from these different lines of investigation in order to better understand the neurodevelopment of skilled hand movements in humans.

In this chapter we suggest that Multiple Motor Channel (MMC) theory has the explanatory breadth needed to organize this wealth of research and provides a useful framework for understanding how skilled hand movements develop in human infants. We start by introducing MMC theory and briefly summarizing the neurobehavioural evidence upon which it is based. Next, we examine some of the evidence for MMC theory obtained from studies on the neurodevelopment of skilled hand movements in humans. Finally, we discuss how MMC theory can be used to organize and advance our understanding of the neurobehavioural development of skilled hand movements in both healthy and developmentally disabled persons.

Multiple Motor Channel Theory

MMC theory has its origins in Jeannerod's (1981, 1999) suggestion that prehension, the act of reaching out to grasp an object with the hand, consists of two separate movements. The reach movement is directed toward the extrinsic properties of the target (i.e. its location and orientation); it transports the hand to the target, is a proximally controlled movement, and is dependent upon egocentric (relative to the body) guidance. The grasp movement is directed toward the intrinsic properties of the target (i.e. its size and shape); it opens, shapes, and closes the hand in proportion to target size for target acquisition, is largely enabled by distal musculature of the forelimb, and is guided by allocentric (relative to the target) cues. In short, the reach is directed to "where" a target is and the grasp is directed to "what" the target is. Anatomical, physiological, and brain imaging studies in both human and non-human primates have provided strong support for Jeannerod's proposal that the neural basis of the reach and the grasp depend upon separable neural pathways that project from visual to motor cortex via the parietal and premotor cortices (for reviews see Andersen, Andersen, Hwang, & Hauschild, 2014; Culham & Valyear, 2006; Kaas & Stepniewska, 2016; Vesia & Crawford, 2012).

Jeannerod's idea has three weaknesses. First, it fails to consider "why". There is no functional utility in just reaching out to grasp an object, and we ordinarily would never do it. When we reach, we have an objective. The simplest "why" for reaching is the intention to place the object in our mouth. This is a primary goal of the reaching movement in many animal species and it is among the first purpose for which reaching is used by human infants. Thus, in addition to the reach and the grasp, MMC theory also includes the movement of bringing the target toward the mouth/body, which is termed "withdraw". Withdraw is an egocentrically directed movement that is dependent upon a third neural pathway from parietal to motor cortex (Kaas & Stepniewska, 2016; Sacrey & Whishaw, 2012). But, unlike the reach and the grasp, withdraw does not have a visual component. Second, Jeannerod's original idea does not account for non-visual sources of sensory guidance such as somatosensory, auditory, olfactory, and cognitive cues, which can also guide our reach, grasp, and withdraw movements. The ability to flexibly modify these movements in response to ever-changing sensory and cognitive demands is indispensable for building and maintaining the impressive repertoire of skilled movements expressed by the human hand. Third, Jeannerod's theory provides neither an evolutionary nor a developmental explanation as to *how* prehensile forelimb movements arose as separate movement components enabled by dissociable neural pathways.

MMC theory proposes that at least three separate movements – a reach, a grasp, and a withdraw – each with its own function, neural pathway, and sensory control accounts for the "where", "what", and "why" of prehension (Figure 3.1). It posits that "the reach" is enabled by circuitry in dorsal motor

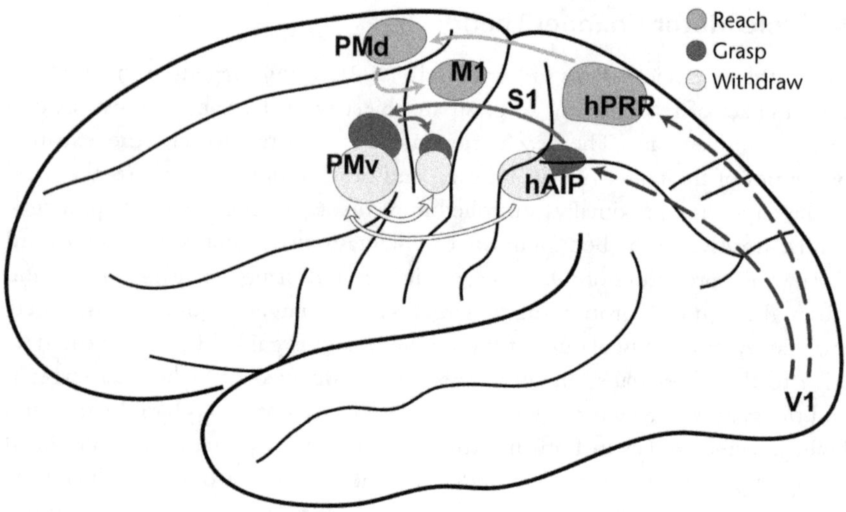

FIGURE 3.1 Proposed reach (medium gray), grasp (dark gray), and withdraw (light gray) pathways from parietal to premotor to motor cortex. Also illustrated are inputs from the cortical visual system (dotted lines)

Notes: PMd = dorsal premotor cortex, PMv = ventral premotor cortex, M1 = primary motor cortex, S1 = primary somatosensory cortex, hPRR = human parietal reach region, hAIP = human intraparietal sulcus, V1 = primary visual cortex.

cortex that originally subserved forelimb stepping movements; "the grasp" is enabled by circuitry in ventrolateral motor cortex that originally enabled manipulatory hand movements for feeding; and "the withdraw" is enabled by circuitry, also in ventrolateral motor cortex, that originally subserved hand-to-mouth behaviours involved in feeding (Karl & Whishaw 2013; Whishaw, 2003; Whishaw & Karl, 2014). MMC theory acknowledges that visual supervision of the hands was not prominent in pre-primates, and as such, proposes that each movement originated under non-visual, primarily somatosensory, control.

According to MMC theory, visual control of action originated for the purpose of executing whole body movements towards visual stimuli, like orienting the mouth towards food, and was enabled by subcortical visuomotor pathways. Throughout evolution, these subcortical visual pathways gained subsequent access to the cortical "reach" circuit, thereby enabling our ancestors to launch ballistic forelimb movements towards nearby visual targets. Still, visual input from these subcortical pathways was insufficient to enable precise online visual monitoring and correction of reach trajectory and hand shaping after movement initiation. It was not until much later in evolution that expansion of the primate visual system resulted in the co-opting of the pre-existing "reach", "grasp", and to a lesser extent "withdraw" circuits by the more sophisticated cortical visual system. This enabled primates to use online visual guidance to efficiently coordinate the reach and grasp into a singular action, in which the hand preshapes to the size of a target while reaching towards it, followed by a somatosensory-guided withdraw movement.

The major implication of MMC theory is that neurobehavioural control of the human hand is best understood by viewing it as a recent coalescence of discrete neurobehavioural systems that served different functions and were enabled by varied sensory control mechanisms throughout their different evolutionary histories. MMC theory is a comprehensive theory of hand use that has implications for understanding how the reach-to-eat movement, and skilled hand movements more generally, develop in human infants.

Neurobehavioural Control of Prehension in Human Adults

Behavioural studies have dissociated what initially appears to be a single fluid action of reaching into its three movement components; the reach, grasp, and withdraw. Whether the movements are integrated or dissociated depends on the sensory information available to guide each movement. The experimental setup is simple; a subject reaches out to take a piece of food and places it in the mouth for eating. Healthy sighted adults can seamlessly direct a hand to a target, shape the fingers to grasp it, and bring the target to the mouth for eating (Figure 3.2A).

The reach involves lifting the hand into an aim position in which the digits are closed and flexed in a "collect" posture, followed by advancing and pronating the hand so that it is juxtaposed to the target. The grasp involves preshaping

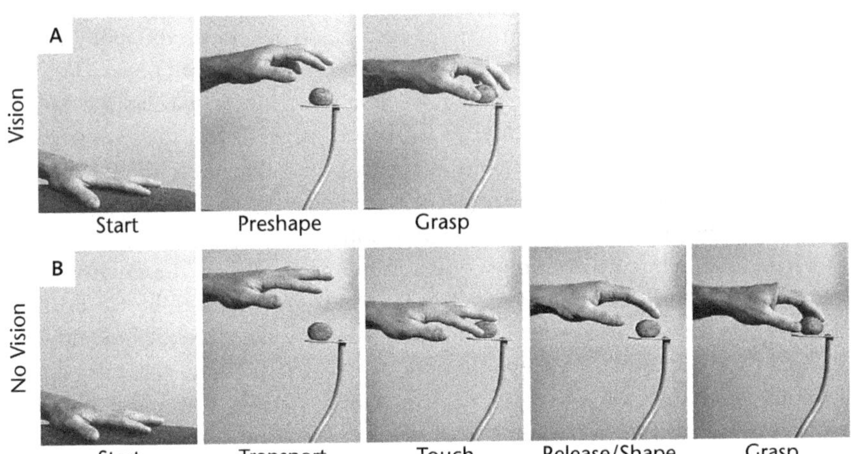

FIGURE 3.2 Reach and grasp movements are (A) integrated in sighted adults and (B) dissociated in unsighted adults. Adapted from Karl & Whishaw, 2013

Notes: Note that sighted adults integrate the reach and grasp into a single hand preshaping action whereas unsighted adults dissociate the two movements by performing them sequentially. The hand first reaches to touch the target and only then shapes to grasp it.

the digits in proportion to the size/shape of the target and closing the digits to grasp. From the point that the reach is initiated until the grasp occurs, visual focus or attention, as measured by eye tracking methods, is "anchored" on the target. When vision is degraded, as occurs when reaching into a peripheral visual field or reaching without vision, the reach and grasp movements dissociate and an open-handed reach is used to locate the target by touching it, after which, touch cues instruct the hand to shape and grasp (Figure 3.2B; Hall, Karl, Thomas, & Whishaw, 2014; Karl, Sacrey, Doan, & Whishaw, 2012a; Karl, Schneider, & Whishaw, 2013). Hand shaping in response to touch cues is remarkably similar to hand shaping guided by vision, a finding that not only demonstrates the autonomy of the grasp but also its multisensory control. The withdraw movement also takes place in two phases; a retract that brings the target toward the body and a transport that supinates the hand and brings the target to the mouth. The withdraw movement is unaffected by visual occlusion revealing that it is mainly dependent upon somatosensory cues (de Bruin, Sacrey, Brown, Doan, & Whishaw, 2008).

This rather simple experimental model of skilled reaching reveals that the main features of the reach, the grasp, and the withdraw are relatively autonomous movements that can be guided by either visual or somatosensory cues and it is within this general framework that we describe the development of reaching in infants.

The Development of Prehension in Infants

From birth to approximately 4 months of age, infants generate semi-spontaneous body, leg, arm, and hand movements sometimes referred to as "general movements" or "motor babbling". Although these movements may appear to be random and chaotic, evidence, as reported below, suggests that they are actively generated by the nervous system in order to probe the limits of the infant's motor abilities and to learn about the sensory and functional consequences of activating different muscle groups in various patterns.

These semi-spontaneous movements can be modulated by the environment (Gibson, 1988) and may be generated, at least in some instances, by the cortical motor systems that will eventually produce voluntary movements in adulthood (Kanazawa et al., 2014). At the same time, Reed (1982) argues that they could be considered subcomponents of distinct action systems that serve different evolutionarily adaptive functions such as feeding, exploring, communicating, or locomoting. From this perspective, these semi-spontaneous movements would be at least partially innate and determined by the evolutionary history of the organism. As the reach, grasp, and withdraw movements of feeding were critical to the survival of our evolutionary ancestors, MMC theory predicts that it should be possible to observe developmental precursors to functional reach, grasp, and withdraw movements among the spontaneous movements of human infants and that these precursors should be influenced by body senses to a greater extent than visual inputs. For clarity, we call these precursor movements pre-reach, pre-grasp, and pre-withdraw movements.

Pre-Withdraw Movements

Pre-withdraw movements, where the hand is withdrawn towards the face resulting in contact with the face, mouth, or inside of the mouth, are observed in human fetuses (Hepper, Shahidullah, & White, 1991; Ianniruberto & Tajani, 1981; Sparling, Van Tol, & Chescheir, 1999; de Vries, Visser, & Prechtl, 1982, 1985). By 12 weeks gestational age, human fetuses generate hand-to-face contacts and by 18 weeks they will regularly insert a thumb inside the mouth. Across this developmental age, the movements become increasingly accurate at reaching towards the mouth (Reissland, Francis, Aydin, Mason, & Schaal, 2014).

Once born, non-feeding neonates spend up to 20% of their waking hours with their hands in contact with the oral region of the face (Kessen, Williams, & Williams, 1961; Korner, Chuck, & Dontchos, 1968). Piaget (1952) described these neonatal hand-to-mouth contacts as fortuitous and brief, but subsequent studies by Butterworth and Hopkins (1988), as well as Rochat and colleagues (Rochat & Hespos, 1997), show that the tactile self-stimulation that occurs when a newborn brings their hand into contact with their face does not elicit rooting reflexes to the same extent as when an external object contacts the same facial location suggesting that they have some intrinsic function. Furthermore, newborns often bring the hand directly to the mouth during ongoing variations in mouth location due to turning of the head and in doing so open the mouth in anticipation of hand-to-mouth contact. Thus, from very soon after birth, infants can generate hand-to-face and hand-to-mouth movements that are flexible and they can discriminate the resulting self-generated somatosensory stimulation from external stimulation.

Studies by Rochat and colleagues (Blass, Fillion, Rochat, Hoffmeyer, & Metzger, 1989; Rochat, Blass, & Hoffmeyer, 1988; Rochat & Senders, 1991) further reveal that hand-to-mouth movements in 1- to 3-day-old infants are closely related to feeding. They find that administration of a sucrose solution into the mouth of neonates immediately elicits mouthing behaviour that is more frequent, intense, and prolonged than the mouthing movements elicited in response to water. Importantly, the sucrose-induced mouthing movements are frequently followed by withdrawal of the infant's hand into the mouth, whereas water-induced mouthing movements are less so. They report that the hand moves towards the face and when perioral contact is established, movement of the hand slows. The hand will then either move toward the mouth, which often opens before actual contact with the hand, or the hand will remain still and the head will turn to allow the mouth to apprehend the hand. Placement of a pacifier into the mouth of the infant prevents hand-to-mouth movements in response to sucrose administration. Together, these findings demonstrate that pre-withdraw movements are not automatic reflexes but are intentional. They are likely one of a variety of ways, along with placement of a soft object such as the mother's breast or a pacifier into the mouth for sucking, that infants obtain feeding-related satisfaction.

That the hand-to-mouth movements of 1- to 3-day-old infants are specifically elicited in response to feeding-related mouth movements is significant as it is not until the second month of life that infants will begin to voluntarily bring an object that has been placed into their hand by an experimenter to the mouth for oral exploration (Rochat, 1987, 1989; Rochat & Senders, 1991). Thus, as predicted by MMC theory, the earliest pre-withdraw movements displayed by human infants are selectively activated in response to feeding-related mouth movements and can be flexibly modified in response to somatosensory feedback from the mouth. As such, we suggest that the hand-to-mouth movements of very young infants are a developmental precursor to the withdraw movements of older infants, which first facilitate oral exploration of retrieved objects and then serve to bring food to the mouth for eating.

Pre-Reach Movements

Before young infants successfully reach out and grasp a visual target with the hand, which they do not do consistently until 6 months of age, they display a variety of pre-reach movements. These pre-reach movements include extending the arm and hand to establish self-contact with their own body and extending the arm and hand in order to touch distal targets.

Awake newborn infants generate a plethora of arm and hand movements that result in contact with their own body (Lobo & Galloway, 2013; Thomas, Karl, & Whishaw, 2015). Early self-touching movements may serve as developmental precursors for mature reach movements during which the infant learns about the size, shape, and organization of their own body and limbs via somatosensory feedback which helps to refine their own body representation (Granmo, Petersson, & Schouenborg, 2008; Yamada, Fujii, & Kuniyoshi, 2013). In other words, self-touching movements appear to help the infant to learn about the sensory and functional consequences of moving their own limbs. This idea is supported by Rapoport (1972), who describes an infant that lacked somatosensory feedback from oral and body surfaces and also displayed significant delays in motor development.

An ethological description of self-touching movements over the first 5 months of life suggests that these self-touching movements undergo significant developmental change (Thomas et al., 2015). Initially, arm movements are generated largely by the shoulder with the elbow stiffly bent and the digits closed and fisted. Self-contacts are initially most prominent between the dorsum of the fisted hand and either the face, head, or upper body (Figure 3.3A). As infants

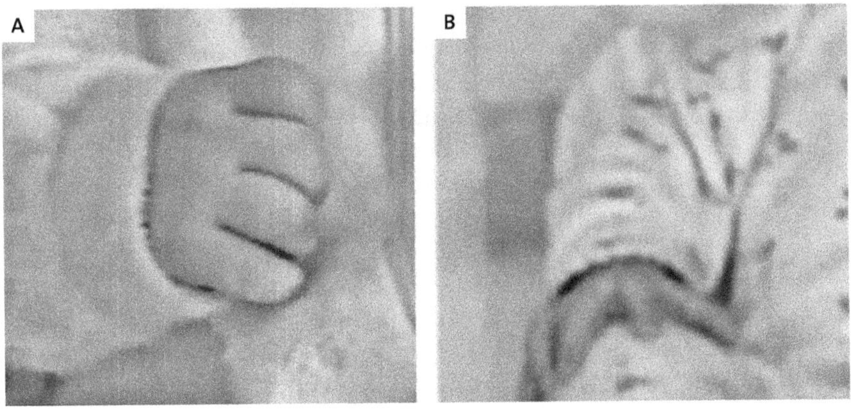

FIGURE 3.3 Differences in digit configurations during self-touching movements in (A) a 1-month-old infant and (B) a 3-month-old infant

age, the movements come to include movements of the trunk, all forelimb joints, and the digits. Consequently, from 2 months of age onwards infants increasingly make palmar and fingertip contacts with the lower torso, legs, and eventually feet (Figure 3.3B). At the same time, forelimb movements become more coordinated with other body movements resulting in frequent self-contacts with the moving legs and feet. Eventually, hand contacts with the lower body begin to feature exploratory movements in which the hand caresses the body part that it contacts. The self-touching movements of 4- to 5-month-old infants are reminiscent of the functional reach movements of unsighted human adults, which feature extension of the arm and opening of the hand to make palmar contact with a distal target. In both cases, if successful contact with a target is not immediate, the hand often palpitates and caresses the contacted surface until it establishes tactile contact with the desired target. As such, we suggest that the forelimb extension, palmar self-contact, and caressing movements observed in 4- to 5-month-old infants serve as a developmental precursor to non-visually and eventually visually guided reach movements in adults.

The proposition that pre-reach movements towards the body are precursors to functional non-visual and visual reach movements in adulthood is further supported by Pellijeff and colleagues (2006). They show that when adults reach toward their own hand, positioned near their own torso, fMRI activity increases in the anterior precuneus and medial intraparietal sulcus of the superior parietal lobe. These same areas are activated when adults perform both non-visually and visually guided reaching movements toward distal objects (Filimon, Nelson, Huang, & Sereno, 2009). Thus, we suggest that when infants use their arm and open hand to contact their own body and torso they are displaying a developmental precursor, in both neural and behavioural terms, to the reach movements that healthy adults direct towards both bodily and non-bodily targets.

By 3–4 months of age, infants also begin to reach out to touch, but not grasp, distal objects (Clifton, Muir, Ashmead, & Clarkson, 1993; Gonçalves et al., 2013; von Hofsten, 1984; Thelen et al., 1993). Because infants of this age generate so many semi-spontaneous arm movements, it is possible that initial contact with distal targets is incidental rather than intentional. In the early stage of their development these reach-to-touch movements are stiff, display limited arm control, and involve multiple corrections in arm trajectory (Berthier, Clifton, McCall & Robin, 1999; Berthier & Keen, 2006; Spencer & Thelen, 2000; Thelen et al., 1993).

It was initially thought that infants used vision to compare the distance between their hand and the target in order to correct errors in reach trajectory and to ultimately bring the hand to the location of the target. More recent research indicates, however, that infants do not need to see their own hand in order to bring it into contact with a distal target. Thelen et al. (1993) find that errors in reach trajectory result from the infant's limited ability to control the speed and direction of their arm rather than as a consequence of using visual feedback to adjust the location of their hand in space. These findings were confirmed by Lee and Newell (2012), who find that infants' reach movements are similar whether or not they have visual feedback of their hand. Meanwhile, Corbetta, Thurman, Wiener, Guan, and Williams (2014) find that young infants learn to direct their visual attention to the location of a hand after it contacts a target instead of learning to direct their hand to where they are looking. Together, this work suggests that young infants may or may not use vision to identify the location of a target in space, but use proprioception to direct the hand toward the target and tactile feedback to determine whether or not they made contact with the target. Through this successful contact, infants learn to constrain and refine their reach movements to distal targets through repeated practice and reinforcement learning (Williams & Corbetta, 2016; Williams, Corbetta, & Guan, 2015).

A number of recent studies have confirmed that haptic contact with the target is important for reinforcing and refining pre-reach movements, as well as for increasing the likelihood that infants will repeat these reach movements in the future. Work by Libertus and Needham (2010) suggests that the use of "sticky mittens", which simulate grasping of a touched object in infants at this stage of development, results in increased reach attempts after a 2-week training period. Williams et al. (2015) extend this work by including an additional group of infants that wore "sticky mittens" that had openings at the tips of the fingers so that infants could still make direct tactile contact with the target at the end of their reach. Their results reveal that infants that made direct finger contact with the target showed greater increases and more improvement in pre-reach behaviour by the end of the experiment compared to the group that did not (Corbetta, Williams, & Haynes, 2016). These results support the view that direct tactile feedback is an important determining factor in the selection and refinement of early pre-reach movements (Schlesinger, Parisi, & Langer, 2000).

Taken together, these studies on the development of pre-reach movements seem to suggest that: (1) early self-touching movements may serve as a developmental precursor for reaching to distal targets during which the infant learns about the size, shape, and organization of their own body and limbs; (2) initially spontaneous arm movements result in contact with either the body or a distal object and this haptic contact provides feedback to the infant about the sensory and functional consequences of moving their forelimbs, which in turn increases infant motivation to continue performing reach-to-touch movements; and (3) with repeated practice, the arm movements that result in successful contact with the target are gradually reinforced, refined, and more likely to be executed in the future. Thus, as predicted by MMC theory, early pre-reach movements, and the neural circuitry that controls them, are gradually selected and refined based on somatosensory feedback from the infant's own body and hand long before feedforward visual control of the reach movement emerges.

Pre-Grasp Movements

Young infants produce a variety of pre-grasp movements before they reliably preshape the hand and digits to the size and shape of a visual target. These pre-grasp movements include hand babbling, self-grasping, and the oral and manual exploration of grasped objects that are brought to the mouth. Initially, infants do not look at their hands when performing these pre-grasp movements, suggesting that the movements originate under tactile and proprioceptive control.

A newborn infant often clench the resting hand and digits in a fist, but will reflexively close the hand if something touches the palm of the hand. This resting posture becomes more open and relaxed by about one month of age, transitioning to a posture called "collection" (Sacrey & Whishaw, 2010). Once infants reach 2 months of age, they start "hand babbling" (Wallace & Whishaw, 2003). They generate a variety of hand and digit movements, including extending and flexing individual digits, performing sequential digit movements, and pressing their fingertips together to form pre-precision and pincer grips (Figure 3.4). Early hand babbling generally consists of fisted configurations between 1 and 2 months, but this transitions to pre-precision configurations (the thumb touches the side of the index or middle finger) between 2 and 3 months, and finally pincer configurations (the tip of the thumb touches the tip of the index or middle finger) between 4 and 5 months. At 5 months of age, hand babbling becomes self-grasping movements, in which the infant grasps and manipulates their own body or clothes (Figure 3.5; Thomas et al., 2015). Self-grasping

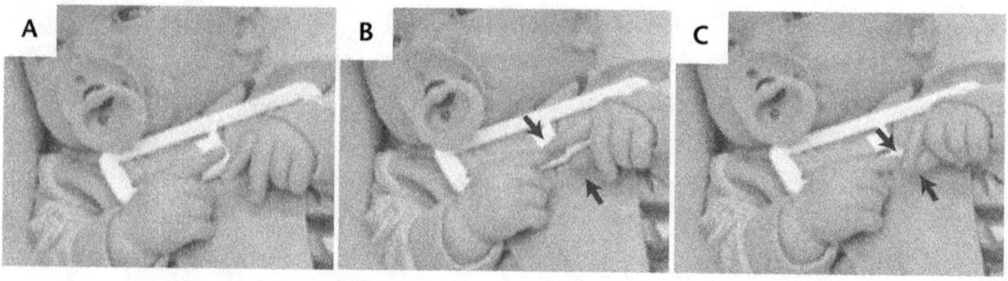

FIGURE 3.4 A 2-month-old infant performs a pincer grip with the left hand during spontaneous hand babbling

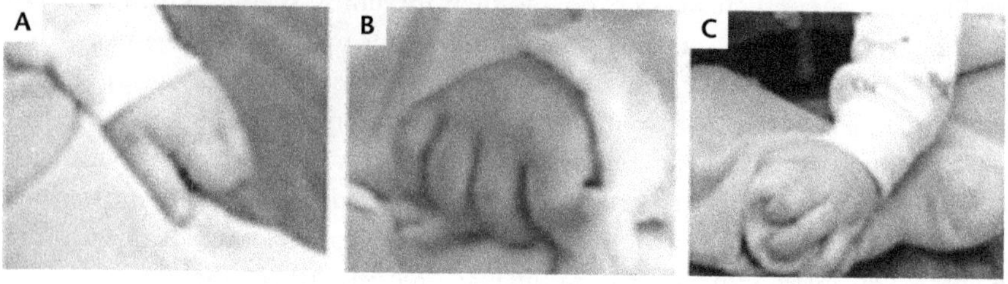

FIGURE 3.5 Typical self-grasping movements in (A) and (B) 5-month-old human infants and (C) a 6-month-old human infant

movements include pre-precision grasps, involving only a few digits, and whole hand grasps, involving all digits, followed by a range of manipulatory hand movements. Thus, long before the infant nervous system can use vision to shape the hand to match the configuration of a distal target, it practises forming different grip configurations on the infant's own body under the auspices of proprioception and touch.

The contributions of tactile feedback to the formation of early grasp patterns is supported by the observation that 2-month-old infants are also able to modify their grasping pattern to reflect the haptic features of an object that is placed into their hand by an experimenter (Case-Smith, Bigsby, & Clutter, 1998). It is at this age that infants will also begin to withdraw such an item towards the mouth and engage in perioral exploration (Rochat, 1987, 1989). Importantly, the hands grasp the object and support it so that the mouth can engage with it. Desmurget et al. (2014) recently described specific sites in human motor cortex that process somatosensory inputs from the mouth while also generating forelimb motor actions. Depending on how early these cortical areas develop, they could contribute to the development of early grasping movements by processing haptic feedback from the mouth which is then used to refine grasp configurations made by the hand. It is not until approximately 4 months of age that infants will visually inspect a grasped object before bringing it to the mouth for perioral exploration. Thus, as predicted by MMC theory, from soon after birth, the infant nervous system actively generates a large repertoire of grasp configurations that are almost entirely refined and reinforced through haptic contact, either with their own body or mouth.

Summary

To summarize, young infants generate a wide variety of semi-spontaneous body, arm, and hand movements that likely reflect attempts by the nervous system to actively explore the sensorimotor consequences of generating different patterns of muscle activity. The presence of discrete pre-withdraw, pre-reach, and pre-grasp movements within these spontaneous movements suggests that discrete muscle synergies for withdraw, reach, and grasp movements are at least partially innate and subject to differential refinement via somatosensory feedback from the infant's own body, mouth, and environment throughout development. The initial expression of these precursor movements are likely enabled by the activation of the discrete withdraw, reach, and grasp areas in motor cortex in relative isolation. Different types of somatosensory feedback – contact with a distal object for the reach, apprehension of an object in the hand for the grasp, and placement of an object in the mouth for the withdraw – help to refine initially crude precursor muscle synergies into increasingly precise muscle synergies for each movement component separately.

Haptic Guidance of Separate Reach, Grasp, and Withdraw Movements

Between 4 and 6 months of age, infants begin to organize the reach, grasp, and withdraw movements into a useful behavioural sequence. That is, to first reach, then grasp, then withdraw the hand in order to bring a grasped target to the mouth. Yet, the structural organization of the reach-to-eat movement continues to differ substantially from that of healthy sighted adults. According to MMC theory, this is due largely to differences in sensory control: developing infants continue to rely heavily on somatosensory feedback to signal successful completion of each movement whereas healthy adults are more reliant on feed-forward visual control, which allows them to integrate their reach and grasp movements into a single cohesive action followed by withdraw of the hand to the mouth.

To test this hypothesis we have compared the reach and grasp movements of 4- to 24-month-old infants to that of sighted and unsighted adults. Our previous work had shown that healthy adults flexibly restructure their reach and grasp movements depending on the availability of visual feedback (Hall, Karl, Thomas, & Whishaw, 2014; Karl et al., 2012a, 2013; Karl, Sacrey, Doan, & Whishaw 2012b). When vision is available, adults organize the movement in a temporally efficient manner: the hand lifts, collects, forms a peak aperture, orients to the target, and begins to close before touching the target.

Consequently, the target is securely grasped almost immediately after contact. In contrast, in unsighted adults, the hand lifts and forms a large peak aperture, contacts the target, and only then orients and closes to grasp. Thus, unsighted adults spend a significant amount of time using haptic feedback to adjust the orientation and aperture of the hand after target contact until they are able to achieve a secure grasp. Our aim was to determine whether the early reach and grasp movements of infants more closely resemble that of unsighted adults, which would suggest that they rely more heavily on somatosensory guidance, or that of sighted adults, which would suggest that the movements have come under visual guidance.

We found that the reach and grasp movements of 4- to 8-month-old infants more closely resemble the dissociated and sequential reach-then-grasp movements of unsighted adults (Figure 3.6; Karl & Whishaw, 2014). The reach and grasp movements of 4- to 8-month-old infants differ from those of sighted adults in a number of ways. First, the adult reach contains one movement unit, which consists of a single acceleration phase and one deceleration phase. This produces a single smooth reach movement. The infant reach movement consists of multiple movement units resulting in a fractionated reach trajectory

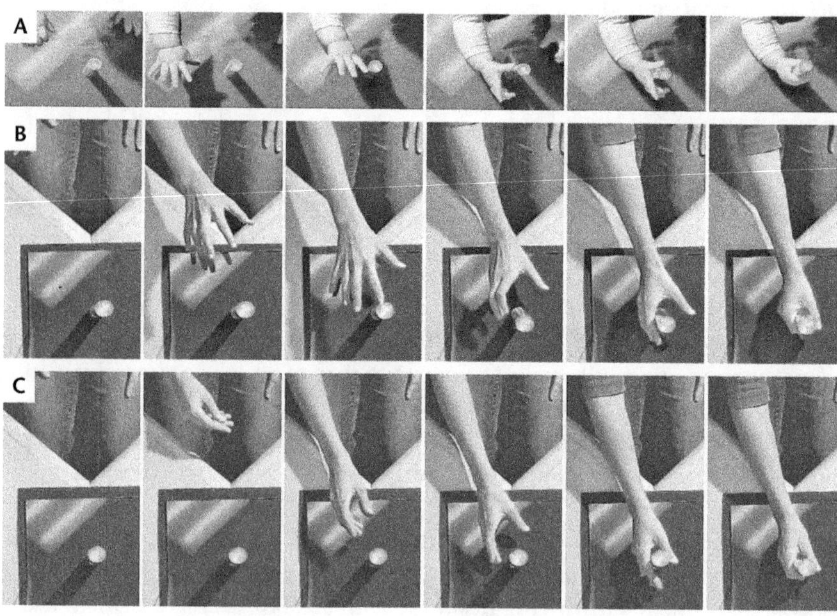

FIGURE 3.6 The reach and grasp movements of (A) a 7-month-old infant, (B) an unsighted adult, and (C) a sighted adult. Adapted from Karl & Whishaw, 2014

Note: Note the similar sequential organization of the reach and grasp movements in the 7-month-old infant and unsighted adult.

(Sacrey, Karl, & Whishaw, 2012a; von Hofsten, 1991). Second, young infants don't pre-emptively orient the hand before they touch the target. Instead, they tend to make imprecise digit-to-target contact (Karl & Whishaw, 2014; McCarty, Clifton, Ashmead, Lee, & Goubet, 2001; Schum, Jovanovic, & Schwarzer, 2011). Third, young infants do not close the digits to the size and shape of the target before contacting it. Rather, they reach out with an open hand, touch the target, and then use tactile cues shape and close the hand on the target (Karl & Whishaw, 2014; Witherington, 2005). Thus, even though 4- to 8-month-old infants can reach and grasp for objects in their environment, they do not display the anticipatory hand shaping characteristic of adult reach-to-grasp movements. Instead, they dissociate their reach and grasp movements, similar to unsighted adults, suggesting that they continue to rely heavily on tactile and proprioceptive cues to guide their movements (Karl & Whishaw, 2014).

In addition to dissociating the reach and the grasp, young infants and unsighted adults use a number of hand adjustments associated with tactile contact to successfully reach and grasp distal targets. First, they tend to contact the target, a vertical rod, with the tips of their digits instead of the palm of their hand. Second, they often make extra contacts, releasing and then re-establishing target contact before they grasp it. Third, the release after contact is often associated with re-positioning of the hand so that a more appropriate target contact can be made. Fourth, young infants delay the onset of grasp closure until they have established contact with the target using an appropriate part of the hand. These results indicate that 4- to 8-month-old infants likely use vision to localize the target in space, but rely heavily on somatosensory cues from the arm and hand to complete separate reach and then grasp movements. Subsequent withdraw of the hand back to the mouth may initially be supervised by vision, as infants at this age often look at objects that they are bringing to the mouth (Sacrey, Karl, & Whishaw, 2012b), but true visual guidance of the withdraw movement is highly unlikely due to the infant's inability to see their own mouth. Nonetheless, future research directed towards this question will be helpful.

In summary, 4- to 8-month-old infants continue to rely on somatosensory feedback to guide separate reach, grasp, and withdraw movements. This early dependence on haptic, rather than visual, inputs may enable the infant's success in acquiring distal targets despite immature postural control, incomplete maturation of the dorsal visual stream, and significant kinematic inconsistencies within individual reach and grasp components. In addition, behavioural dissociation of the reach, grasp, and withdraw under haptic control likely facilitates further refinement of discrete reach, grasp, and withdraw areas in parietal, premotor, and motor cortex through differential somatosensory consequences. This would help to prevent overly tight coupling of the three movements, in both neural and behavioural terms, during early development, which would have the unfortunate consequence of reducing the adaptability and functionality of the human hand in adulthood.

Visual Harnessing of Separate Reach and Grasp Movements

From 8 months of age onwards, infants must learn to use feedforward visual control to integrate the reach and grasp into a single movement whereby the hand opens, shapes, and closes before target contact and then withdraw the hand to the mouth using somatosensory guidance. But how exactly does visual takeover of the separate reach and grasp components occur? MMC theory predicts that this developmental transition is facilitated by the infant's ongoing practice of reach-to-eat behaviour coupled with increasing visual supervision of their own movements. In doing so, the infant's nervous system begins to learn which types of sensory information are most useful for guiding each movement component across a variety of environmental contexts. This in turn, facilitates the strengthening of corticocortical connections between the cortical visual system in occipitoparietal cortex and the premotor and motor areas that produce separate reach and grasp movements.

As it turns out, young infants do display what appear to be some visually elicited movements from very early in development. For example, from soon after birth, infants can orient the eyes and head to a visual target (Greenman, 1963; von Hofsten and Rosander 1997; Kremenitzer, Vaughan, Kurtzberg, & Dowling, 1979), "reach" for an object with the mouth by thrusting the head forward and flexing the abdominals (Foroud & Whishaw, 2012), and, if properly supported, will extend a single arm and sometimes contact moving objects in their field of view (von Hofsten, 1984). These observations appear to contradict the postulate of MMC theory that reach, grasp, and withdraw movements undergo substantial refinement under the auspices of proprioception and somatosensation before they come under visual control. MMC theory predicts, however, that the earliest function of vertebrate vision is to elicit whole body movements towards visual targets, and in particular, to orient the mouth towards food as a fish might capture a prey item. These visual orienting movements are proposed to be mediated by subcortical visual systems involving the tectum or superior colliculus. MMC theory further proposes that these subcortical visual pathways eventually gained access to the cortical motor circuit that generates stepping ("reach") movements, long before the expansion of the cortical visual system enabled true feedforward visual control of reach and grasp movements in primates. Thus, we might expect that one of the very first visually elicited movements in human development should involve orienting the mouth to a visual stimulus that can potentially be eaten. We would also expect that from very early in development, young infants should be able to use subcortical visual pathways to produce ballistic reach-type movements to salient visual stimuli in their environment. Furthermore, the ability to perform reach and grasp movements towards visual targets guided by the cortical geniculostriate system should require a prolonged developmental period and visual guidance of the two movements may not necessarily develop at the same rate.

In relation to the first prediction, a recent study by Futagi and colleagues (2016) confirms that hungry infants as young as 2 days of age open the mouth in response to the visual approach of a pointed fingertip. That this mouth opening behaviour is selectively elicited in hungry infants and in response to a pointed fingertip (which resembles a nipple), but not an open palm, suggests that it is closely related to feeding. By 3 months of age, this behaviour can no longer be elicited, but at the same time, if a visible target is presented within reaching distance, infants will attempt to capture the object with the mouth by thrusting the head forward, flexing the abdominals, and generating active jaw, lip, and tongue movements (Figure 3.7; Foroud & Whishaw, 2012; von Hofsten, 1984). Infants will continue to demonstrate this visually elicited "mouth capture" behaviour even after they begin using their arms to reach for visual targets (between 4 and 6 months of age) if their arms are restricted (Rochat & Senders, 1991). Thus, as predicted by MMC theory, orientation of the mouth towards external objects does appear to be one of the very first visually elicited actions to develop during infancy.

In relation to the second prediction, that subcortical visual inputs should be capable of eliciting ballistic arm movements to salient visual stimuli in very young infants, von Hofsten (1982) has shown that when the head, neck, and body are properly supported, young infants are capable of extending a single hand to swipe at visual targets (Figure 3.8), however, their aim is quite poor, they hardly ever contact the target, and they never actually grasp it. We propose that these ballistic pre-reach movements are enabled, not by the cortical visual inputs that produce precise visually guided reach movements in healthy adults, but by earlier developing subcortical visual inputs to the parietofrontal reach circuit.

In relation to the third prediction, that cortical geniculostriatal visual guidance of the reach and grasp will develop slowly and at different rates; recent work has revealed that over the first year of life, infants show significant

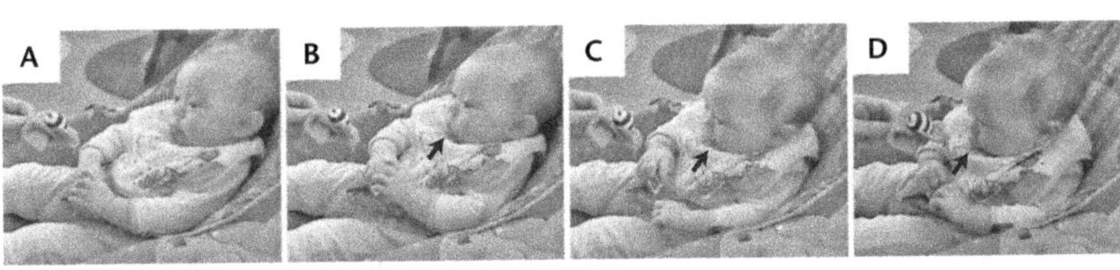

FIGURE 3.7 Visually elicited mouth movements in a 2-month-old infant involve (A) visual fixation on the target, (B) mouth opening, (C) flexion of the abdominals, and (D) puckering of the lips and mouth . Adapted from Karl & Whishaw, 2013

FIGURE 3.8 When properly supported, a young infant performs a pre-reach movement by extending a single fisted hand towards a visual target

improvements in transporting and orienting the hand to the target prior to contact (Corbetta, Guan, & Williams, 2012; von Hofsten & Fazel-Zandy, 1984; Karl & Whishaw, 2014; Schum et al., 2011); by 11–12 months consistently contact the target with an appropriate part of the hand (Karl & Whishaw, 2014); and by 12 months the jerkiness of the reach is greatly reduced due to a gradual decrease in the number of movement units over time (von Hofsten, 1991; Sacrey et al., 2012b; Thelen, Corbetta, & Spencer, 1996). Thus, both visual guidance and motor control of the reach appears to be approaching maturity by 12 months of age.

Despite this, 1-year-old infants show only partial improvement in using visual guidance to prepare the hand for grasping prior to target contact. Prior to 1 year of age, infants use vision to select an appropriate grip configuration. They will, for example, use a pincer grip to grasp a cheerio versus a palmar grip to grasp a plastic rod (Butterworth, Verweij, & Hopkins, 1997; Sacrey et al., 2012b). They will also begin to close their hand on approach to the target. Nonetheless, the extent of hand closure is greatly reduced compared to that of adults and the majority of grasp closure continues to occur after contact (Figure 3.9; von Hofsten & Rönnqvist, 1988; Karl & Whishaw, 2014). Thus, mature visual guidance of the grasp continues to develop beyond 2 years of age.

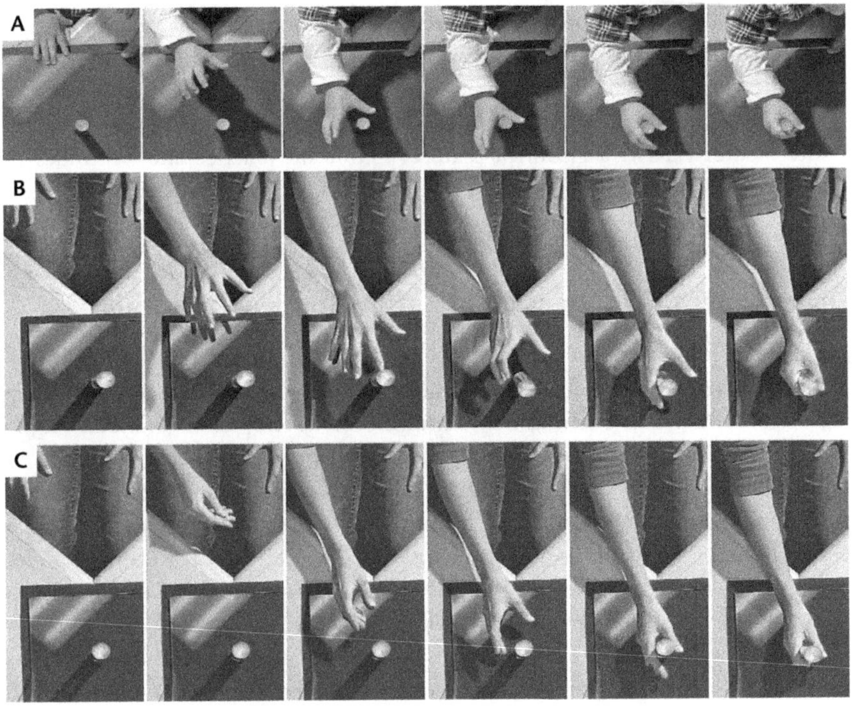

FIGURE 3.9 The reach and grasp movements of (A) a 24-month-old infant, (B) an unsighted adult, and (C) a sighted adult. Adapted from Karl & Whishaw, 2014

Note: Note the reach and grasp movements of the 24-month-old infant more closely resemble that of the sighted adult, with the exception that the 24-month-old infant delays the majority of grasp closure until after target contact.

In summary, as predicted by MMC theory, newborn infants do display a number of visually elicited behaviours, which include orienting the mouth to visual targets or swiping a single hand at a visible target. These movements appear to be ballistic, are likely mediated by earlier developing subcortical visual pathways, and largely dissipate between the second and third month of life. By 6 months of age, infants begin to reach, grasp, and withdraw target objects to the mouth, but each movement component remains highly reliant on somatosensory cues. It is not until 9 months of age that infants begin to display effective feedforward visual control of functional reach movements by orienting the hand to match the orientation of a visual target prior to touching it. Although feedforward visual control of the reach appears largely functional by this age, precise feedforward visual control of the grasp is still attenuated compared to adults, indicating that development of feedforward visual control of the reach precedes that of the grasp. Differences in the rate of development of visual guidance for the reach and grasp is proposed to be related to the ongoing maturation of cortical visual inputs to the separate reach and grasp movements in motor cortex from the dorsal visual stream (Loenneker et al., 2010). Additional studies in 10-year-old children confirm that adult-like integration of the reach and grasp under feedforward visual control requires significant practice and continues to develop over the first decade of life (Kuhtz-Buschbeck, Stolze, Jöhnk, Boczek-Funcke, & Illert, 1998).

Development of Prehension in Autism Spectrum Disorder

It has been suggested that a lack of variance in early spontaneous movements along with overly tight coupling of different movement components may be indicative of abnormal neural and sensorimotor development (Hadders-Algra, 2007; Piek, 2002; Touwen, 1995). There is debate as to whether this lack of movement variability and flexibility is due to abnormal motor development, abnormal sensory development, or both. MMC theory provides a possible explanation. That is, abnormal somatosensory inputs to the motor cortices during early development could impede the proper reinforcement and refinement of separate reach, grasp, and withdraw movements. Subsequent maturation of cortical visual inputs to abnormally developing reach and grasp regions of motor cortex would further impede one's ability to adaptively coordinate the movements under feedforward visual control. If many higher level perceptual, cognitive, and social abilities are enabled by the same neural substrates that initially evolved to subserve skilled motor behaviour, then abnormal separation, refinement, and coordination of these three movement components would also result in abnormalities in higher-level behaviour.

As an example, Autism Spectrum Disorder (ASD) is one of the most prevalent forms of developmental disability (Christensen et al., 2016). It is characterized by impairments in social-communication and the presence of repetitive or restricted behaviours (American Psychiatric Association, 2013). A growing literature suggests that early sensory and motor differences form a "prodrome" of ASD that manifests in the first year of life (Baranek, 1999; Kaur, Srinivasan, & Bhat, 2015; Libertus, Sheperd, Ross, & Landa, 2014; Trevarthen & Delafield-Butt, 2013). These include impairments in basic motor control (Teitelbaum, Teitelbaum, Nye, Fryman, & Maurer, 1998), difficulty performing skilled motor gestures (Mostofky et al., 2006), and disturbances in coordinating reach and grasp movements (Mari, Castiello, Marks, Marraffa, & Prior, 2003; Noterdaeme, Mildenberger, Minow, & Amorosa, 2002). At the neural level, ASD is characterized by abnormal structural as well as functional brain connectivity with localized brain areas displaying hyperconnectivity and more distal brain areas displaying hypoconnectivity (e.g., Kana, Uddin, Kenet, Chugani, & Müller, 2014).

Interestingly, early behavioural work revealed that infants as young as 12 months of age who are later diagnosed with ASD display abnormal oral and manual exploration of objects (Palmer, 1989; Ruff, 1984). More recent studies have identified similar abnormalities beginning even earlier at 6 months of age (Kaur et al., 2015; Koterba, Leezenbaum, & Iverson, 2012). In addition, at-risk infants show a prolonged visual fixation on objects that continues into early childhood (Zwaigenbaum et al., 2005). For example, Sacrey, Bryson, & Zwaigenbaum (2013) filmed at-risk and low-risk infants between 6 and 36 months of age as they played with toys and recorded the amount of time they spent looking at the toy both before and after it was grasped. Group differences began to emerge at 12 months of age, with high-risk infants continuing to look at the toy after it was grasped. These group differences continued until 24 months of age. Together, these studies suggest that early impairments in somatosensory processing that normally serve to reinforce and refine reach, grasp, and withdraw movements during object exploration could lead to a higher incidence of visual supervision of the hands in infants at-risk of developing ASD.

Yet, there is also evidence that children with ASD have difficulty using vision to guide their arm and hand movements. Masterton and Biederman (1983) trained children with ASD, intellectual disability (ID), and control children to place a wooden block onto a target while wearing prism lenses that displaced vision of their environment. The ASD and ID groups took almost twice as long to adapt to the prisms compared to the control children. Interestingly, adjustments made using the reaching hand were transferred to the non-reaching hand, but only for the ASD group. The authors suggest that this slower learning combined with transfer to the non-reaching hand is evidence that children with ASD rely on proprioceptive rather than visual feedback to complete the task. Similarly, when completing a grasp and place-in-a-hole task, children with ASD take longer to complete the movement, have higher terminal velocities, and make corrections *at least once* after entering the area of the hole compared to controls. Comparatively, *fewer than half* the controls made such corrections (Forti et al., 2011). Together these results suggest that older ASD children have difficulty using visual guidance to control their reach movements so they revert to relying on proprioceptive and tactile cues to assist in guiding them.

Abnormalities in the somatosensory and visual guidance of discrete reach, grasp, and withdraw movements could lead to abnormalities in more complex movements built up from these simpler movement components. For example, Mari et al. (2003) found that low-functioning ASD children are not able to execute the reach and grasp components in a simultaneous manner. Rather, low-functioning ASD children separate the reach and the grasp movement by only beginning to open the digits for grasping at 72% of movement duration. In contrast, the high-functioning ASD children demonstrated a closely integrated and overlapping movement, with the onset of grasping occurring at 15% of movement duration. In addition, children with ASD show irregular muscle activity associated with mouth opening during eating tasks. Cattaneo et al. (2007) compared electromyography (EMG) recorded muscle activity involving the mouth during an eating task in children with ASD and age-matched controls. In the control group, EMG activity began before the hand grasped the object, whereas EMG activity did not begin until much later in the ASD group, when the hand was withdrawing the food to the mouth. Together, these results suggest that children with ASD are impaired in using both online visual and somatosensory guidance to efficiently coordinate reach, grasp, and withdraw movements into more complex reach-to-eat actions.

When one considers that a primary feature of ASD is an abnormality in brain connectivity, one can begin to see how MMC theory could provide insights into the neurobehavioural basis of ASD. First, MMC theory would predict that early impairments in oral and manual interactions with objects likely result from abnormal somatosensory inputs into the reach, grasp, and withdraw circuits in the motor cortices. Support for this idea is provided by a recent diffusion tensor imaging study by Thompson et al. (2017), which reveals hypoconnectivity between the hand areas of primary somatosensory cortex and primary motor cortex in individuals with ASD. This reduction in somatosensory input to primary motor cortex would likely contribute to abnormal refinement and reinforcement of separate reach, grasp, and withdraw movements during early development. Second, MMC theory would predict that impairments in coordinating reach, grasp, and withdraw movements into more complex actions would result from an inability of cortical visual areas to coordinate the separate reach and grasp movements into a singular hand preshaping action. Again, support for this prediction is derived from a number of neuroimaging studies that have found reduced structural and functional connectivity within the dorsal visual stream in individuals with ASD (e.g. Villalobos, Mizuno, Dahl, Kemmotsu, & Müller, 2005). Third, MMC theory would predict that because individuals with ASD have difficulties using both somatosensory and visual information to predict and execute their own hand movements, they would also have difficulties using somatosensory and visual information to predict and interpret the hand movements of others, a view espoused by the mirror neuron hypothesis of ASD (Saffin & Tohid, 2016).

Conclusion

We have noted throughout this review that we have some understanding of the neural circuitry that supports reach-to-eat behaviour in healthy human adults. Nevertheless, from a developmental perspective, the neural underpinnings of the reach-to-eat movement must change as the behaviour itself changes during development. MMC theory provides a useful way to organize the extensive behavioural findings on the development of hand movements in human infants, and as such can provide insight into the underlying changes in neural organization that likely accompany and enable these behavioural changes from infancy to adulthood. Future work should continue to examine the many subcortical and cortical changes that support this development. We also note that primate reaching is special in its visual guidance and future work could examine the relationship between the development of visually guided hand movements in humans relative to the development of specialized primate neuroanatomy, such as the fovea and its visual projections as well as the direct corticospinal projections from the arm and hand areas of motor cortex to the spinal cord. We suggest that the development of the human reach and grasp is closely related to the maturation of the extended visual pathway from the retina through visual cortex to motor cortex and ultimately from motor cortex through the corticospinal tracts to motor neurons. Finally, we began our description of the three components of prehension – the reach, grasp, and withdraw – by observing that each serves a different behavioural function. The reach represents "where" an object is, the grasp represents "what" an object is, and the withdraw represents "why" the reach is being performed. Future work can examine whether these cognitive correlates of prehension also shape our cognitive functions more generally.

References

American Psychiatric Association. (2013). *Diagnostic and statistical manual, 5th edition (DSM-5)*. Washington, DC: American Psychiatric Association.

Andersen, R.A., Andersen, K.N., Hwang, E.J., & Hauschild, M. (2014). Optic ataxia: from Balint's syndrome to the parietal reach region. *Neuron, 81*, 967–983.

Baranek, G.T. (1999). Autism during infancy: a retrospective video analysis of sensory-motor and social behaviours at 9–12 months of age. *J Autism Dev Disord, 29*, 213–224.

Berthier, N.E., Clifton, R.K., McCall, D.D., & Robin, D.J. (1999). Proximodistal structure of early reaching in human infants. *Exp Brain Res, 127*, 259–269.

Berthier, N.E., & Keen, R. (2006). Development of reaching in infancy. *Exp Brain Res, 169*, 507–518.

Blass, E.M., Fillion, T.J., Rochat, P., Hoffmeyer, L.B., & Metzger, M.A. (1989). Sensorimotor and motivational determinants of hand-mouth coordination in 1–3 day old human infants. *Dev Psychol, 25*, 963–975.

de Bruin, N., Sacrey, L.A., Brown, L.A., Doan, J., & Whishaw, I.Q. (2008). Visual guidance for hand advance but not hand withdrawal in a reach-to-eat task in adult humans: reaching is a composite movement. *J Mot Behav, 30*, 337–346.

Butterworth, G., & Hopkins, B. (1988). Hand-mouth coordination in the newborn baby. *British J Dev Psychol, 6*, 303–314.

Butterworth, G., Verweij, E., & Hopkins, B. (1997). The development of prehension in infants: Halverson revisited. *Br J Dev Psychol, 15*, 223–236.

Case-Smith, J., Bigsby, R., & Clutter, J. (1998). Perceptual-motor coupling in the development of grasp. *Am J Occup Ther, 52*, 102–110.

Cattaneo, L., Fabbri-Destro, M., Boria, S., Pieraccini, C., Monti, A., Cossu, G., et al., (2007). Impairments of actions chains in autism and its possible role in intention understanding. *PNAS USA, 104*, 17825–17830.

Christensen, D.L., Baio, J., Braun, K.V., Bilder, D., Charles, J., Constantino, J.N., et al. (2016). Prevalence and characteristics of autism spectrum disorder among children aged 8 years – Autism and Developmental Disabilities Monitoring Network, 11 sites,

United States, 2012. *Morbidity and Mortality Weekly Report Surveillance Summaries, 65*, 1–23.

Clifton, R.K., Muir, D.W., Ashmead, D.H., & Clarkson, M.G. (1993). Is visually guided reaching in early infancy a myth? *Child Dev, 64*, 1099–1110.

Corbetta, D., Guan, Y., & Williams, J.L. (2012). Infant eye-tracking in the context of goal-directed actions. *Infancy, 17*, 102–125.

Corbetta, D., Thurman, S.L., Wiener, R.F., Guan, Y., & Williams, J.L. (2014). Mapping the feel of the arm with the sight of the object: on the embodied origins of infant reaching. *Front Psychol, 5*, 576.

Corbetta, D., Williams, J.L., & Haynes, J.M. (2016). Bare fingers, but no obvious influence of "prickly" Velcro! In the absence of parents' encouragement, it is not clear that "sticky mittens" provide an advantage to the process of learning to reach. *Infant Behav Dev, 42*, 168–178.

Culham, J.C., & Valyear, K.F. (2006). Human parietal cortex in action. *Curr Opin Neurobiol, 16*, 205–212.

Desmurget, M., Richard, N., Harquel, S., Baraduc, P., Szathmari, A., Mottolese, C., & Sirigu, A. (2014). Neural representations of ethologically relevant hand/mouth synergies in the human precentral gyrus. *Proc Natl Acad Sci U S A, 111*, 5718–5722.

Filimon, F., Nelson, J.D., Huang, R.S., & Sereno, M.I. (2009). Multiple parietal reach regions in humans: cortical representations for visual and proprioceptive feedback during on-line reaching. *J Neurosci, 29*, 2961–2971.

Foroud, A., & Whishaw, I.Q. (2012). The consummatory origins of visually guided reaching in human infants: a dynamic integration of wholebody and upper-limb movements. *Behav Brain Res, 231*, 343–355.

Forti, S., Valli, A., Perego, P., Nobile, M., Crippa, A., & Molteni, M. (2011). Motor planning and control in autism: a kinematic analysis of preschool children. *Research in Autism Spectrum Disorders, 5*, 834–842.

Futagi, Y., Ozaki, N., Matsubara, T., Futagi, M., Suzuki, Y., & Kitajima, H. (2016). Eye-mouth associated movement in the human newborn and very young infants. *Pediatr Neurol, 58*, 75–82.

Gibson, E.J. (1988). Exploratory behavior in the development of perceiving, acting and the acquiring of knowledge. *Ann Rev Psychol, 39*, 1–41.

Gonçalves, R.V., Fibueiredo, E.M., Mourão, C.B., Colosimo, E.A., Fonseca, S.T., & Mancinin, M.C. (2013). Development of infant reaching behaviours: kinematic changes in touching and hitting. *Infant Behav Dev, 36*, 825–832.

Granmo, M., Petersson, P., & Schouenborg, J. (2008). Action-base body maps in the spinal cord emerge from a transitory floating organization. *J Neurosci, 28*, 5494–5503.

Greenman, G.W. (1963). Visual behavior of newborn infants. In A.J. Solnit & S.A. Provence (Eds.), *Modern perspectives in child development*. New York: Hallmark.

Hadders-Algra, M. (2007). Putative neural substrate of normal and abnormal general movements. *Neurosci Biobehav Rev, 31*, 1181–1190.

Hall, L.A., Karl, J.M., Thomas, B.L., & Whishaw, I.Q. (2014). Reach and grasp reconfigurations reveal that proprioception assists reaching and hapsis assists grasping in peripheral vision. *Exp Brain Res, 232*, 2807–2819.

Hepper, P.G., Shahidullah, S., & White, R. (1991). Handedness in the human fetus. *Neuropsychologia, 29*, 1107–1111.

von Hofsten, C. (1982). Eye-hand coordination in the newborn. *Dev Psychol, 18*, 450–461.

von Hofsten, C. (1984). Developmental changes in the organization of prereaching movements. *Dev Psychol, 3*, 378–388.

von Hofsten, C. (1991). Structuring of early reaching movements: a longitudinal study. *J Mot Behav, 23*, 280–292.

von Hofsten, C., & Fazel-Zandy, S. (1984). Development of visually guided hand orientation in reaching. *J Exp Child Psychol, 38*, 208–219.

von Hofsten, C., & Rönnqvist, L. (1988). Preparation for grasping an object: a developmental study. *J Exp Psychol, 14*, 610–621.

von Hofsten, C., & Rosander, K. (1997). Development of smooth pursuit tracking in young infants. *Vision Res, 37*, 1799–1810.

Ianniruberto, A., & Tajani, E. (1981). Ultrasonographic study of fetal movements. *Semin Perinatol, 5*, 175–181.

Jeannerod, M. (1981). Intersegmental coordination during reaching at natural visual objects. In J. Long & A. Badeley (Eds.), *Attention, performance IX* (pp. 153–169). Hillsdale, NJ: Lawrence Erlbaum Associates.

Jeannerod, M. (1999). Visuomotor channels: their integration in goaldirected prehension. *Hum Mov Sci, 18*, 201–218.

Kaas, J.H., & Stepniewska, I. (2016). Evolution of posterior parietal cortex and parietal-frontal networks for specific actions in primates. *J Comp Neurol, 524*, 3123–3140.

Kana, R.K., Uddin, L.Q., Kenet, T., Chugani, D., & Müller, R.A. (2014). Brain connectivity in autism. *Front Hum Neurosci, 8*, 349.

Kanazawa, H., Kawai, M., Kinai, T., Iwanaga, K., Mima, T., & Heike, T. (2014). Cortical muscle control of spontaneous movements in human neonates. *Eur J Neurosci, 40*, 2548–2453.

Karl, J.M., Sacrey, L.A., Doan, J.B., & Whishaw, I.Q. (2012a). Hand shaping using hapsis resembles visually guided hand shaping. *Exp Brain Res, 219*, 59–74.

Karl, J.M., Sacrey, L.A., Doan, J.B., & Whishaw, I.Q. (2012b). Oral hapsis guides accurate hand preshaping for grasping food targets in the mouth. *Exp Brain Res, 221*, 223–240.

Karl, J.M., Schneider, L.R., & Whishaw, I.Q. (2013). Nonvisual learning of intrinsic object properties in a reaching task dissociates grasp from reach. *Exp Brain Res, 225*, 465–477.

Karl, J.M., & Whishaw, I.Q. (2013). Different evolutionary origins for the reach and the grasp: an explanation for dual visuomotor channels in primate parietofrontal cortex. *Front Neurol, 4*, 208.

Karl, J.M., & Whishaw, I.Q. (2014). Haptic grasping configurations in early infancy reveal different developmental profiles for visual guidance of the reach versus the grasp. *Exp Brain Res, 232*, 3301–3316.

Kaur, M., Srinivasan, S.M., & Bhat, A.N. (2015). Atypical object exploration in infants at-risk for autism during the first year of life. *Front Psychol., 6*, 798.

Kessen, W., Williams, E.J., & Williams, J.P. (1961). Selection and test of response measures in the study of the human newborn. *Child Dev, 32*, 7–24.

Korner, A.F., Chuck, B., & Dontchos, S. (1968). Organismic determinants of spontaneous oral behaviour in neonates. *Child Dev, 39*, 1145–1157.

Koterba, E.A., Leezenbaum, N.B., & Iverson, J.M. (2012). Object exploration at 6 and 9 months in infants with and without risk for autism. *Autism, 18*, 97–105.

Kremenitzer, J.P., Vaughan, H.G., Kurtzberg, D., & Dowling, K. (1979). Smooth-pursuit eye movements in the newborn infant. *Child Dev, 50*, 442–448.

Kuhtz-Buschbeck, J.P., Stolze, H., Jöhnk, K., Boczek-Funcke, A., & Illert, M. (1998).

Development of prehension movements in children: a kinematic study. *Exp Brain Res, 122*, 424–432.

Lee, M.H., & Newell, K.M. (2012). Visual feedback of hand trajectory and the development of infant prehension. *Infant Behav Dev, 35*, 273–279.

Libertus, K., & Needham, A. (2010). Teach to reach: the effects of active vs passive reaching experience on action and perception. *Vision Res, 50*, 2750–2757.

Libertus, K., Sheperd, K.A., Ross, S.W., & Landa, R.J. (2014). Limited fine motor and grasping skills in 6-month-old infants at high risk for autism. *Child Development, 85*, 2218–2231.

Lobo, M.A., & Galloway, J.C. (2013). The onset of reaching significantly impacts how infants explore both objects and their bodies. *Infant Behav Dev, 36*, 14–24.

Loenneker, T., Klaver, P., Bucher, K., Lichtensteiger, J., Imfeld, A., & Martin, E. (2010). Microstructural development: organizational differences of the fiber architecture between children and adults in dorsal and ventral streams. *Hum Brain Mapp, 32*, 935–946.

Mari, M., Castiello, U., Marks, D., Marraffa, C., & Prior, M. (2003). The reach-to-grasp movement in children with autism spectrum disorder. *Philosophical Transactions of the Royal Society of London. Series B Biological Sciences, 358*, 393–403.

Masterton, B.A., & Biederman, G.B. (1983). Proprioceptive versus visual control in autistic children. *Journal of Autism and Developmental Disorders, 13*, 141–152.

McCarty, M.E., Clifton, R.K., Ashmead, D.H., Lee, P., & Goubet, N. (2001). How infants use vision for grasping objects. *Child Dev, 72*, 973–987.

Mostofsky, S.H., Dubey, P., Jerath, V.K., Jansiewicz, E.M., Goldberg, M.C., & Denckla, M.B. (2006). Developmental dyspraxia is not limited to imitation in children with autism spectrum disorders. *Journal of the International Neuropsychological Society, 12*, 314–426.

Noterdaeme, M., Mildenberger, K., Minow, F., & Amorosa, H. (2002). Evaluation of neuromotor deficits in children with autism and children with a specific speech and language disorder. *European Journal of Child and Adolescent Psychology, 11*, 219–225.

Palmer, C.F. (1989). The discriminating nature of infants' exploratory actions. *Dev Psychol, 25*, 885–893.

Pellijeff, A., Bonilha, L., Morgan, P.S., McKenzie, K., & Jackson, S.R. (2006). Parietal updating of limb posture: an event-related fMRI study. *Neuropsychologia, 44*, 2685–2690.

Piaget, J. (1952). *The origins of intelligence in children.* New York: International University Press.

Piek, J. (2002). The role of variability in early motor development. *Infant Behav and Dev, 25*, 452–465.

Rapoport, J.L. (1972). Normal psychological development in the absence of oral and body surface sensation: a case report. In J.F. Bosma (Ed.), *Third symposium on oral sensation and perception: the mouth of the infant*. Springfield, IL: Charles C. Thomas.

Reed, E.S. (1982). An outline of a theory of action systems. *J Mot Behav, 14*, 98–134.

Reissland, N., Francis, B., Aydin, E., Mason, J., & Schaal, B. (2014). The development of anticipation in the fetus: a longitudinal account of human fetal mouth movements in reaction to and anticipation of touch. *Dev Psychobiol, 56*, 955–963.

Rochat, P. (1987). Mouthing and grasping in neonates: evidence for the early detection of what hard or soft substances afford for action. *Infant Behav and Dev, 10*, 435–449.

Rochat, P. (1989). Object manipulation and exploration in 2- to 5-month old infants. *Dev Psychol, 25*, 871–884.

Rochat, P., Blass, E.M., & Hoffmeyer, L.B. (1988). Oropharyngeal control of hand-mouth coordination in newborn infants. *Devel Psychol, 24*, 459–463.

Rochat, P., & Hespos, S.J. (1997). Differential rooting response by neonates: evidence for an early sense of self. *Early Dev Parent, 6*, 179–192.

Rochat, P., & Senders, S.J. (1991). Active touch in infancy: action systems in development. In M.J. Weiss & P.R. Zalazo (Eds), *Infant attention: biological constraints and the influence of experience* (pp. 412–442). Norwood, NJ: Ablex.

Ruff, H.A. (1984). Infants' manipulative exploration of objects: effects of age and object characteristics. *Developmental Psychology, 20*, 9–20.

Sacrey, L.R., Bryson, S.E., & Zwaigenbaum, L. (2013). Prospective examination of visual attention during play in infants at high-risk of autism spectrum disorder: a longitudinal study from 6 to 36 months of age. *Behav Brain Res, 256*, 441–450.

Sacrey, L.A., Karl, J.M., & Whishaw, I.Q. (2012a). Development of visual and somatosensory attention of the reach-to-eat movement in human infants aged 6 to 12 months. *Exp Brain Res, 223*, 121–136.

Sacrey, L.A., Karl, J.M., & Whishaw, I.Q. (2012b). Development of rotational movements, hand shaping, and accuracy in advance and withdrawal for the reach-to-eat movement in human infants aged 6–12 months. *Infant Behav Dev, 35*, 543–560.

Sacrey, L.A., & Whishaw, I.Q. (2010). Development of collection precedes targeted reaching: resting shapes of the hands and digits in 1–6-month-old human infants. *Behav Brain Res, 214*, 125–129.

Sacrey, L.A., & Whishaw, I.Q. (2012). Subsystems of sensory attention for skilled reaching: vision for transport and pre-shaping and somatosensation for grasping, withdrawal and release. *Behav Brain Res, 231*, 356–365.

Saffin, J.M., & Tohid, H. (2016). Walk like me, talk like me: the connection between mirror neurons and autism spectrum disorder. *Neurosciences, 21*, 108–119.

Schlesinger, M., Parisi, D., & Langer, J. (2000). Learning to reach by constraining the movement search space. *Dev Science, 3*, 67–80.

Schum, N., Jovanovic, B., & Schwarzer, G. (2011). Ten- and twelve-month-olds' visual anticipation of orientation and size during grasping. *J Exp Child Psychol, 109*, 218–231.

Sparling, J.W., Van Tol, J., & Chescheir, N.C. (1999). Fetal and neonatal hand movement. *Phys Ther, 79*, 24–39.

Spencer, J.P., & Thelen, E. (2000). Spatially specific changes in infants' muscle coactivity as they learn to reach. *Infancy, 1*, 275–302.

Teitelbaum, P., Teitelbaum, O., Nye, J., Fryman, J., & Maurer, R.G. (1998). Movement analysis in infancy may be useful for early diagnosis of autism. *PNAS USA, 95*, 13982–13987.

Thelen, E., Corbetta, D., Kamm, K., Spencer, J.P., Schneider, K., & Zernicke, R.F. (1993). The transition to reaching: mapping intention and intrinsic dynamics. *Child Dev, 64*, 1058–1098.

Thelen, E., Corbetta, D., & Spencer, J.P. (1996). The development of reaching during the first year: the role of movement speed. *J Exp Psychol: HPP, 22*, 1059–1076.

Thomas, B.L., Karl, J.M., & Whishaw, I.Q. (2015). Independent development of the reach and the grasp in spontaneous self-touching by human infants in the first 6 months. *Front Psychol, 5*, 1526.

Thompson, A., Murphy, D., Dell'Acqua, F., Ecker, C., McAlonan, G., Howells, H., Baron-Cohen, S., Lai, M.C., Lomgbardo, M.V., MRC AIMS Consortium, & Catani, M. (2017). Impaired communication between the motor and somatosensory homunculus is associated with poor manual dexterity in autism spectrum disorder. *Biol Psychiatry, 81*, 211–219.

Touwen, B.C. (1995). The neurological development of prehension: a developmental neurologist's view. *Int J Psychophysiol, 19*, 115–127.

Trevarthen, C., & Delafield-Butt, J.T. (2013). Autism as a developmental disorder in intentional movement and affective engagement. *Frontiers of Integrated Neuroscience, 17*, 49.

Vesia, M., & Crawford, J.D. (2012). Specialization of reach function in human posterior parietal cortex. *Exp Brain Res, 221*, 1–18.

Villalobos, M.E., Mizuno, A., Dahl, B.C., Kemmotsu, N., & Müller, R.A. (2005). Reduced functional connectivity between V1 and inferior frontal cortex associated with visuomotor performance in autism. *Neuroimage, 25*, 916–925.

de Vries, J.I., Visser, G.H., & Prechtl, H.F. (1982). The emergence of fetal behaviour. I. Qualitative aspects. *Early Hum Dev, 7*, 301–322.

de Vries, J.I., Visser, G.H., & Prechtl, H.F. (1985). The emergence of fetal behaviour. II. Quantitative aspects. *Early Hum Dev, 12*, 99–120.

Wallace, P.S., & Whishaw, I.Q. (2003). Independent digit movements and precision grip patterns in 1–5-month-old human infants: hand babbling, including vacuous then self-directed hand and digit movements, precedes targeted reach. *Neuropsychologia, 41*, 1912–1918.

Whishaw, I.Q. (2003). Did a change in sensory control of skilled movements stimulate the evolution of the primate frontal cortex? *Behav Brain Res, 146*, 31–41.

Whishaw, I.Q., & Karl, J.M. (2014). The contribution of the reach and the grasp to shaping brain and behaviour. *Can J Exp Psychol, 68*, 223–235.

Williams, J.L., & Corbetta, D. (2016). Assessing the impact of movement consequences on the development of early reaching in infancy. *Front Psychol, 7*, 587.

Williams, J.L., Corbetta, D., & Guan, Y. (2015). Learning to reach with "sticky" or "non-sticky" mittens: a tale of developmental trajectories. *Infant Behav Dev, 38*, 82–96.

Witherington, D.C. (2005). The development of prospective grasping control between 5 and 7 months: a longitudinal study. *Infancy, 7*, 143–161.

Yamada, Y., Fujii, K., & Kuniyoshi, Y. (2013). Impacts of environment, nervous system and movements of preterms on body map development: fetus simulation with spiking neural network. *IEEE Third Joint International Conference on Development and Learning and Epigenetic Robotics (ICDL)*, Osaka, pp. 1–7.

Zwaigenbaum, L., Bryson, S., Rogers, T., Roberts, W., Brian, J., & Szatmari, P. (2005). Behavioural manifestations of autism in the first year of life. *International Journal of Developmental Neuroscience, 23*, 143–152.

4

THE DEVELOPMENT OF ANTICIPATORY PLANNING SKILLS IN 3- TO 12-YEAR-OLD CHILDREN

Matthias Weigelt

Introduction

In their review article on the anticipatory control of human behavior, Rosenbaum and colleagues (Rosenbaum, Chapman, Weigelt, Weiss, & van der Wel, 2012) proposed the *concept of orders of planning* to account for a large body of research on object manipulation. The concept is based on a simple assumption: If reaching and grasping actions depend on anticipatory planning processes, then the same object will be grasped differently, depending on what a person intends to do with the object after grasping has occurred. Hence, from examining how people grasp objects, it would be possible to infer their mental states, if one observes reliable changes in object manipulation in task n as a function of task $n + j$, $j > 0$, where the order of planning is given by $j > 1$. Accordingly, *first-order planning* addresses rudimentary perceptual-motor coupling based on information directly available from the scene, as a particular grip in task n is selected as a function of task $n+0$. Reach and grasp formation is driven by object affordances, such as the object's size, shape, and orientation. In *second-order planning*, actions are organized as a function of task $n+1$. Grip selection is driven by what the person intends to do with the object in the next step. Therefore, second-order planning must be based on some form of cognitive representation, because this kind of information is not directly available at the beginning of the task. Of course, the sequence of actions can be extended further and one can also look at *third-order planning*, as a function of task $n+2$, and for even higher orders of planning.

From a cognitive science perspective, second- and higher-order planning processes are especially interesting, because here, object manipulation relies on cognitive representations of future states. Thus, investigating motor planning skills in object manipulation tasks provides a window into the general organization of human behavior. An observation that has been taken as a paradigmatic approach to study motor planning skills, is the so-called *end-state comfort* (ESC) effect (Rosenbaum et al., 1990). Rosenbaum et al. (1990) demonstrated the ESC effect for the first time for the bar-transport-task (BTT). In the original study, the participant's task was to reach for a horizontal bar and to place either the left or right bar end down onto a target on the left or right side. When placing the right end down, they selected an overhand grasp (palm facing down) and rotated the bar by 90° into a vertical orientation. An interesting observation was made, however, when they placed the left end down. Now, participants changed their grip to an underhand grasp (palm facing up) and performed the rotation. This strategy enabled them to finish the manipulation in a comfortable position of the arm and hand (i.e., with a thumb-up posture). In doing so, participants sacrificed initial-state comfort for end-state comfort, which was taken as a signification of more sophisticated motor planning skills (*posture-based planning account*, Rosenbaum, Meulenbroek, Vaughan, & Jansen, 2001).

In the past 25 years, much research has been dedicated to the ESC effect and it has been observed in adult participants reliably across a large variety of different task scenarios (cf. Rosenbaum et al., 2012). Of importance for the present chapter are reports of the ESC effect to be present in a number of non-human primate species, such as chimpanzees (Frey & Povinelli, 2012), rhesus macaques (Nelson, Berthier, Metevier, & Novak, 2010), cotton-top tamarin (Weiss, Wark, & Rosenbaum, 2007), and lemurs (Chapman, Weiss, & Rosenbaum, 2010). The latter are prosimian primates, who are the most evolutionarily distant primate relatives to the *Homo sapiens* (Yoder, 2007), having diverged from the hominid line roughly 65 million years ago. The presence of the ESC effect in these monkeys suggests some phylogenetic roots of second-order planning skills. How these are represented in the ontogenetic development across childhood will be the focus of the remainder of this chapter.

State of Research on the Development of Anticipatory Planning Skills in Object Manipulation

In 2013, Wunsch, Henning, Aschersleben, and Weigelt published a systematic review on the development of anticipatory planning skills (as signified by the ESC effect) for object manipulation across childhood. These authors identified a total of 13 studies, beginning with the first study on children by Hughes in 1996. Since this review, the developmental pathway of the ESC effect has drawn much attention in the research community and a considerable number of new studies have been published. Performing a quick (and not very systematic) search myself (in PubMed), revealed that at least 18 new studies came out within the last 4 years, testing the ESC effect in children. This highlights the attractiveness of studying the ESC effect as a paradigmatic approach to examine the development of more complex anticipatory planning skills.

When taking a closer look at the presence of the ESC effect in children of different ages across all of these studies, it becomes clear that such sophisticated planning skills do not seem to be rooted as a phylogenetic trait in humans, as could be assumed from the presence of the ESC effect in the different animal species, but rather mature over the course of ontogenetic development. In fact, the ESC effect was found to be completely absent in children younger than 3 years old (Jovanovic & Schwarzer, 2011) and the majority of children shows the ESC for the first time, when they are about 5–6 years of age (cf. Wunsch et al., 2013). Strikingly, even when closely replicating the task design previously used in two of the studies on monkeys (Chapman et al., 2010; Weiss et al., 2007), 4-, 6-, and 8-year-old children do not demonstrate the ESC effect (Wunsch, Weiss, Schack, & Weigelt, 2014). With regard to second-order planning skills in children, previous research therefore points to a rather protracted development of anticipatory planning skills for object manipulation (as signified by the ESC effect). Below, I will summarize the state of research, focusing on the developmental pathway for different ESC tasks in children across the age span of 3 to 12 years.

The Bar-Transport-Task (BTT)

As mentioned above, the bar-transport-task (BTT) was originally developed by Rosenbaum et al. (1990). Weigelt and Schack (2010) tested pre-school children across three age groups and found that 18% of 3-year-old, 45% of 4-year-old, and 67% of 5-year-old children demonstrated the effect. Thus, the ESC effect in children increases (steadily) during pre-school years (for similar results using a 180°-rotation-version of the BTT, see Jovanovic & Schwarzer, 2011). This developmental trend was further examined for the BTT across a larger age span, including pre-school and primary-school children (as well as adults), by Thibaut and Toussaint (2010) and by Wunsch, Henning, Pfister, Aschersleben, and Weigelt (2016). Both studies replicated the pattern of results for pre-school children and found that primary-school children reach adult-like proficiency in the BTT, when they are about 10 years old (for similar results across this age span using a 180°-rotation-version of the BTT, see Knudsen, Henning, Wunsch, Weigelt, & Aschersleben, 2012). Hence, the developmental pathway for the BTT appears to be positively accelerated between the ages of 3 and 10 years, with the critical age for the ESC effect to be present in the majority of children (>50%) being 5–6 years old and adult-like behavior around the age of 10.

Handle-Rotation-Task (HRT)

The handle-rotation-task (HRT) was constructed by Rosenbaum, Vaughan, Jorgensen, Barnes, and Stewart (1993). In this task, participants were presented with a handle, which was fixed at the center of a disk, and they were asked to rotate the marked end of the handle to one of eight target locations, being spaced equidistant around the disk. Although participants were not told how to perform the manipulation, ESC was considered to be satisfied, when they grasped the handle in a way that ensured a comfortable hand posture at the final location.

The HRT has been used in very different versions to test children. A child-like version represents the sword-rotation-task (SRT), originally developed by Crajé, Aarts, Nijhuis-van der Sanden, and Steenbergen (2010), in which children played pirates and reached for a wooden sword that was presented in six different orientations. In each trial, the task's goal was to insert the blade of the sword into a tight hole of a wooden block. Results revealed a pattern of gradual development for pre-school children from 3 to 6 years. While only about 10% of the 3-year-old and 30% of the 4-year-old children demonstrated the effect, it was present in the majority (>50%) of the 5- and 6-year-old children. This developmental trend was further examined for the SRT by Wunsch et al. (2016), again examining children across a larger age span (3–10 years), including pre-school and primary-school children (as well as adults). Although their youngest children were a little more proficient, the results for pre-school children replicated the earlier findings of Crajé et al. (2010), with the ESC effect being present in the majority of 5- and 6-year-old children. In primary-school children, however, the development seems to stagnate, with the effect being present in 70–75% of 7-, 8-, 9-, and 10-year-old children (for a similar pattern of results, see Jongbloed-Pereboom, Nijhuis-van der Sanden, Saraber-Schiphorst, Crajé, & Steenbergen, 2013). Much like for the BTT, the developmental pathway of the SRT is positively accelerated between the ages of 3 and 10 years, with the critical age for the ESC effect to be present in the majority of children (>50%) being 5–6 years. Adult-like proficiency, however, seems to be even further protracted, somewhere beyond the age of 10.

Overturned-Glass-Task (OGT)

The overturned-glass-task (OGT) is based on the anecdote of a waiter, who picked up an inverted glass from a shelf in the restaurant using a thumb-down grasp, in order to hold the glass comfortably with a thumb-up grasp, when pouring water into it (Rosenbaum et al., 1990). This observation was brought into the laboratory by Fischman (1997). Two task versions were realized in the original study: (1) pick up a plastic cup, turn it over by 180°, and put it on a coaster, and (2) pick up a plastic cup, turn it over by 180°, and pour water from a pitcher into it. Like the waiter, Fischman's participants reached for the plastic cup with a thumb-down grasp, a strategy representing the ESC effect in the OGT.

A number of studies used the OGT to test children, but the presence of the ESC effect seems to depend highly on the specific task version. Knudsen et al.

(2012) examined the first task version of the OGT ("pick up and place") in children from 3 to 8 years (and adults). They found the ESC effect to be already present in 60% of the 3-year-old children, and an increase of the effect until children's grip strategy reached adult-like proficiency at the age of 8. Less proficient were the 2–3 and the 5–6 year-old children in the study of Adalbjornsson, Fischman, and Rudisill (2008). They used the second task version ("pick up, place, and pour water"), for which the ESC effect was only present in 20% of 2–3 year olds and only in 35% of the 5–6 year olds. Finally, Scharoun and Bryden (2014) extended these earlier task versions: In one version, 3- to 12-year-old children (and adults) were asked to pick up a plastic cup with one hand and hold on to it, while pouring water with the other hand (bimanual, "pick-up and pour water"). In the other version, the same children picked up the glass, but this time passed it to another person (unimanual, "pick-up and pass"). In the "pick up and pour water" version, the majority of children (>50%) demonstrated the ESC effect for the first time at age 5–6, whereas in the "pick up and pass" version, the effect was further delayed, with only about half of the children showing the effect at age 7–8. Children reached adult-like proficiency in both task versions only when they were 11–12 years old. Thus, and quite interestingly, when the task was self-directed ("pick up and pour"), the level of proficiency was higher, than when the task was other-directed ("pick up and pass").

Hammering Task (HT)

Most recently, Comalli et al. (2016) introduced the hammering task as a novel task to investigate the ESC effect in children. In Experiment 1, 4-, 8-, and 12-year-old children (as well as adults) were asked to grasp a hammer, whose handle pointed either to the right (easy trials) or to the left (difficult trials), and to pound a peg with the hammer until it was flat. Children were free to choose their preferred, right hand or their non-preferred, left hand. On difficult trials, requiring an underhand, radial grip with the preferred hand, 38% of 4-year-old, 76% of 8-year-old, and 91% of 12-year-old children (and 100% of adults) showed the ESC effect. Interestingly, 4 year olds changed their grip in almost 50% and 8 year olds in almost 25% of the trials, before implementing the action, indicating a competition between different grip strategies in these children. In Experiment 2, only 4-year-old children were examined for hammering and the hand used was restricted to the preferred hand, reducing planning demands, because children did not have to choose which hand to use in the task. Now, 68% of these children selected a grip in accordance with the ESC effect. Thus, depending on the task constraints (free hand choice vs. no choice), the critical age for the ESC effect to show up in the majority of children was between 4 and 8 years, but only 12-year-old children displayed adult-like behavior.

Grasp-Height-Task (GHT)

The GHT was originally developed by Cohen and Rosenbaum (2004). It represents a simple placing task: Participants stand in front of a book shelf and are required to reach for a plunger, resting on a home base, and place it on one of several target platforms varying in height (home-to-target moves). Results revealed that adults grasp the plunger lower along its shaft, when moving it to a high target platform, and higher along the shaft, when placing it at a low target platform (Cohen & Rosenbaum, 2004). This adaptation of grasp height (so-called *grasp-height effect*) has been interpreted to signify second-order planning skills, somewhat similar to the ESC effect. The GHT has only been recently used in children (Jovanovic & Schwarzer, 2017; Wunsch et al., 2016). Jovanovic and Schwarzer (2017) found the grasp-height effect to be already present in 3 year olds, but it was less pronounced than in 5-year-old children, who behaved similar to adults. Very differently, Wunsch et al. (2016) found (1) that 3-year-old children grasp the plunger higher for high target platforms and lower for low target platforms (thus, matching grasp height with target height), (2) that this effect reverses in children between 3 and 10 years, (3) that there is a developmental spurt for children between 4 and 8 years, and (4) that even the 10-year-old children were far from being as proficient as adults were in this task. The latter study thus suggests that the grasp-height effect develops beyond the age of 10.

Factors Constraining the ESC Effect Across Different ESC Tasks

The factors that may have potentially constrained the ESC effect in some of the studies above, have been addressed previously (Scharoun & Bryden, 2014; Weigelt & Schack, 2010; Wunsch et al., 2013, 2016), and I will therefore only mention them briefly here. The first collection of factors relates to the task design. Among them are the *precision requirements* (high vs. low precision), the *degree of manual rotation* (90° vs. 180°), the *number of action steps* (one, two, or three action steps), the *number of limbs to be coordinated* (uni-manual vs. bimanual control), and the *familiarity with the object* to be manipulated (new object vs. everyday object). The second collection of factors deals with differences in task procedure. Here, *familiarity with the task* (number of practice trials or the opportunity to play with the object before testing), the *specific test set-up* (whether children are sitting vs. standing), the *total number of test trials*, and the *number of trial repetitions* in a particular condition (assembling trials blocked vs. randomly) should be considered. A third collection of factors can be classified in terms of possible motivational effect. These may arise from the *self-directedness of the action goal* (self-directed vs. other-directed action goal), *specific action effects* (action effects vs. no effect; rewards for participating), and possible *effects of social facilitation* (child performs alone vs. in the presence of other children).

A last (but important) factor, which has not been attended to previously, may be the difference of when (i.e., at which age) children enter school in different countries. In some countries a 6-year-old child will still be in Kindergarten (falling into a pre-school sample), whereas in other countries, it is in 1st grade of primary school (falling into a primary-school sample). This should be kept in mind, especially when looking at the age of 5–6 years, which seems to be a critical age for the ESC effect to develop. Together, these factors should be considered, when designing experiments on the ESC effect in the future, and these constraining variables themselves (their relative contributions to the presence or absence of the ESC effect) may become the focus of research.

A Three-Stage Developmental Model on the Acquisition of Anticipatory Planning Skills

To account for the previous findings revealing the protracted developmental pathway of the end-state comfort effect in young children, Wunsch and Weigelt (2016) proposed a three-stage developmental model on the acquisition of anticipatory planning skills for grip selection in object manipulation across childhood. The outline of the three-stage model is presented in Figures 4.1–4.3. Thereby, the general assumptions of the model and the basic construction of each of the three stages draw on two theoretical concepts, which both address the control of complex behavior, but from different perspectives.

The first theoretical concept has been introduced by the famous Russian physiologist Nikolas A. Bernstein in his seminal book on *The Coordination and Regulation of Movements* (Bernstein, 1967). To him, the major challenge of the control of voluntary movements is posed by the so-called *degrees of freedom* (DOF) problem. The DOF problem can be easily demonstrated by the fact that there are always multiple ways for the sensory-motor systems to solve a particular movement task. For example, when reaching for an object, (1) the reaching path can divert from a straight trajectory by moving along any trajectory connecting the initial position of the hand with the location of the object, (2) during the approach phase, the orientation of the hand can be pronated or supinated by various degrees of rotation within the biomechanically possible range of hand and arm motion, and (3) during the grasping phase, the fingers (and finger digits) can form many different grip types to securely grasp the object. Hence, the space of motor solutions for a (seemingly simple) object grasping task is manifold and poses an almost infinite number of ways to solve this task. This example shows that there is no simple motor solution for a particular task, because there is no simple one-to-one correspondence between movement and task. Therefore, the question central to the DOF problem is, how the nervous system "chooses" a (particular) set of DOF from the almost infinite number of possible motor solutions (Kelso, 1995).

In order to account for the DOF problem, Bernstein's concept addresses motor control in two ways: First, the concept can be used to describe the restriction of movements, whenever people perform novel motor actions, by limiting the range of motion for individual joints. Initially "freezing" individual DOF comes with the benefit of lower movement variability and better task control in the early motor learning stage. As the learning process continues and the actor becomes more dexterous, more DOF will be exploited, which results in a larger space of motor solutions for a particular movement task. This most often results in a reorganization of (previous) motor behavior and in new task solutions. Second, the concept addresses the strategy to limit the variety of motor actions used to accomplish a certain task. For example, limiting the infinite posture space in the object reaching task to two grasp postures (i.e., thumb-up and thumb-down grasp) reduces planning costs (which otherwise arise from computing many different grasping actions) and supports the automatic grip selection as the behavioral control becomes more efficient and movement execution more skillful. As it is argued by Wunsch and Weigelt (2016), both strategies are employed during the acquisition of anticipatory planning skills for object manipulation in young children.

The second theoretical concept relevant to the three-stage model has been put forward by the German psychologist Joachim Hoffmann (Hoffmann, 2003; Hoffmann et al., 2007). His so-called Anticipatory Behavioral Control (ABC) framework is based on associative learning and focuses on two learning mechanisms, which provide the cognitive structure for the voluntary control of anticipatory behavior. The first learning mechanism realizes the acquisition of action-effect associations in the early motor learning stage. Thereby, the strength of these action-effect associations relies on the contingency with which a particular motor action will produce a certain action effect (in the environment). For example, if a particular grasp (e.g., a thumb-down grasp) always (in other words reliably) results in a comfortable position of the arm and hand at the end of the object manipulation (e.g., holding the cup comfortably with a thumb-up posture), then a strong association between initial grasp (action) and final posture (effect) is established.

The second learning mechanism contextualizes these action-effect associations to specific situational conditions (in the environment), which define the behavioral context. For example, if a thumb-down grasp always results in a comfortable end posture, whenever the object grasping task requires to manipulate an inverted cup, then the previously established action-effect association between the initial grasp (thumb-down grasp) and the final posture (thumb-up posture) will be contextualized to the situational condition featuring the inverted cup. In other words, the contextualization of action-effect associations to specific situational conditions is the learning mechanism that systematically modulates the contingencies between actions and effects as the learning process continues. Thus, the ABC theory takes both "the primacy of action-effect learning as well as the conditionalization of action-effect relations" (Hoffmann et al., 2007, p. 134) into account. With regard to object manipulation, once established, both learning mechanism enable the flexible selection of a particular grip (e.g., a thumb-down grasp) by anticipating the intended action effect (e.g., holding the cup comfortable at the final position) based on the situational condition, which defines the behavioral context (e.g., inverted cup on the shelf).

So far, the DOF problem introduced by Bernstein (1967) and the ABC framework put forward by Hoffmann (Hoffmann, 2003; Hoffmann et al., 2007) have not been viewed together. Considering the combination of these two concepts (or aspects thereof), however, may benefit the understanding of the acquisition of object manipulation skills in young children. In the following, I will present each of the three stages of the developmental model by Wunsch and Weigelt (2016) separately, for a more detailed illustration of the different skill acquisition stages. Thereby, the organization of skill acquisition into three distinct developmental stages was actually inspired by the writings of Bernstein (1967), who also assumed that the acquisition of complex skills develops through three stages, which can be characterized by (1) the "freezing" of all DOF through simultaneous contractions of the agonist and antagonist muscles at the joints, that are not completely necessary to solve the task, (2) by the "freeing" of additional DOF, to overcome these reactive forces, by reducing the simultaneous contraction of agonist and antagonist muscles at the same joint and the emergence of so-called coordination structures (i.e., temporary assemblies of muscular synergies), and (3) by the "exploitation" of the neuro-mechanical properties of these coordinative structures to reach dynamic stability.

The learning mechanisms affecting the acquisition of object manipulation skills at each stage are taken from the ABC framework by Hoffmann (Hoffmann, 2003; Hoffmann et al., 2007). Thus, the three-stage developmental model by Wunsch and Weigelt (2016) is also an associative learning model. In line with Hoffmann et al.'s (2007, pp. 134–135) assumptions, it is also argued (1) that voluntary behavioral control is mainly driven by the anticipation of intended future goal-states (i.e., action effects), instead of being driven by the stimulus itself (i.e., the object), (2) that whenever a certain behavioral outcome satisfies the effect anticipation, the action selected is being reinforced, (3) that action selection is thereafter determined by action-effect associations, instead of stimulus-response association, and (4) that strong action-effect associations can turn into habits (meaning that actions are habitually selected), when they have been contextualized to a situational condition.

The development of children's level of proficiency in object manipulation tasks will be further explained by using the OGT as a paradigmatic example of the acquisition of motor planning skills for object manipulation. To this end, consider the situation in which a child would like to have a cup of milk and, therefore, is going into the kitchen to pick up an empty cup, which is standing upside down on the shelf. According to Hoffmann's ABC framework, the inverted cup on the shelf serves as the *situational condition*. The grasping action to be performed, namely to take the inverted cup from the shelf to have a cup of milk, is the *voluntary action*. The (body-internal) *effect anticipation* is the intended end-posture and the *real effect* is the re-afferent perception of the actual end-posture, which is attained when the manual rotation is completed. This defines the main components of each of the three stages of the developmental model. In order to better illustrate the learning mechanisms that underlie each stage, dark gray boxes and black arrows, as well as black ink for text, mark those associations between any two or more components that have already been established at a certain age. Empty boxes and white arrows, as well as light gray ink for text, depict a lack of associations between any two or more components.

Stage 1 – Habitual Grip Selection

The control of behavior in Stage 1 is fully driven by habitual grasp selection (see Figure 4.1). When reaching for the inverted cup in the shelf, children younger than (approximately) 3–4 years will most likely (habitually) select a default thumb-up grasp. This behavioral strategy can explain the complete absence of the ESC effect in the large majority of children at this age (e.g., Adalbjornsson et al., 2008; Jovanovic & Schwarzer, 2011; Weigelt & Schack, 2010). Thus, at Stage 1, the habitual grasping system dominates the goal-directed grasping system (Herbort & Butz, 2011) and children will not adapt their grasps relative to the *situational condition* (i.e., the orientation of the object). According to the concept of orders of planning by Rosenbaum et al. (2012), such grasping behavior is an indication of first-order planning (i.e., selecting a grasp relative to the immediate task demands). At the same time, children of this age may not be able to anticipate an action outcome other than ending the movement in a thumb-up posture (as a default posture). This is most likely due to the

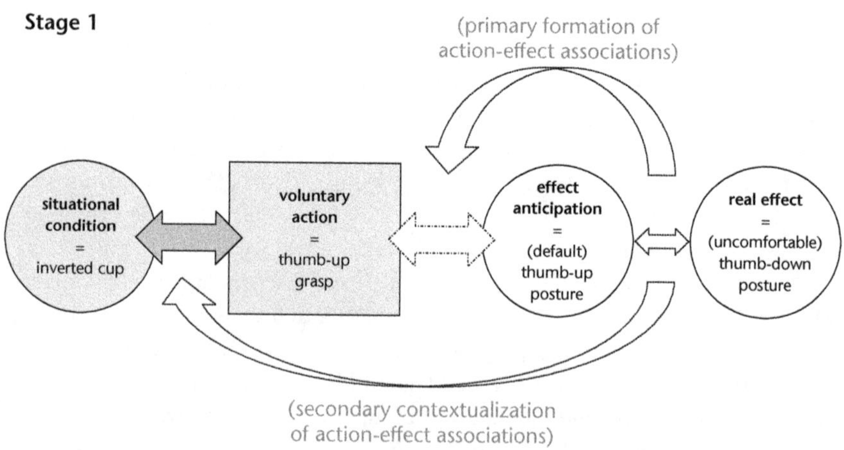

FIGURE 4.1 In Stage 1 (grayed areas), children rely on habitual grasp selection and display only first-order planning

"freezing" of additional DOF in order to constrain movement variability and gain better task control in the first learning stage, as was originally proposed by Bernstein (1967) and later on confirmed in young children by Steenbergen, van der Kamp, Smitsman, and Carson (1997). By "freezing" some additional DOF for the reaching phase, children select the habitual grasp posture for object manipulation in terms of first-order planning (i.e., the thumb-up posture). As a consequence, they will finish the manipulation in an uncomfortable thumb-down posture.

Here, it is important to note that there is most likely no comparison between the *real effect* (i.e., uncomfortable thumb-down posture), which children inevitably experience at the end of their maneuvers, and the *effect anticipation* (default thumb-up posture), which will always be the posture dictated by the habitual grasping system. At least the comparison does not result in a change of behavior, although children finish the action most often in an awkward position. This can be concluded from previous research showing that experience with a particular ESC task, either gained through familiarization and/or trial repetitions, does not seem to have a great influence on grasp selection in children of this age (Wunsch et al., 2013). This means that children at this age may not gain much experience in (otherwise) comfortable end postures. Therefore, neither the primary formation of action-effect associations nor the secondary contextualization of these associations is realized in Stage 1.

Stage 2 – Variability in Grip Selection

The control of behavior in Stage 2 is characterized by a large variability in grasp selection (see Figure 4.2). At this stage, children are between 5 and 10 years of

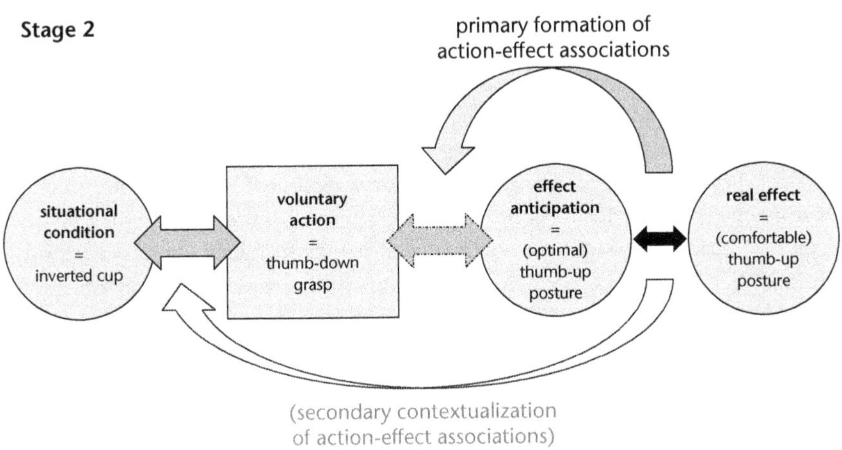

FIGURE 4.2 In Stage 2 (added grayed areas), new action-effect associations are formed and children show second-order planning for the first time

age and begin to "free" additional DOF, which is accompanied by higher movement variability and less task control. But most importantly, as a positive consequence, the freeing of additional DOF results in the exploration of a larger space of motor solutions for the object manipulation task (Bernstein, 1967). In other words, children of this age "practice" with different sets of DOF, becoming more and more available from the freeing of additional DOF. This is accompanied by eminent processes of motor reorganization, as has been reported in a number of studies for children at this age (e.g., Bard, Hay, & Fleury, 1990; Meulenbroek & van Galen, 1988; Thibaut & Toussaint, 2010).

From various experiences with different motor solutions for the object manipulation task, children are now able to anticipate different action outcomes, such as a thumb-up posture, which represents the optimal posture to finish the OGT. Accordingly, they are starting with second-order planning and many of the children experience the ESC effect for the first time. Whenever this is the case, the *real effect* matches the *effect anticipation* and new action-effect associations are formed (i.e., initial thumb-down grasp results in a final thumb-up posture). These new action-effect associations, however, are not yet contextualized to the situational condition (i.e., inverted cup). Therefore, the contingency between the most efficient grip selected (i.e., initial thumb-down grasp) and the desired action effect (i.e., comfortable end-posture) is still weak and unstable. It is most likely, that at this stage, the habitual grasping system competes with the goal-directed system for grip selection (Herbort & Butz, 2011). As a consequence, children will select the habitual grasp in some of the trials, while choosing a goal-directed grasp in the other trials. This notion can be supported by a number of previous research studies, demonstrating large variability in children's grasp selections during Stage 2 (e.g., Comalli et al., 2016; Joengbloed-Pereboom et al. 2013; Wunsch et al., 2014).

Stage 3 – Flexibility in Grip Selection

The control of behavior in Stage 3 reflects the strategy to flexibly select grasp postures in order to reach any desired goal state. This behavior is in line with the notion of second-order planning (Rosenbaum et al., 2012). Accordingly, the most proficient space of motor solutions is exploited and the optimal set of DOF is chosen in order to control the movement (Bernstein, 1967). Children (typically) older than 8–10 years and adults are able to reliably anticipate (body-internal) action effects (i.e., to end comfortably with a thumb-up grasp) (Knudsen et al., 2012; Stöckel, Hughes, & Schack, 2012; Wunsch et al., 2016), based on the strong action-effect associations acquired in Stage 2. These children experience over many trials that the *real effect* consistently matches the *effect anticipation* and they therefore contextualize these strong action-effect associations to the *situational condition* (i.e., inverted cup in the shelf). The contextualization of different action-effect associations to specific situational conditions allows for the flexible selection of grasping actions, which enables the child to

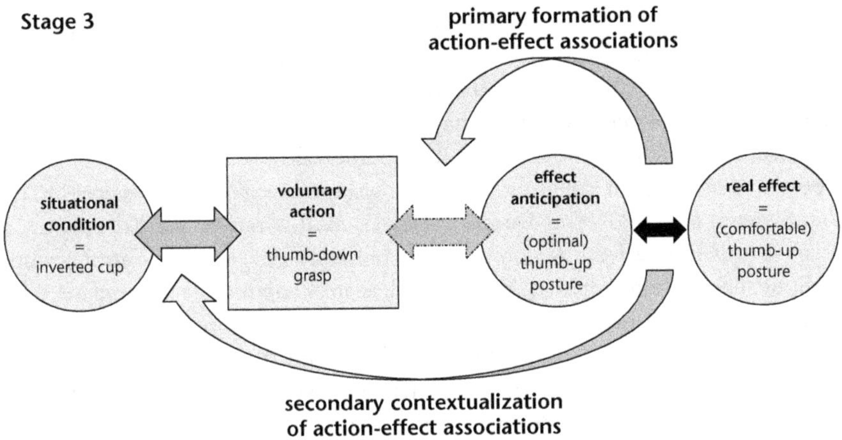

FIGURE 4.3 In Stage 3, children's flexible grasp selection is based on the activation of action-effect associations, which are triggered by the situational condition

choose the optimal grip to reach comfortable end postures (Rosenbaum et al., 2001). Hence, anticipatory planning skills for object manipulation (as signified by the ESC effect) are fully developed in Stage 3. Finally, it may be speculated that these contextualized action-effect associations turn into (new) habits, in the sense, that the most proficient grasping action is habitually selected in the future, for as long as it has been "practiced" enough.

Some Final Considerations

The protracted developmental pathway of the ESC effect across childhood has been taken as an indication of the development of motor planning skills for object manipulation (cf. Wunsch et al., 2013). Consequently, the absence of the ESC effect in young children has been interpreted as a lack of planning skills in these children (e.g., Weigelt & Schack, 2010). This interpretation has not been undisputed, however. Recently, Rosenbaum, Herbort, van der Wel, and Weiss (2014) raised the argument that children may not perceive extreme joint angles as uncomfortable as adults do, because of their limber limbs and more flexible joints. There would simply be no reason for young children to adapt their grasps according to the task demands, because there is no need to avoid (otherwise) uncomfortable body postures, postponing the presence of the ESC effect. After all, children are almost always able to solve the task (that is, they complete the task, although finishing in an uncomfortable end state). Therefore, it may be that the need for end-posture anticipation becomes more important as children mature and their joints and limbs become stiffer. If this argument were true, it would explain the absence of the ESC effect in young children.

At the moment, this argument cannot be refuted easily. Comfort ratings have not been carried out with children (cf. Wunsch et al., 2013), and it would be questionable if the subjective data from comfort ratings would be reliable. Two other lines of research, however, may shed more light on this question. The first examines children with different developmental disorders. Here, it has been reported that children suffering from *developmental coordination disorder* (DCD, Van Swieten et al., 2010) or from *cerebral palsy* (CP, Crajé et al., 2010), show a delay in the ESC effect, as compared to normally developing control groups without such planning deficits. Both disorders are known to be accompanied by (somewhat severe) planning deficits. The second line of research investigates anticipatory planning skills in older people. Quite interestingly, it has been demonstrated that the developmental pattern reverses at the other end of the life span, as indicated by the decline of the ESC effect at old ages (Scharoun, Gonzalez, Roy, & Bryden, 2016; Wunsch, Weigelt, & Stöckel, 2017, Experiment 1). Most interestingly, the decline is stronger, when the movement task becomes more complex and different goal-states have to be anticipated, which suggests that decreasing planning skills are responsible for the decline (Wunsch et al., 2017, Experiment 2). Now, if one flips the argument of limber limbs and more flexible joints, raised for children by Rosenbaum et al. (2014), then we must expect stronger ESC effects for older people, because their limbs and joints become more rigid and stiffer as they get older. That this is not the case reassures us that anticipatory planning skills drive the ESC effect. Therefore, the ESC effect can be taken as a paradigmatic way to investigate the development of anticipatory planning skills across childhood (and beyond). In any case, considering both ends of the life span in the future will provide a more complete picture of the developmental constraints on the voluntary control of object manipulation.

Regarding the different stages of skill acquisition presented in the three-stage developmental model by Wunsch and Weigelt (2016), the age ranges provided are only estimates or corridors based on previous findings (cf. Wunsch et al., 2013). The exact age at which children pass from one skill acquisition stage to another appears to depend on the specific ESC task under investigation. When the same children are tested in two (Jovanovic & Schwarzer, 2011; Knudsen et al., 2012) or even three (Wunsch et al., 2016) different ESC tasks, these children may demonstrate the effect for one particular task, but not for another. For example, children may show the ESC effect earlier in the OGT than in the BTT (Knudsen et al., 2012). Also, ESC performance across different tasks does not seem to be correlated (Wunsch et al., 2016), which poses another challenge for researchers, as it suggests that these kinds of anticipation skills are highly task-dependent. It is also not known for how long children remain in each acquisition stage and also, how long it takes to transit from a lower stage to the next higher stage. It appears, however, that several strategies compete with each other at the first transitional period, before children become sensitive to more goal-directed strategies (Comalli et al., 2016). These issues should be another focus of future research.

There may also be a difference in the developmental pathway between self-directed and other-directed actions (e.g., Keen, Lee, & Adolph, 2014; McCarty, Clifton, & Collard, 2001; Scharoun & Bryden, 2014). For example, children show the ESC effect earlier in the OGT when they reach for the glass to pour water into it (self-directed), as compared to when reaching for the glass to pass it to another person (other-directed) (Scharoun & Bryden, 2014). In a similar vein, McCarty, Clifton, and Collard (2001) found 24-month-old children (but not younger ones) to use more efficient grasp strategies, when manipulating familiar objects (spoon, hairbrush) in self-directed actions (i.e., grasping a spoon to eat or using a hairbrush), as compared to other-directed actions (i.e., grasping to feed or hairbrush a stuffed animal). At least for the spoon-feeding task, this self-other distinction can also be observed in older children of 4 and 8 years (Keen et al., 2014). More research is certainly needed to understand the role of the self-directedness of action goals (self- vs. other-directed action goal) as a motivational variable affecting motor planning skills.

The present considerations are limited to examining grip selection during object manipulation in ESC tasks. Some intriguing observations by Keen and colleagues suggest that children show advanced planning skills for object manipulation at much earlier ages (e.g., Clifton, Rochat, Litovsky, & Perris, 1991; McCarty et al., 1999; McCarty, Clifton, Ashmead, Lee, & Goubet, 2001). For example, Clifton et al. (1991) presented children either a small or a large object in the dark (after a familiarization phase in the light), which afforded either a one-handed or a two-handed grip configuration. Children as young as 6 months old selected the grip configuration that fitted the object's size. That is, they selected a one-handed grip for small objects and a two-handed grip for large objects. In another study by McCarty, Clifton, Ashmead et al. (2001), children were presented a horizontally or vertically oriented rod in the light. When the light was switched off, 9-month-old children already oriented their grasp in a way to adapt to the rod's orientation. Finally, McCarty et al. (1999) presented 9-, 14-, and 19-month-old children with spoons loaded with a toy or with food. Like in the BTT, the spoon's handle pointed either to the left or to the right side. Whereas the 9- and 14-month-old children predominantly selected the overhand grasp with their preferred hand, being unaffected by the spoon's orientation, the 19-month-old children reached for the spoon with either their preferred or non-preferred hand, depending on whether the spoon pointed to the left or to the right. Despite the fact that the older children selected the most proficient hand for the task, a closer look at the data reveals that they still always used an overhand grasp. Together, these three studies are examples of the emergence of first-order planning in toddlers and young children (i.e., rudimentary perceptual-motor coupling), where grip selection is driven by object affordances, such as the object's size, shape, and orientation.

Evidence for second-order planning at such young ages (toddlers) is rare. One exception is the study by Claxton, Keen, and McCarty (2003). In their study, 10-month-old children were asked to reach for a ball and then either to throw it into a tub (low-precision requirements) or to fit it into a tube (high-precision requirements). The children reached for the ball faster, when subsequently throwing it, as compared to when they used it in the precision task (e.g., fitting it down a tube). Hence, in principle, children are capable of advanced planning at young ages, and they are certainly able to perform movements of much greater complexity before they enter primary school (such as bicycling or skiing), but it still remains unclear why these planning skills do not seem to generalize to more sophisticated ESC tasks.

A new perspective on the development of anticipatory behavior, which has drawn considerable interest lately, addresses the question of whether the level of proficiency in object manipulation depends on the level of cognitive development. Accordingly, efficient grip selection may rely on a child's stage of cognitive development, a notion which is currently under a vivid debate (e.g., Stöckel & Hughes, 2015; Van Swieten et al., 2010; Wunsch et al., 2016). In short, this debate revolves about whether performance in object manipulation tasks is driven by executive planning (i.e., actively planning ahead to solve actions correctly or to avoid mistakes, for example) or motor planning (i.e., planning motor actions in advance in order to solve these tasks correctly or most economically), or both (Stöckel et al., 2012). Quite interesting in this regard is the fact that the transition from Stage 1 to Stage 2 in between the age of 5 and 6 years, as proposed in the three-stage model by Wunsch and Weigelt (2016), is a time during which children experience a major growth spurt for a number of important cognitive skills (Zelazo, Craik, & Booth, 2004). For example, they gain control over their reflexive behavior in go/no-go tasks (Bell & Livesey, 1985; Dowsett & Livesey, 2000), are able to inhibit pre-activated responses and show fewer perseveration effects in the Wisconsin Card Sorting Task (WCST) (Chelune & Baer, 1986; Zelazo, Frye, & Rapus, 1996), can follow action rules that are contrary to their intuitive behavior in tapping tasks (Diamond & Taylor, 1996), and they are better able to switch from previously acquired stimulus-triggered response mappings to new action-effect associations (Eenshuistra, Weidema, & Hommel, 2004). Hence, the interdependency between the development of higher cognitive control functions and the maturation of sensory-motor functions (as signified by anticipatory planning abilities) may be stronger than has been assumed in the past.

A strong claim, which could be made from such an interdependency between cognitive and motor functions, would be that, for as long as certain cognitive functions are not in place, efficient grip selection will not be possible (see also Scharoun & Bryden, 2014 for a discussion on the influence of cognitive development on motor planning skills). At least, if these cognitive functions empower the motor functions underlying anticipatory object manipulation skills. This claim is intriguing, because if the notion of a closer link between cognitive and motor functions proves to be true, then this could help to understand the inter-individual differences in motor planning skills between children of the same age.

Therefore, most recent studies, investigating this potential interdependency directly, have focused on a collection of different executive functions, which are thought to play a role in anticipatory planning (Stöckel & Hughes, 2016; Wunsch et al., 2016). At this point in time, however, the findings are not conclusive. Stöckel and Hughes (2016) investigated a group of 5- to 6-year-old children and only found correlations between children's performance in the BTT and their planning and problem-solving skills in the Tower-of-London (TOL) task. But no correlations were detected between the BTT and their working memory capacity in the Corsi-Block-Tapping (CBT) test, as well as their inhibition control in the Animal Stoop (AS) task. Also, Wunsch et al. (2016) found no correlation between any of three ESC tasks (BTT, SRT, and GHT) and any of three tasks testing executive functions (Tower-of-Hanoi [TOH] task, Mosaic-task, and D2-attention-endurance-test) for a large group of children between 3 and 10 years old. Given the results of these two studies, it may be that motor anticipation skills develop fairly independent from other cognitive skills, as has been suggested by Van Swieten et al. (2010). At this point in time, however, there are simply not enough studies to decide on this issue and, therefore, more research is warranted to examine the potential link between cognitive and motor development more systematically.

Summary

The aim of the present chapter was to provide an overview of the state of research on the development of anticipatory planning skills in object manipulation for children from the age of 3 to 12. Across five different object manipulation tasks, the research findings presented suggest a rather protracted developmental pathway, with the critical age of second-order planning skills to effect grip selection being 5–6 years, while adult-like behavior may only be reached at age 10 and beyond. The three-stage developmental model by Wunsch and Weigelt (2016) was introduced to account for the different levels of proficiency across childhood, providing an associative learning framework for the acquisition of object manipulation skills. Future research should consider exploring motor planning skills at both ends of the life span, examine effects of self-other distinction in self-directed and joint-action tasks, and address parallels in the development of cognitive and motor functions.

References

Adalbjornsson, C.F., Fischman, M.G., & Rudisill, M.E. (2008). The end-state comfort effect in young children. *Research Quarterly for Exercise and Sport, 79*(1), 36–41.

Bard, C., Hay, L., & Fleury, M. (1990). Timing and accuracy of visually directed movements in children: Control of direction and amplitude components. *Journal of Experimental Child Psychology, 50*, 102–118.

Bell, J.A. & Livesey, D.J. (1985). Cue significance and response regulation in 3- to 6-year-old children's learning of multiple choice discrimination tasks. *Developmental Psychobiology, 18*, 229–245.

Bernstein, N.A. (1967). *The coordination and regulation of movements*. Oxford: Pergamon Press.

Chapman, K.M., Weiss, D.J., & Rosenbaum, D.A. (2010). Evolutionary roots of motor planning: The end-state comfort effect in lemurs. *Journal of Comparative Psychology, 124*(2), 229–232.

Chelune, G.J., & Baer, R.A. (1986). Developmental norms for the Wisconsin Card Sorting Test. *Journal of Clinical and Experimental Neuropsychology, 8*, 219–228.

Claxton, L.J., Keen, R., & McCarty, M.E. (2003). Evidence of motor planning in infant reaching behavior. *Psychological Science, 14*, 354–356.

Clifton, R.K., Rochat, P., Litovsky, R.Y., & Perris, E.E. (1991). Object representation guides infants' reaching in the dark. *Journal of Experimental Psychology: Human Perception & Performance, 17*, 323–329.

Cohen, R.G., & Rosenbaum, D. (2004). Where grasps are made reveals how grasps are planned: Generation and recall of motor plans. *Experimental Brain Research, 157*(4), 486–495.

Comalli, D.M., Keen, R., Abraham, E.S., Foo, V.J., Lee, M.-H., & Adolph, K.E. (2016). The development of tool use: Planning for end-state comfort. *Developmental Psychology, 52*(11), 1878–1892.

Crajé, C., Aarts, P., Nijhuis-van der Sanden, M., & Steenbergen, B. (2010). Action planning in typically and atypically developing children (unilateral cerebral palsy). *Research in Developmental Disabilities, 31*(5), 1039–1046.

Diamond, A., & Taylor, C. (1996). Development of an aspect of executive control: Development of the abilities to remember what I said and to "do as I say not as I do". *Developmental Psychobiology, 29*, 315–334.

Dowsett, S.M., & Livesey, D.J. (2000). The development of inhibitory control in preschool children: Effects of "executive skills" training. *Developmental Psychobiology, 36*, 161–174.

Eenshuistra, R.M., Weidema, M.A., & Hommel, B. (2004). Development of the acquisition and control of action-effect associations. *Acta Psychologica, 115*, 185–209.

Fischman, M.G. (1997). End-state comfort in object manipulation [Abstract]. *Research Quarterly for Exercise and Sport, 68(Suppl.)*, A-60.

Frey, S.H., & Povinelli, D.J. (2012). Comparative investigations of manual action representations: Evidence that chimpanzees represent the costs of potential future actions involving tools. *Philosophical Transactions of the Royal Society B, 367*, 48–58.

Herbort, M.V., & Butz, M. (2011). Habitual and goal-directed factors in (everyday) object handling. *Experimental Brain Research, 213*(4), 371–382.

Hoffmann, J. (2003). Anticipatory behavioral control. In M.V. Butz, O. Sigaud, & P. Gérard (Eds.), *Anticipatory behavior in adaptive learning systems* (pp. 44–65). Heidelberg: Springer.

Hoffmann, J., Berner, M., Butz, M.V., Herbort, O., Kiesel, A., Kunde, W., & Lenhard, A. (2007). Explorations of anticipatory behavioral control (ABC): A report from the cognitive psychology unit of the University of Würzburg. *Cognitive Process, 8*, 133–142.

Hughes, C. (1996). Planning problems in autism at the level of motor control. *Journal of Autism and Developmental Disorders, 26*, 99–107.

Jongbloed-Pereboom, M., Nijhuis-van der Sanden, M.W.G., Saraber-Schiphorst, N., Crajé, C., & Steenbergen, B. (2013). Anticipatory action planning increases from 3 to 10 years of age in typically developing children. *Journal of Experimental Child Psychology, 114*, 295–305.

Jovanovic, B., & Schwarzer, G. (2011). Learning to grasp efficiently: The development of motor planning and the role of observational learning. *Vision Research, 51*, 945–954.

Jovanovic, B., & Schwarzer, G. (2017). The development of the grasp-height effect as a measure of efficient action planning in children. *Journal of Experimental Child Psychology, 153*, 74–82.

Keen, R., Lee, M.-H., & Adolph, K. (2014). Planning an action: A developmental progression in tool use. *Ecological Psychology, 26*, 98–108.

Kelso, J.A.S. (1995). *Dynamic patterns: The self-organization of brain and behavior.* Cambridge, MA: MIT Press.

Knudsen, B., Henning, A., Wunsch, K., Weigelt, M., & Aschersleben, G. (2012). The end-state comfort effect in 3- to 8-year-old children in two object manipulation tasks. *Frontiers in Psychology, 3*, 1–10.

McCarty, M.E., Clifton, R.K., Ashmead, D.H., Lee, P., & Goubet, N. (2001). How infants use vision for grasping objects. *Child Development, 72*, 973–987.

McCarty, M.E., Clifton, R.K., & Collard, R.R. (1999). Problem solving in infancy: The emergence of an action plan. *Developmental Psychology, 35*, 1091–1101.

McCarty, M.E., Clifton, R.K., & Collard, R.R. (2001). The beginnings of tool use by infants and toddlers. *Infancy, 2*, 233–256.

Meulenbroek, R.G., & van Galen, G.P. (1988). The acquisition of skilled handwriting: Discontinuous trends in kinematic variables. *Advances in Psychology, 55*, 273–281.

Nelson, E.L., Berthier, N.E., Metevier, C.M., & Novak, M.A. (2010). Evidence for motor planning in monkeys: Rhesus macaques select efficient grips when transporting spoons. *Developmental Science, 14*, 822–831.

Rosenbaum, D.A., Chapman, K.M., Weigelt, M., Weiss, D.J., & Van der Wel, R. (2012). Cognition, action, and object manipulation. *Psychonomic Bulletin, 138*(5), 924–946.

Rosenbaum, D.A., Herbort, O., van der Wel, R., & Weiss, D.J. (2014). What's in a grasp? *American Scientist, 102*(5), 366–373.

Rosenbaum, D.A., Marchak, F., Barnes, H.J., Vaughan, J., Slotta, J.D., & Jorgensen, M.J. (1990). Constraints for action selection: Overhand versus underhand grip. In M. Jeannerod (Ed.), *Attention and performance XIII: Motor representation and control* (pp. 321–342). Hillsdale, NJ: Lawrence Erlbaum Associates.

Rosenbaum, D.A., Meulenbroek, R.G., Vaughan, J., & Jansen, C. (2001). Posture-based motion planning: Applications to grasping. *Psychological Review, 108*, 709–734.

Rosenbaum, D.A., Vaughan, J., Jorgensen, M.J., Barnes, H.J., & Stewart, E. (1993). Plans for object manipulation. In D.E. Meyer & S. Kornblum (Eds.), *Attention and performance XIV – a silver jubilee: Synergies in experimental psychology, artificial intelligence and cognitive neuroscience* (pp. 803–820). Cambridge, MA: MIT Press, Bradford Books.

Scharoun, S.M., & Bryden, P.J. (2014). The development of end- and beginning-state comfort in a cup manipulation task. *Developmental Psychobiology, 56*(3), 407–420.

Scharoun, S.M., Gonzalez, D.A., Roy, E.A., & Bryden, P.J. (2016). How the mode of action affects evidence of planning and movement kinematics in aging: End-state comfort in older adults. *Developmental Psychobiology, 58*(4), 439–449.

Steenbergen, B., van der Kamp, J., Smitsman, A.W., & Carson, R.G. (1997). Spoon handling in two- to four-year-old children. *Ecological Psychology, 9*(2), 113–129.

Stöckel, T., & Hughes, C.M.L. (2015). Effects of multiple planning constraints on the development of grasp posture planning in 6- to 10-year old children. *Developmental Psychology, 51*(9), 1254–1261.

Stöckel, T., & Hughes, C.M.L. (2016). The relation between measures of cognitive and motor functioning in 5- to 6-year old children. *Psychological Research, 80*(4), 543–554.

Stöckel, T., Hughes, C.M.L., & Schack, T. (2012). Representation of grasp postures and anticipatory motor planning in children. *Psychological Research, 76*(6), 768–776.

Thibaut, J.-P., & Toussaint, L. (2010). Developing motor planning over ages. *Journal of Experimental Child Psychology, 105*(1–2), 116–129.

Van Swieten, L.M., van Bergen, E., Williams, J.H.G., Wilson, A.D., Plumb, M.S., Kent, S.W., & Mon-Williams, M.A. (2010). A test of motor (not executive) planning in developmental coordination disorder and autism. *Journal of Experimental Psychology: Human Perception and Performance, 36*, 493–499.

Weigelt, M., & Schack, T. (2010). The development of end-state comfort planning in pre-school children. *Experimental Psychology, 57*, 476–782.

Weiss, D.J., Wark, J.D., & Rosenbaum, D.A. (2007). Monkey see, monkey plan, monkey do: The end-state comfort effect in cotton-top tamarins. *Psychological Science, 18*, 1063–1068.

Wunsch, K., Henning, A., Aschersleben, G., & Weigelt, M. (2013). A systematic review of the end-state comfort effect in normally developing children and in children with developmental disorders. *Journal of Motor Learning and Development, 1*, 59–76.

Wunsch, K., Henning, A., Pfister, R., Aschersleben, G., & Weigelt, M. (2016). No interrelation of motor planning and executive functions across young ages. *Frontiers in Psychology, 7*, Article 1031.

Wunsch, K., & Weigelt, M. (2016). A three-stage model for the acquisition of anticipatory planning skills for grip selection during object manipulation in young children. *Frontiers in Psychology, 7*, Article 958.

Wunsch, K., Weigelt, M., & Stöckel, T. (2017). Anticipatory motor planning in older adults. *Journal of Gerontology: Psychological Science, 72*(3), 373–382.

Wunsch, K., Weiss, D., Schack, T., & Weigelt, M. (2014). Second-order motor planning in children: Insights from a cup-manipulation-task. *Psychological Research, 79*(4), 669–677.

Yoder, A. (2007). Lemurs. *Current Biology, 17*, R866–R868.

Zelazo, P.D., Craik, F.I.M., & Booth, L. (2004). Executive functions across the life span. *Acta Psychologica, 115*, 167–183.

Zelazo, P.D., Frye, D., & Rapus, T. (1996). An age-related dissociation between knowing rules and using them. *Cognitive Development, 11*, 37–63.

PART II

Neurophysiological Bases of Reaching, Grasping, and Action Selection

5

NEURAL CIRCUITS FOR ACTION SELECTION

Paul Cisek and David Thura

Introduction

Imagine that you're driving a car, and need to decide on the best route to reach your destination. As you approach a potential exit on the freeway, you consider the relevant information at your disposal: the road signs, information on a map, radio traffic reports, etc. Importantly, as the exit is getting closer, there is an increasing urgency to decide one way or the other. After some deliberation, you commit to a choice and your decision is expressed through a movement – a turn of the wheel in the chosen direction.

In this simple example, there are several kinds of decisions problems. There is the medium-term decision on whether to remain on the highway or take the exit. Within that, there are the immediate and concrete decisions involved in executing your choice, such as how to get around another car that is driving in your chosen lane. Finally, all of these lie embedded within an abstract longer-term decision, taken earlier, to embark on the journey in the first place.

These different kinds of decisions face us with different challenges, and it is likely that they are implemented through different neural mechanisms. Some, like the long-term decision to take the trip, may involve abstract cognitive representations entirely divorced from the sensorimotor contingencies of execution, and may be implemented through a serial model that governs "economic" decisions (Padoa-Schioppa, 2011). In contrast, such a serial model is presumably not well-suited to arbitrate between the moment-to-moment choices that ultimately determine which muscles contract at which time (Cisek & Pastor-Bernier, 2014). For such "embodied" decisions, some have proposed a parallel architecture in which different potential actions compete for execution within the same sensorimotor circuits that guide their execution (Cisek, 2007; Gold &

Shadlen, 2007). These extremes and all of the intermediate kinds of decisions must ultimately be coordinated into a coherent behavior that accomplishes some relevant goal. How does the nervous system manage to achieve this coordination, and which circuits are involved?

Here, we focus primarily on the question of how the nervous system makes decisions about actions. As noted by Gibson (1979), the world at all times presents animals with many potential actions, called "affordances," whose metrics are specified by geometric information sampled by the sensors. Furthermore, in most natural circumstances, new opportunities and dangers constantly present themselves and evolve, even during ongoing activity. In these cases, the decision might be considered not as an abstract computation but rather as the explicit intention to initiate (or avoid) a particular course of action. It has been proposed that during such interactive behavior, the brain simultaneously specifies the affordances currently available in its environment, and selects among them through a competition that is biased by the desirability of their predicted outcomes (Cisek, 2007; Cisek & Kalaska, 2010). The "affordance competition hypothesis" proposes that action specification engages reciprocally interconnected sensorimotor maps in parietal and frontal cortical regions, each of which represents potential actions as peaks of activity in tuned neural populations. These potential action representations compete against each other through mutual inhibition, via direct cortico-cortical connections or competing cortico-striatal loops, until one suppresses the others and the decision is made. This distributed competition is continuously influenced by a variety of biasing inputs, including rule-based inputs from prefrontal regions, reward predictions from basal ganglia, or any variable pertinent to making a choice. While these biases may contribute their votes to different loci along the distributed fronto-parietal sensorimotor competition, their effects are shared across the network owing to its reciprocal connectivity. Consequently, the decision is not determined by any single central executive, but simply depends upon which regions are the first to commit to a given action strongly enough to pull the rest of the system into a "distributed consensus" (Cisek, 2007, 2012).

Below, we summarize some of the neural evidence for this kind of distributed, action-centric model of decisions, focusing mostly on data from nonhuman primates. In particular, we address three important questions: Is sensorimotor cortex involved in selecting between actions? What role is played by the basal ganglia? And finally, is this action-based architecture at all related to the kinds of abstract decisions that dominate so much of our human behavior?

Action Selection and Sensorimotor Cortex

Parietal cortical areas whose activity varies during preparation and execution of movements are strongly and reciprocally connected with frontal regions involved in movement control (Figure 5.1): the lateral intraparietal area (LIP) is

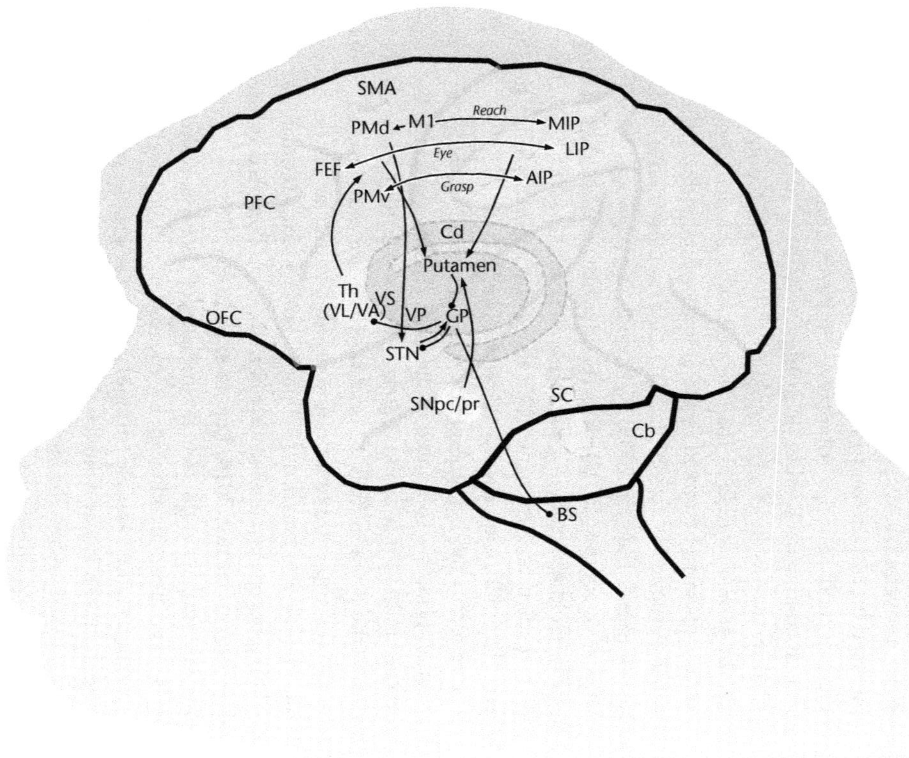

Lateral view of a macaque monkey brain

AIP: Anterior intraparietal area
BG: Basal ganglia
BS: Brainstem
Cb: Cerebellum
Cd: Caudate nucleus
FEF: Frontal eye field
GP: Globus pallidus
LIP: Lateral intraparietal area
M1: Primary motor cortex
MIP: Medial intraparietal area
OFC: Orbitofrontal cortex
PFC: Prefrontal cortex
PMd: Dorsal premotor cortex
PMv: Ventral premotor cortex
SC: Superior colliculus
SMA: Supplementary motor area
SN: Substantia nigra
STN: Subthalamic nucleus
Th: Thalamus
VP: Ventral pallidum
VS: Ventral striatum
VA: Ventroanterior nucleus of the thalamus
VL: Ventrolateral nucleus of the thalamus

FIGURE 5.1 Localization of some cortical and sub-cortical brain areas involved in the selection and the regulation of action in the macaque monkey

interconnected with the frontal eye field (FEF) and both together control eye movements; the medial intraparietal area (MIP) is interconnected with the dorsal premotor cortex (PMd) and primary motor cortex (M1), together controlling arm reaching; the anterior intraparietal cortex (AIP) is interconnected with ventral premotor cortex (PMv) and controls grasping (Matelli & Luppino, 2001). Within each of these systems, distinct potential actions can be specified in parallel (Baumann, Fluet, & Scherberger, 2009; Cisek & Kalaska, 2005; Kim & Shadlen, 1999; Klaes, Westendorff, Chakrabarti, & Gail, 2011; McPeek, Han, & Keller, 2003; Platt & Glimcher, 1997).

There is strong neurophysiological evidence that movement-related fronto-parietal activity is modulated by decision variables. For example, LIP activity correlates not only with sensory and motor variables related to eye movements, but also with decision variables such as expected utility (Platt & Glimcher, 1999), local income (Sugrue, Corrado, & Newsome, 2004), hazard rate (Janssen & Shadlen, 2005) and relative subjective desirability (Dorris & Glimcher, 2004). For control of arm movements, recent studies demonstrate that activity in PMd and M1 is influenced by rewards and PMd activity related to the representation of action choice depends on the robustness of information about potential targets (Dekleva, Ramkumar, Wanda, Kording, & Miller, 2016; Klaes et al., 2011; Pastor-Bernier & Cisek, 2011; Ramkumar, Dekleva, Cooler, Miller, & Kording, 2016). In general, variables that are considered as sensory, cognitive, or motor in nature appear to be mixed in the activity of individual cells in many regions traditionally assigned as action planning and execution areas, such as the premotor cortex (Cisek & Kalaska, 2005; Nakayama, Yamagata, & Hoshi, 2016; Romo, Hernandez, & Zainos, 2004), M1 (Ramkumar et al., 2016; Thura & Cisek, 2014), FEF (Gold & Shadlen, 2000; Kim & Shadlen, 1999), LIP (Platt & Glimcher, 1997; Shadlen & Newsome, 2001) and the superior colliculus (SC) (Basso & Wurtz, 1998; Horwitz, Batista, & Newsome, 2004).

These data have led to the hypothesis that when a decision entails the simple selection of an action plan, it is computed in the same circuits that guide the preparation and the execution of that action (Cisek & Kalaska, 2010; Gold & Shadlen, 2007). In this view, the decision itself is best thought of not as an abstract computation but rather as the explicit intention to execute a particular course of action (Shadlen, Kiani, Hanks, & Churchland, 2008; Tosoni, Galati, Romani, & Corbetta, 2008). This concept is sometimes referred to as "embodied" decision-making (Cisek & Pastor-Bernier, 2014; Lepora & Pezzulo, 2015).

While pioneering studies on the neural correlates of decision-making in non-human primates have been conducted by means of simple, well-controlled experimental paradigms (i.e., simple stimulus-response trials), natural actions during interactive behavior in the wild are determined by a constantly changing and unpredictable environment. This motivates studies that use dynamic and/or probabilistic stimuli, multimodal sensory stimuli, complex geometrical arrangements of targets, manipulation of reward size and/or probability, and manipulation of the speed/accuracy trade-off context in which a decision has to be made, etc.

For example, if the competition that determines decisions occurs within sensorimotor maps that reflect the geometrical properties of the environment, then this predicts that the strength of the competition between options should depend on their geometric relationship. Thus, if an animal is faced with two sets of actions that are similar, the decision formation can be gradual, first computing the properties shared by the different options and only later diverting toward one option versus another. By contrast, when faced with very different response options, an all-or-none decision has to be made. Pastor-Bernier and Cisek (2011) tested these predictions by recording PMd activity while monkeys made decisions between potential targets with different reward values and various geometrical arrangements. They found that neural activity in PMd was modulated by relative reward values when two targets were presented. Crucially, the gain of the interaction between targets was the strongest when they were furthest apart. This is consistent with the prediction that the competition between options depends on their geometric relationship, as predicted by the hypothesis that action decisions evolve within a representation of potential action space.

More recently, we investigated the dynamic nature of action selection and the speed-accuracy trade-off faced in many natural decisions (Figure 5.2). We trained two monkeys to perform the "tokens" task (Figure 5.2A), during which they had to decide between two options based on constantly changing sensory evidence and were free to commit whenever they felt ready. The task allowed us to dissociate in time the two key stages of the decision-making process: the deliberation period during which a choice is made; and the commitment to that choice. We also manipulated the timing parameters of the task in blocks of trials, encouraging either slow, accurate decisions or fast, risky decisions. At the behavioral level, we found that animals need less evidence to commit to their choice as time passes in the trial, in agreement with the proposal that an "urgency signal" pushes subjects to make decisions as time is running out, to maximize reward rate (Cisek, Puskas, & El-Murr, 2009; Thura, Beauregard-Racine, Fradet, & Cisek, 2012). Critically, we observed that the level of urgency at decision time significantly influenced the kinematics of the movements performed by the animals to report their decisions (Figure 5.2B): early decisions (usually made on the basis of strong sensory evidence but low urgency) were followed by long duration movements whereas later decisions (relying on weak sensory evidence but stronger urgency) were followed by shorter and faster movements, as if they were influenced by the strong level of urgency encountered when a significant amount of time has elapsed during a trial. When we encouraged animals to make earlier and less accurate decisions in blocks of trials (fast speed-accuracy trade-off regime), we observed faster movements compared to a condition where accuracy was emphasized, implying a similar adaptation of the speed-accuracy trade-off during both decision and action

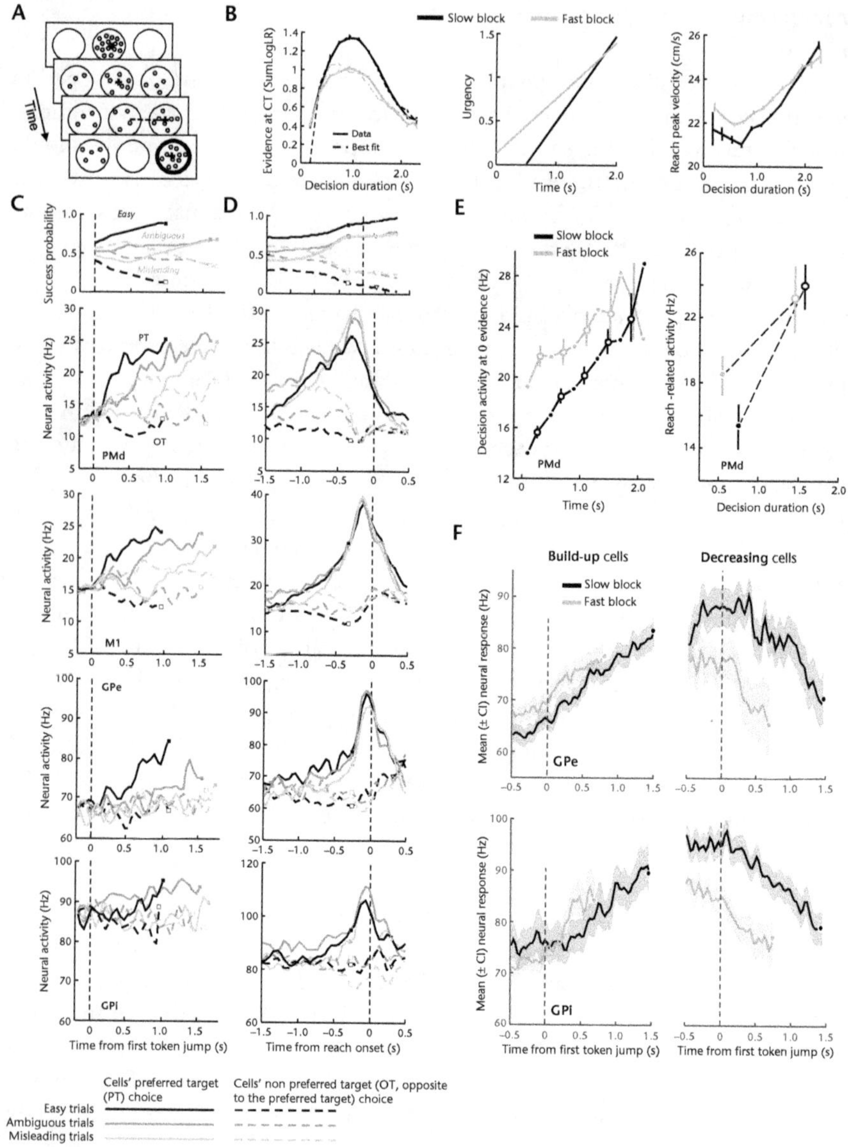

FIGURE 5.2 Neural substrates of dynamic decisions between reaching movements in premotor and motor cortices

Note: A. The "tokens" task. During each trial, tokens jump from the center to one of the outer targets every 200 ms. The monkey's task is to move the cursor (cross) to the target which he believes will ultimately receive the majority of tokens (thick black circle). B. The left panel shows the quantity of sensory evidence available to the monkey at time of commitment (CT) as a function of decision duration and speed-accuracy trade-off (SAT) context, i.e., trials in which slow and accurate decisions are favored (black) or trials encouraging fast and risky decisions (gray). The middle panel shows the estimated shapes of the urgency functions in the two SAT conditions (slow and fast) computed by

(Thura et al., 2014). We recorded single neurons in animals' premotor (PMd) and primary motor cortex (M1) and showed that both areas are strongly involved in both deliberation and choice commitment (Figure 5.2C, D). Both areas combine sensory information with an urgency signal, as predicted by the "urgency-gating" model (Thura & Cisek, 2014). Moreover, in agreement with the behavioral observations, we observed that activity of both decision and movement related cells in PMd and M1 was amplified in blocks of trials in which speed was encouraged over accuracy (Figure 5.2E; Thura & Cisek, 2016) and varied as a function of the outcome of previous trials (Thura, Guberman, & Cisek, 2017). In summary, we found evidence for a "continuous flow" (Coles, Gratton, Bashore, Eriksen, & Donchin, 1985) of sensory information into sensorimotor cortical areas, whereby activity related to specific movements is continuously biased by the sensory evidence in favor of performing that movement. This sensory information is further combined with a global "urgency signal" that gradually grows over the course of each trial. This signal provides unified control of speed-accuracy trade-off adjustments by energizing both the urgency of decisions and the vigor of the selected action. This apparent link between decision urgency and movement vigor led us to hypothesize that both are controlled by a unified signal emanating from the basal ganglia (Thura et al., 2014).

---

fitting the urgency-gating model to the monkey's behavior (dotted curves in the left panel) using different values of urgency slope and intercept. The right panel shows the peak velocity of reaching movements performed by the monkey to report his choices as a function of decision duration and SAT condition. Reproduced with permission from Thura, Cos, Trung, and Cisek (2014). C. The top panel shows the success probability of the preferred target (PT) of recorded cells during easy (black), ambiguous (dark gray), and misleading (light gray) trials, in which the monkey correctly chose the cells' PT (solid lines) or the opposite target (OT, dotted lines). Below, we show the average activity, during those same trials, of spatially tuned neurons recorded in PMd, M1, GPe, and GPi. Activity is aligned on the first token jump and truncated 280ms before movement onset (squares) to avoid averaging artifacts. Modified with permission from Thura and Cisek (2017). D. Same as C but aligned on movement onset. Squares represent our estimate of the monkey's commitment time, inverted triangles mark movement offset. E. The left panel shows the evolution of the average activity of the PMd population calculated for the condition when the evidence is equal for each target, plotted as a function of time in either the slow (black) or the fast blocks (gray). The right panel illustrates the average movement time response of PMd cells involved in movement execution during the tokens task. Data are sorted according to the duration of decisions preceding the reach that reports them, either the shortest (filled circles) or the longest (open circles), and as a function of the SAT condition, the slow (black) and the fast (gray) SAT regime. Reproduced with permission from Thura and Cisek (2016). F. The top left panel shows the average activity (with 95% confidence intervals) of 19 build-up GPe cells aligned on the first token jump during the fast (gray) and slow blocks (black). The top right panel shows the average activity of 5 GPe decreasing cells. The bottom left shows the average activity of 11 build-up GPi cells. The top right panel shows the average activity of 11 GPi decreasing cells. Activity is truncated before decision commitment (circles). Reproduced with permission from Thura and Cisek (2017).

Action Selection and the Basal Ganglia

The basal ganglia (BG) are a group of subcortical nuclei that are interconnected with the cerebral cortex through a series of loops. The input structures are together called the striatum, and include the caudate (Cd), putamen, and ventral striatum (VS). They receive direct projections from the cerebral cortex. The output structures are together called the pallidum, and include the globus pallidus pars interna (GPi), substantia nigra pars reticulata (SNpr), and ventral pallidum (VP). They project to the brainstem movement generators as well as back to the cerebral cortex via the thalamus (Figure 5.1). Intrinsic nuclei such as the globus pallidus pars externa (GPe), the substantia nigra pars compacta (SNpc), and the subthalamic nucleus (STN) activate and regulate these input-output regions. The motor functions of the BG appear to be represented in limited parts in the basal ganglia, specifically, the dorsal part of the striatum (caudate and putamen) and related areas that receive strong projections from the motor areas in the cerebral cortex. In the context of limb movement control, all cortical areas involved in the planning and execution of movements project to the putamen (the caudate being more involved in the control of eye movements) and striatal neurons receiving these cortical inputs then project directly or indirectly (see below) to the GPi (whereas the SNpr is more involved in the control of eye movements), which projects to the brainstem circuitry as well as to ventrolateral (VL) and ventroanterior (VA) nuclei of the thalamus. VL/VA in turn project to premotor, supplementary motor, and primary motor cortices (Middleton & Strick, 2000; Nakano, 2000; Romanelli, Esposito, Schaal, & Heit, 2005).

While it is well-established that the basal ganglia are involved in some aspects of motor control, their exact role during voluntary behavior is still under debate. Influential models suggest a role in motor decision-making (Frank, 2011; Mink, 1996; Redgrave, Prescott, & Gurney, 1999), whereby desired actions are selected and competing ones suppressed through the "direct" and "indirect" dopamine dependent pathways (Cox et al., 2015; DeLong, 1990; Leblois, Boraud, Meissner, Bergman, & Hansel, 2006). This general theory is supported by observations of decision-related modulation of neural activity in the striatum (Ding & Gold, 2010; Samejima, Ueda, Doya, & Kimura, 2005) and pallidum (Arimura, Nakayama, Yamagata, Tanji, & Hoshi, 2013; Pasquereau et al., 2007), and by reinforcement signals in dopamine neurons (Dayan & Daw, 2008; Schultz, 1997). Overall, these models describe information flow in the BG nuclei as follows: Dopamine signals from the SNpc together with cortical projections of glutamate activate the medium spiny neurons of the striatum (MSNs, with D1-type receptors of dopamine, which increase intracellular calcium concentration), which project to the SNpr and the GPi. This first cascade of activation represents the so-called "direct" pathway. Because MSNs are GABAergic cells (i.e., they produce γ-Aminobutyric acid, an *inhibitory* neurotransmitter), they exert an inhibitory action on neurons of the SNpr/GPi complex, which are also GABAergic. This inhibition of the SNpr/GPi leads to a disinhibition of the thalamo-cortical and brainstem circuitry, promoting movements. Conversely, activation of other striatum MSNs (with D2-type receptors of dopamine, which inhibit production of intracellular calcium) projecting indirectly to the SNpr/GPi via the GPe and the STN (the so-called "indirect" pathway), inhibits the GABAergic neurons of the GPe, leading to a disinhibition of the glutamatergic neurons of the STN. The increased discharge of these excitatory STN neurons activates the SNpr/GPi neurons projecting to the brainstem and the thalamus, resulting in the reduction of downstream activity and the suppression of movements (Albin, Young, & Penney, 1989; DeLong, 1990). Finally, a third, hyper-direct pathway connecting several cortical areas (including M1, the supplementary motor area [SMA], PMd and PMv) to the STN represent the fastest way for action to access the BG output nuclei (Kitai & Deniau, 1981).

Based on this functional organization, the model of action selection in the BG proposes that the cortex generates ensembles of possible actions and the striatum selects among these actions by activating specific downstream circuits either promoting or suppressing movements (Gurney, Prescott, & Redgrave, 2001; Mink, 1996; Redgrave et al., 1999). Some theories suggest that reinforcement learning is the process by which the striatum learns to select appropriate actions within a given behavioral context (Frank, 2005).

Despite its influence, however, the hypothesis that the basal ganglia are a system for selection and generation of actions is still under debate (Calabresi, Picconi, Tozzi, Ghiglieri, & Di Filippo, 2014). As mentioned above, the predominant "direct/indirect pathways" model proposes an opposing scheme of action selection primarily based on the antagonist neural transmissions in these two pathways and their final convergence onto basal ganglia output nuclei (Kravitz et al., 2010). But recent data, based in part on advances in optogenetic technology, suggest instead a coordinated activation of both pathways during goal-directed behavior. For example, one study found that activation of the direct pathway with optical stimulation in mice both activates and inhibits the response of the GABAergic output neurons of the BG (Freeze, Kravitz, Hammack, Berke, & Kreitzer, 2013). Another experiment conducted on mice conditioned to perform an operant task (a two-lever free choice task), showed that neural activity in both pathways transiently increases when animals initiated an action, suggesting that coordinated activity of direct and indirect pathway is critical for the appropriate timing of basal ganglia circuits during movement (Cui et al., 2013). Arguing against a strict functional dichotomy between the direct and indirect pathways, it has been shown that neurons of the two pathways collateralize far more than proposed in classical models (Cazorla et al., 2014).

Other data further challenge the role of the BG in selecting the appropriate action while suppressing competing ones. First, electrophysiological studies in behaving monkeys trained to execute reaching or wrist movements found that movement-related activities in the GPi consist of an increase in firing in the majority of the recorded neurons (Mink & Thach, 1991a; Turner and Anderson, 1997). Because an increase in GPi firing strengthens the inhibition on the recipient motor control circuits, this observation supports one of the proposed role of the BG output nuclei in suppressing motor plans that would be inappropriate with the movement being performed. However, it has been also shown that during the performance of a reaching choice reaction time task, movement-related activities initiate later in the striatum and globus pallidus than in connected regions of cortex (Anderson & Horak, 1985; Turner & Anderson, 1997). These patterns also reflect choices significantly later than cortical regions (Arimura et al., 2013; Seo, Lee, & Averbeck, 2012) and crucially, they are largely simultaneous with muscle contraction (Anderson & Horak, 1985; Turner & Anderson, 1997). Thus, the timing of movement-related activity in GPi is not compatible with a role in any processes that are completed prior to the initial activation of a movement, like the selection of the appropriate action and the suppression of competing ones. Furthermore, GPi inactivation or ablation does not lengthen reaction times and does not produce deficits in reach target selection (Desmurget & Turner, 2008, 2010; Mink & Thach, 1991b). Once again, these observations are not consistent with the idea that the BG contributes to the selection or to the initiation of movements.

Our own recordings in the globus pallidus of monkeys performing the tokens task show that during deliberation, information pertinent to selection of reaching movements is continuously influencing activity in reach-related regions of PMd and M1, but is much weaker and significantly delayed in the BG, particularly in its output via the GPi (Figure 5.2C). Then, as the cortical bias grows in favor of one of the targets, it begins to influence activity in the GPe, producing a gradual emergence of tuning before commitment. We propose that eventually, action-specific activity becomes strong enough to engage tuning in the GPi, leading to a positive feedback that constitutes commitment to the action choice (Figure 5.2D). This argues against the role of the BG in the deliberation process that determines which reach target is selected but instead suggests that these nuclei are involved in confirming a choice that is determined in the cortex (Thura & Cisek, 2017).

Moreover, during deliberation we found build-up and decreasing activities, in both GPe and GPi, which are strongly modulated by the speed-accuracy trade-off condition in which the task is performed. Crucially, build-up neurons are often more active in blocks of trials encouraging fast and risky decisions compared to blocks in which monkeys make safe and conservative choices, and decreasing cells almost always show the opposite pattern of modulation (Figure 5.2F). BG activity thus appears to reflect an urgency signal and its adjustment between SAT policies. Together, our data suggest that BG output is not involved in deliberation at all, but instead participates in the regulation of urgency and commitment to the choice made in the cortex (Thura & Cisek, 2017).

Many studies support a role of the basal ganglia in regulating the execution of movements, *after* they have been selected and initiated. For example, in monkeys trained to execute reaching movements, responses in GPi are strongly influenced by kinematic properties of a movement such as its direction, amplitude, and speed (Turner & Anderson, 1997). Moreover, inactivation or ablation of the GPi reduces the velocity and extent of movements (Desmurget & Turner, 2010; Horak & Anderson, 1984a; Mink & Thach, 1991b), and stimulation during movement planning and during the early stages of execution had the largest effects on kinematics (Horak & Anderson, 1984b). A recent study (Yttri & Dudman, 2016) found that targeted optogenetic stimulation of MSNs in the direct (or indirect) pathway can be used to train mice to either increase (or decrease) their movement speed. Given that GPi neurons are also often influenced by the context in which a task is performed, including reward expectation (Pasquereau et al., 2007), these data support a role of the BG in the regulation of motivation and response vigor (Dudman & Krakauer, 2016; Turner & Desmurget, 2010). This is consistent with the similarity of speed-accuracy tradeoff regulation we observed in decision-making and movement (Figure 5.2B,E; Thura & Cisek, 2017; Thura et al., 2014), and with the symptoms of Parkinson's Disease (Mazzoni, Hristova, & Krakauer, 2007).

Decisions Beyond Simple Action Selection

As reviewed above, there is still no consensus on how the cerebral cortex and basal ganglia share the labor of action selection. This may be attributable to differences between the experimental paradigms used in different studies. Indeed, what may seem like relatively trivial variation in behavioral methodologies (e.g., a right versus left reach or a go versus no go to report a decision) can potentially result in the recruitment of distinct mechanisms, and alter the neurobiological processes that regulate choice. For example, until now we have focused on situations in which animals make decisions about different specific responses that are defined from the start of the trial. But obviously, humans and animals can make decisions that are not at all about specific actions. It is thus necessary to clarify the type of decision we are talking about. Although every decision is ultimately expressed through an action, the relationship between the two is sometimes very indirect. For example, when deciding on a house to buy, one presumably does not choose between movements to open the door, but more likely between abstract quantities like cost, location, and the commute to work. In this case, the decision is abstract in nature and computed outside of the motor domain. By contrast, when playing tennis, the sensory evidence provided by the moving ball's trajectory affords the player multiple motor options. In this scenario, an embodied model is much more relevant as it allows a very effective way to compute the decision directly in the effector-relevant sensorimotor areas of the brain, ready for release in a fraction of a second. Any situation in-between these extreme scenarios probably involves the sensorimotor structures as well as the more abstract executive centers through some distributed mechanisms (Cisek, 2012).

These considerations have motivated some researchers to investigate activity of sensorimotor neurons during behavioral tasks in which the motor response used to report a decision was dissociated from the stimulus informing the subject about the choice. In most cases, these studies showed that sensorimotor areas did not exhibit decision-related activity when information about actions was not yet provided (Bennur & Gold, 2011; Cisek & Kalaska, 2005; Filimon, Philiastides, Nelson, Kloosterman, & Heekeren, 2013; O'Connell, Dockree, & Kelly, 2012). Does this imply that decisions are always made in cognitive regions, and just "spill into" sensorimotor cortex when conditions permit? We think it is more likely that decision-making is not a single, unified phenomenon with a single neural substrate, but rather a distributed process that emerges differently across the nervous system depending on the information that finally determines the choice (Cisek, 2012). Moreover, we believe that many discrepancies between data could be reconciled if we take into account the evolution of brain structures and the accompanying ethological motor repertoires of animals. Indeed, we believe that evolutionary history can be a guide not only for providing answers to the still outstanding questions about the neural mechanisms of decision-making, but for defining better questions to start with.

The Evolution of Decision-Systems

As noted above, there are many ways to classify decisions, and it is likely that different kinds of decisions involve different neural circuits. Many studies have aimed to achieve a correspondence between specific circuits and specific types of decisions. A complementary approach to finding that correspondence is to start with the specific neural circuits themselves, consider how and when they emerged in evolution, and then define different kinds of decisions on the basis of the particular behavioral innovations made possible by the emergence of those specific circuits. This is a useful approach because of the highly conservative nature of neural evolution, which can only develop new mechanisms through gradual modifications of existing ones. Thus, here we take a brief detour into describing, in chronological order, a putative sequence of the evolution of different types of decision systems. What follows is necessarily simplified, and rather speculative, but we propose it leads to a potentially useful new perspective that may shed light on some of the unanswered issues about decision-making described above.

Figure 5.3 shows a schematic summary of the evolutionary history of the lineage leading to humans, starting from the earliest multicellular animals. Along the way, different taxa diverged from the human lineage and led to lineages that

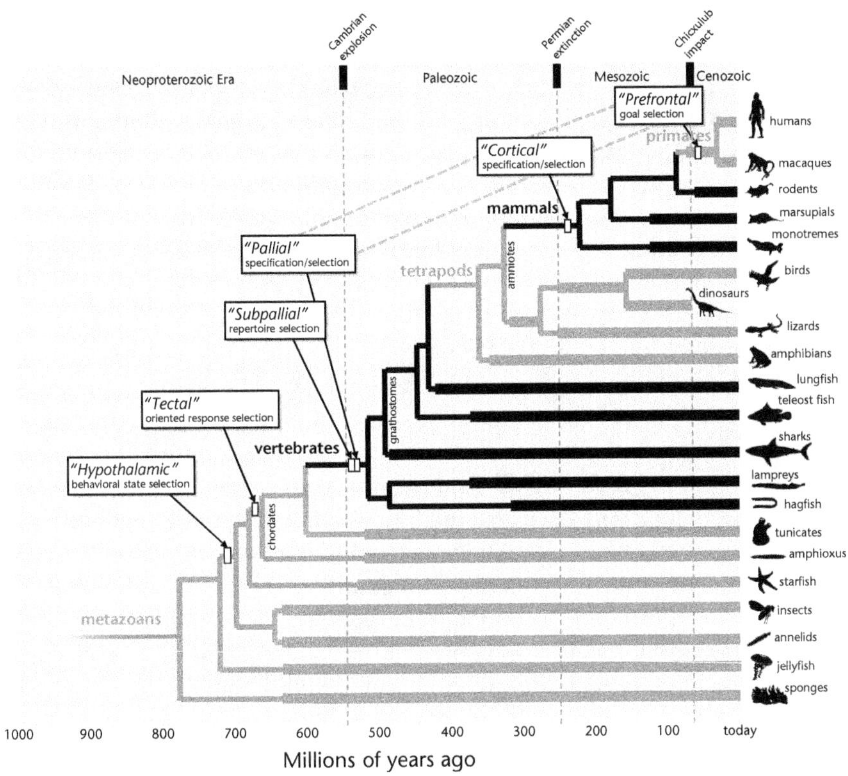

FIGURE 5.3 The phylogenetic tree of animals, emphasizing the lineage leading to humans

Note: Vertical lines indicate times of divergence between different lineages, estimated on the basis of molecular clock analyses (Erwin et al., 2011). Horizontal lines indicate evolution of a lineage toward the specific species pictured on the right, with thicker lines indicating the timing of relevant fossil evidence (from paleobiodb.org). Key innovations in decision-making systems are indicated as small white squares placed at their latest putative appearance, with definitions provided in the boxes. Dashed lines connect the "pallial," "cortical," and "prefrontal" systems to emphasize that they are innovations of each other. Silhouettes are from phylopic.org.

are now represented by a variety of living species. The comparison of the nervous systems of those species allows us to determine when in evolution specific circuits appeared and make inferences about the behavioral innovations they conferred. Here, we focus on the circuits involved in specific kinds of decision-making processes.

Neurons appeared between 700 and 800 million years ago, after our lineage diverged from the one leading to sponges (Brunet & Arendt, 2016; Mackie, 1970). Soon after, the simple ancestral nerve net specialized into an "apical nervous system" (ANS) and a "blastoporal nervous system" (BNS), utilizing hormonal and synaptic transmission, respectively (Arendt, Tosches, & Marlow,

2016). In all animals with neurons, the legacy of the ANS still acts as a high-level controller governing general physiological functions, from thermoregulation to changes in the life cycle, as well as general behavioral state, such as the sleep/wake cycle or decisions about resting, feeding, or seeking food. We can consider that as an ancient system for selecting the animal's behavioral state, such as sleep/wake, explore/exploit, etc. Meanwhile, the legacy of the BNS has evolved into circuits which accomplish those general goals by mediating sensorimotor interactions with the environment. In some groups of animals (chordates, arthropods, annelids, and some mollusks) the ANS and BNS became integrated into a central brain (Northcutt, 2012). Within our specific lineage (chordates), the ANS evolved into the rostral hypothalamus while the BNS evolved into the caudal hypothalamus and the rest of the brain, spinal cord, and peripheral nervous system (Tosches & Arendt, 2013). Therefore, we can define the earliest type of decision-making system as one that selects the behavioral state, and in vertebrates we can call that the "hypothalamic system."

After the evolution of the neural tube, the early chordate brain consisted of a rudimentary hypothalamus (including homologues of the paraventricular nucleus, anterior pituitary, and the pineal gland) attached to a basal locomotor midbrain and a spinal cord. Homologues of these structures are still found in the lancelet Amphioxus (Lacalli, 2008; Shimeld & Holland, 2005), a small filter-feeding animal whose ancestors diverged from the vertebrate lineage about 650 million years ago and appear to have not changed much since. The Amphioxus also possesses a group of cells that receive tactile input arriving from skin receptors on both the rostral and caudal body, and project to the midbrain locomotor centers (Lacalli, 1996). These cells appear to be involved in selecting between forward versus backward locomotion that the lancelet performs in response to physical contact. They also receive input from the simple frontal eye and may be involved in escaping from sudden visual stimulation. Thus, they are anatomically and functionally similar to the "tectum" of vertebrates (Lacalli, 1996), a structure homologous to the superior colliculus (SC) of mammals.

In vertebrates, the tectum is implicated in the selection of oriented reactions to stimuli in the spatial modalities: visual, auditory, and somatosensory. For example, in rodents, stimulation of those regions of the tectum that receive visual input from space above the animal leads to coordinated escape behaviors, while stimulation of the regions that receive input from the lower visual field leads to approach behavior (Comoli et al., 2012; Sahibzada, Dean, & Redgrave, 1986). Analogous results are obtained from stimulation of the tectum in lamprey (Saitoh, Menard, & Grillner, 2007), which diverged from our lineage more than 500 million years ago. Thus, we can define a second type of decision-making system we'll call the "tectal system," which selects the animal's oriented responses with respect to spatial stimuli (e.g., the approach vs. escape decision). It clearly existed in the early vertebrates, and the putative homology with Amphioxus suggests it evolved more than 650 million years ago.

These findings suggest that the early ancestors of vertebrates possessed two decision systems: one, associated with the hypothalamus, which selected between behavior types (e.g., seek food vs. rest); and a second, associated with the tectum, which resolves selection between spatially oriented actions (e.g., turn to the right vs. left).

Around 540 million years ago, the world's fauna underwent a dramatic change called the "Cambrian Explosion," which led to the appearance of all of the major animal phyla. It is believed that this explosion of animal diversity was stimulated by the advent of active predation, initiated by the arthropods, leading to an "arms race" of predators and prey. During those tumultuous times, many species of animals died out, and many new strategies proved themselves valuable for survival. Among our chordate ancestors, three strategies appear to have succeeded, each exemplified by one of the three remaining chordate branches: The cephalochordates, like Amphioxus, survived by spending most of their time hiding buried in the ocean floor. The tunicates, like sea squirts, built a tough shell and returned to a sedentary life. The vertebrates, in contrast, became experts in fast swimming and later, with the advent of jaws, turned the tables and became expert predators as well.

Accompanying the explosion of body types, there was great innovation in nervous system complexity, especially in sensory processing. Among the vertebrates, there was the emergence of paired image-forming eyes (Shu et al., 2003) accompanied by a dramatic expansion of the dorsal portion of the caudal hypothalamus into what is now called the telencephalon. In all vertebrates, it is composed of two parts. The first is a spatially topographic sensorimotor region called the pallium, which in mammals corresponds to the cerebral cortex, hippocampus, septum, and "pallial" amygdala. The second is a modulatory region called the subpallium, which in mammals corresponds to the basal ganglia (BG), along with the "subpallial" amygdala and other subcortical structures (Butler & Hodos, 2005; Puelles, Harrison, Paxinos, & Watson, 2013). All of these systems have become incredibly elaborate along the human lineage, but we can infer much about their original roles by studying animals that are "living fossils," such as lamprey, which diverged from our lineage long ago and have not dramatically altered their way of life (Gess, Coates, & Rubidge, 2006).

Lamprey possess nearly all components of the mammalian BG circuits (Grillner & Robertson, 2015), including input from the pallium through the striatum, and output through the GP to the thalamus and pallium and through the SNpr to the brainstem and tectum. As reviewed above for mammals, lamprey possess direct, indirect, and hyperdirect routes through this circuit, with cell types and neurotransmitters conserved among all vertebrates studied to date. Thus, it is reasonable to suggest that whatever functions the basal ganglia were performing 500 million years ago, they are still performing similar ones today.

Grillner and colleagues propose that the main role of the basal ganglia, in lamprey and other vertebrates, is to select between the different types of species-typical actions available in an animal's behavioral repertoire (Grillner, Robertson, & Stephenson-Jones, 2013). For lamprey, this is relatively limited and includes such behaviors as orienting, locomotion, and feeding. As animals evolved more complex body types capable of a richer set of behaviors, there was a multiplication of BG modules (Stephenson-Jones, Samuelsson, Ericsson, Robertson, & Grillner, 2011) to support the selection among this growing set of behaviors. In primates it includes a great variety of action types, including reaching and grasping, object manipulation, hand-to-mouth movements, defensive responses, several kinds of locomotion (walking, climbing, and leaping), grooming oneself and others, as well as vocalization and other social behaviors. Thus, we can define a third type of decision-making system – the "subpallial system" – which selects between different types of action within an animal's behavioral repertoire.

Deciding whether you want to grasp something within reach or instead walk elsewhere is important, but it doesn't solve the whole problem of goal-directed action. One must also select what it is you'll reach out for, or where you'll transport yourself. These are fundamentally spatial decisions, where the options themselves are defined by the geometrical arrangement of objects around the body. For reaching-and-grasping, the critical information concerns the locations of objects that afford grasping. For locomotor guidance, it is the layout of obstacles that constrain one's movement through space. In each case, information from spatial modalities must be used to specify the potential actions made possible by the current environment, and a selection must be made between these different potential actions. We propose that these kinds of spatial decisions are made by a fourth decision system, which we will call the "pallial system." As noted above, this evolved along with the subpallial system as an integrated telencephalic circuit around the time of the Cambrian Explosion, and its basic topological features are shared by all vertebrates (Butler & Hodos, 2005).

We propose that these two systems, pallial and subpallial, together implement spatial goal-directed action through a simple division of labor: The pallium (e.g., cerebral cortex) includes dedicated sensorimotor streams for different types of species-typical behavior (Ocana et al., 2015), each of which implements a "within behavior" competition between different potential actions of the same type (Cisek, 2007). Meanwhile the subpallium (e.g., the basal ganglia) makes the "between behaviors" decision on which of these sensorimotor streams is released into execution at a given time. For example, in a lizard the subpallium might select the behavior of running (as opposed to feeding), while the pallium selects between different ways to run around objects in its world.

The basic topology of the vertebrate brain, seen already in lamprey, has been remarkably conserved for the last 500 million years. One novel innovation was the evolution of the cerebellum, which appeared between 520 and 490 million years ago. Since then, most changes were morphological rather than topological, though of course many of them were nevertheless quite significant. Along the human lineage, a significant morphological change involved the massive expansion of the dorsal pallium, which occurred 300–250 million years ago in concert with the great diversification of tetrapods during the Carboniferous period. As in the evolution of many neural systems, expansion was accompanied by lamination (Striedter, 2005), leading to the layered structure we call the mammalian cerebral cortex. In the scheme we describe here, the role of the cerebral cortex in decision-making is analogous to that of the pallial system from which it evolved – the specification and selection of spatial actions. However, given the dramatic increase in the complexity and richness of the actions that could now be specified, and the fine resolution of the selection among them, we consider it appropriate to define this as the emergence of a new, fifth decision-making system, which we'll call the "cortical system."

Similar to the lamprey pallium (Ocana et al., 2015), the mammalian cerebral cortex appears to contain within it parallel sensorimotor streams dedicated to different types of species-typical behaviors. As in the lamprey pallium, long-duration microstimulation of specific regions of mammalian sensorimotor cortex can elicit a variety of ethologically relevant action types. For example, Michael Graziano and colleagues showed that stimulation within specific regions of the premotor and motor cortex of rhesus monkeys can elicit reach-to-grasp actions, manipulation in central space, hand-to-mouth, defensive actions, and climbing-like actions (Graziano, 2016; Graziano, Taylor, & Moore, 2002). These findings have been replicated by Jon Kaas and colleagues, who extended them to the posterior parietal cortex of rhesus monkeys (Gharbawie, Stepniewska, Qi, & Kaas, 2011b), prosimian galagos (Stepniewska, Fang, & Kaas, 2005), squirrel and owl monkeys (Gharbawie, Stepniewska, & Kaas, 2011a), and Leah Krubitzer and colleagues who found similar results in tree shrews (Baldwin, Cooke, & Krubitzer, 2017). That similar maps of actions have also been observed in squirrels (Cooke, Padberg, Zahner, & Krubitzer, 2012) and rats (Brown & Teskey, 2014) supports the proposal that the sensorimotor regions of the cerebral cortex are organized in terms of ethological action maps (Graziano, 2016). Most importantly, the topographic consistency of action maps across related species, as well as their existence in diverse mammalian taxa, suggests that they are a fundamental principle of cortical organization. These cortical action maps differentiated and expanded in concert with the expanding repertoire of species-typical behaviors of which animals become capable during their evolution.

As the repertoire of behavioral types expanded in evolution, so did the sophistication and variety of the movements within each type. For example, in anthropoid primates, whose forelimbs are capable of very fine motor skills, the cortical regions associated with manipulation actions are large, permitting great variability and control of specific types of manipulation actions. Nevertheless, you only have two hands, and each hand can only accomplish one action at a given time. Thus, within each region associated with each type of action, a competition between different specific actions must be resolved. As described above, the affordance competition hypothesis (Cisek, 2007; Cisek & Kalaska, 2010) suggests that the dorsal visual system (Milner & Goodale, 1995) specifies, in parallel, several potential actions currently available in the world, and that these actions compete against each other for overt execution through a reciprocal competition biased by a variety of sources of selection-relevant information. Here, we propose that in addition to the "within behavior" competition that plays out within specific fronto-parietal action streams, there is a "between behaviors" competition between different streams, mediated by the basal ganglia (Grillner et al., 2013).

Finally, along the primate branch, innovations in cortical processing made possible new variations of decision-making. Passingham and Wise (2012) review data suggesting that granular regions of the prefrontal cortex (PFC) appear in primates along with advanced foraging strategies necessary for their arboreal niche, in which high-value food items are distributed unevenly between sites separated by regions with high predation risk. The orbitofrontal cortex (OFC) and caudal PFC were the first to appear, and respectively conferred the ability to select between predicted outcomes associated with specific objects and to establish sub-goals between them and the actions that are currently available. Later PFC areas appeared in anthropoid primates and allowed the generation of sub-goals from current context and recent experience. This motivates us to define a final, sixth decision-making system which we'll call the "prefrontal system." It is an extension of the concept of ethological action maps, but to an increasingly abstract domain of sub-goals and outcomes, linked together through the brain's ability to predict the consequences of actions at multiple levels of abstraction (Pezzulo & Cisek, 2016).

A New Perspective

Figure 5.4 illustrates the general scheme at which we have now arrived. It maps the distinction between the subpallial/BG and the pallial/cortical decision systems on the brain of a rhesus monkey. The fronto-parietal cortical regions are organized into parallel streams for sensorimotor guidance of different types of actions (gaze orientation, reach-to-grasp, hand-to-mouth, defensive actions, etc.). Each of these streams receives input appropriate for specifying those types of actions. For example, the reaching system includes MIP and the interconnected PMd (Caminiti, Ferraina, & Johnson, 1996), and receives proprioceptive information about limb configuration as well as visual information about targets in reachable space (Andersen, Snyder, Bradley, & Xing, 1997; Gallivan, Cavina-Pratesi, & Culham, 2009). Meanwhile, the gaze orientation system includes LIP and the interconnected FEF, and receives information about visual targets in retinotopic coordinates as well as proprioceptive information about head position (Andersen et al., 1997). Each of these ethologically defined zones independently resolves the competition between potential actions of that type (e.g., different directions to reach, different directions to saccade), but each also

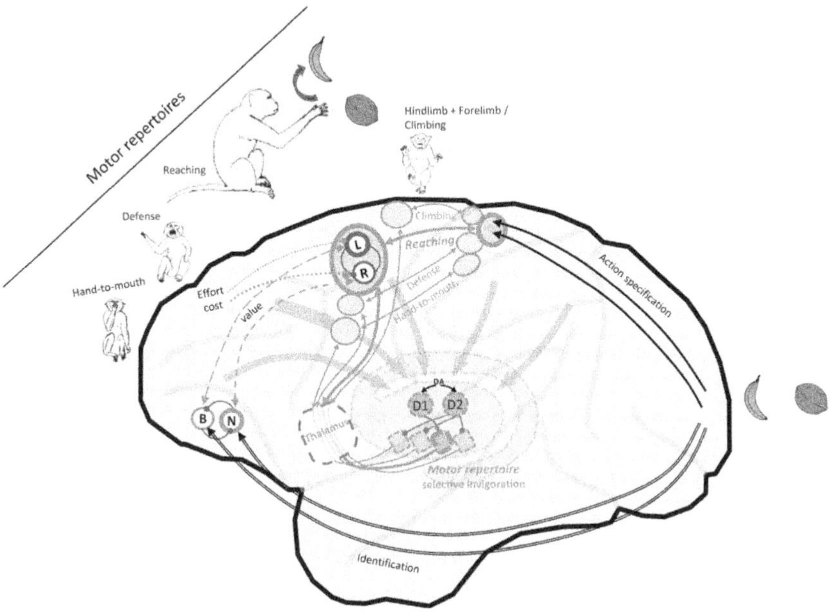

FIGURE 5.4 Schematic illustration of putative mechanisms of between- and within-behavior action selection mapped onto the rhesus monkey brain

Notes: The premotor/motor and rostral posterior parietal cortices are organized in "zones" of ethologically relevant types of motor behaviors (gray ovals), such as reaching, climbing, defensive, or hand-to-mouth movements, defined according to long-train intracortical microstimulation experiments (see Graziano, 2016; Kaas & Stepniewska, 2016). Here, the reaching behavior is selected and invigorated while other behaviors are suppressed through a cortico-basal ganglia (BG)-thalamo-cortical loop. In this loop, the dopamine (DA)-dependent MSNs of the putamen (dotted gray circles) receive massive input from all cortical areas (wide gray arrows) and project to downstream structures of the BG organized as segregated modules (dotted gray squares) (see Grillner et al., 2013) connected to a given cortical zone. The output nuclei of the BG activate the reaching system through projections to the appropriate cortical zone via connections in the thalamus. In this reaching system, available actions specified along the parietal cortex (thick black arrows) compete during the deliberation period (black circles illustrated in the premotor/motor cortex, L: cells voting for a leftward reach; R: cells voting for a rightward reach). The competition is biased by value signals specified in OFC (gray circles, B: cells voting for the banana; N: cells voting for the nut) on the basis of stimulus identification mechanisms in temporal cortex (black double-line arrows) and by effort costs associated with each action computed in anterior cingulate cortex (dotted gray lines). In this example, a banana and a nut afford two potential reaching movements. The competition within the reaching system is biased by the monkey's preference (favoring the nut) and effort cost (favoring the banana). Because the effort cost associated with eating a nut is very high, it prevails over the slight preference for the nut and consequently, reaching neurons in the premotor/motor cortex commit for a leftward movement to reach the banana. Monkey images reprinted with permission from Graziano, 2016.

receives a signal from the basal ganglia that selectively invigorates a given type of action (Figure 5.2F, Thura & Cisek, 2017). In keeping with the old division of labor between subpallium and pallium, the signal from the basal ganglia is specific only with respect to a given action type, leaving the choice of the specific action of that type to be determined by the relevant zone of the cortex.

As an example of how these different decision systems can cooperate to guide goal-directed activity, consider the following example. Suppose a monkey starts to feel hungry and decides to seek food. This is an example of a top-level "hypothalamic" decision, which motivates the activation of telencephalic systems for finding and obtaining food sources. Within the telencephalon, the basal ganglia first selectively disinhibit the locomotion system (a "subpallial" decision to move through space), which presumably includes regions of parietal, premotor, and primary motor cortex involved in the guidance of walking (Andujar, Lajoie, & Drew, 2010) as well as the navigation circuits of the hippocampus (O'Keefe & Nadel, 1978). If the behavior succeeds, and the animal arrives at a site with some graspable food objects, the basal ganglia now inhibit locomotion and instead disinhibit the reach-to-grasp system (MIP, PMd/M1). Once disinhibited, that system now begins to specify the potential reaching movements to several graspable objects, which compete against each other within the cortical reach regions (a "pallial/cortical" decision begins). This competition is biased by a variety of factors, including the value of the different food items, possibly computed in orbitofrontal cortex (Wallis, 2007), as well as the energetic costs of the required actions, possibly computed in the anterior cingulate cortex (Kennerley, Behrens, & Wallis, 2011). In the example shown on Figure 5.4, the nutritional value of the nut on the right is greater than that of the banana on the left, but comes at a higher energetic cost. Consequently, the competition is ultimately resolved in favor of reaching to the left and grasping the banana. This nested process continues, with a subpallial decision to disinhibit hand-to-mouth actions, followed by feeding actions, until finally the banana is consumed, the top-level hypothalamic needs are satisfied, and the animal can rest.

Can this scheme, focused so strongly on embodied decisions, apply to abstract decisions that do not refer to actions at all? It is difficult to say. Perhaps economic choices and other purely abstract decisions are simply different, and engage circuits that have nothing in common with those sketched in Figure 5.4. Perhaps they involve serial architectures closer to those classically defined in traditional cognitive psychology (Newell & Simon, 1972; Pylyshyn, 1984). While this is plausible, evolutionary considerations motivate us to explore the possibility that abstract decisions have themselves emerged as specializations of the circuits underlying embodied choices, and that, following Piaget (1963), cognition is built on a foundation of sensorimotor control.

Pezzulo and Cisek (2016) proposed a way in which the "affordance competition hypothesis" can be extended to address these kinds of more abstract goal-directed behaviors. Key to their proposal is the brain's ability to predict the consequences of actions, using what in sensorimotor control is called "forward models" and often ascribed to the cerebellum (Miall, 1998). A forward model takes the efference copy of the current motor command and together with information about the action context, predicts the sensory consequences of that command. This is useful because it allows effectively "zero-lag" closed loop control, yielding faster and more efficient action execution. That same principle can be applied to higher-order consequences of actions, such as the prediction that moving forward brings objects in front of you into reach. This allows the nervous system to predict how currently available actions can make new actions possible – how a given action choice leads to the creation of new affordances. Crucially, the cerebellum forms closed loops with all parts of the frontal cortex, including motor execution regions like M1 as well as more "abstract" regions such as prefrontal cortex (Middleton & Strick, 2000). If it implements a predictive forward model, then it may provide that for many levels of abstraction – from predicting the proprioceptive consequences of a given muscle contraction to the social consequences of a particular treatment of conspecifics. Based in part on the proposed architecture of primate prefrontal areas described by Passingham and Wise (2012) and the predictive control framework of Friston (2010), Pezzulo and Cisek (2016) proposed that the expansion of the frontal cortex in primates expanded the architecture of brain circuitry allowing the extension of concrete and instantaneous planning to more abstract and more temporally delayed activities. Admittedly, this is still far from abstract decision-making and human cognition, but we propose that it may be a phylogenetically plausible step in the right direction.

Conclusions

A central goal of neuroscience is to understand how the brain chooses and controls actions. This chapter focused on the first part of this fascinating question: how the brain selects an action among multiple alternatives. Major advancements in our understanding of the neural mechanisms of action selection were made by studying decisions about actions in human and non-human primates alongside single-neuron recording to evaluate the underlying mechanisms. By showing a signature of action selection in brain areas traditionally associated with action preparation, execution, and regulation, these neurophysiological data rejected theories from classic cognitive psychology stipulating that all decisions are computed in an abstract form, outside of the sensorimotor network. Instead, a wealth of neuroscientific data now argues that decisions, at least those concerning actions, are made within the same regions and the same circuits that guide movement execution. This includes sensorimotor regions of the frontal and parietal cerebral cortex as well as interconnected parts of the basal ganglia, although the precise role of each region is still under lively debate. Also under debate is whether the neural mechanisms of abstract decisions, divorced from specific actions, have any relation to the circuits underlying sensorimotor decisions.

Here, we have reviewed some of the data relevant to these debates, interpreting them within the context of the evolution of neural systems. We propose that this context can help resolve many outstanding questions, leading to a putative sequence of how simple decisions (approach vs. avoid, exploit vs. explore) may have evolved into the complex and abstract decision-making behavior that is the hallmark of human cognition.

References

Albin, R.L., Young, A.B., & Penney, J.B. (1989). The functional anatomy of basal ganglia disorders. *Trends Neurosci, 12*, 366–375.

Andersen, R.A., Snyder, L.H., Bradley, D.C., & Xing, J. (1997). Multimodal representation of space in the posterior parietal cortex and its use in planning movements. *Annu Rev Neurosci, 20*, 303–330.

Anderson, M.E., & Horak, F.B. (1985). Influence of the globus pallidus on arm movements in monkeys. III. Timing of movement-related information. *J Neurophysiol, 54*, 433–448.

Andujar, J.E., Lajoie, K., & Drew, T. (2010). A contribution of area 5 of the posterior parietal cortex to the planning of visually guided locomotion: limb-specific and limb-independent effects. *J Neurophysiol, 103*, 986–1006.

Arendt, D., Tosches, M.A., & Marlow, H. (2016). From nerve net to nerve ring, nerve cord and brain – evolution of the nervous system. *Nat Rev Neurosci, 17*, 61–72.

Arimura, N., Nakayama, Y., Yamagata, T., Tanji, J., & Hoshi, E. (2013). Involvement of the globus pallidus in behavioral goal determination and action specification. *J Neurosci, 33*, 13639–13653.

Baldwin, M.K., Cooke, D.F., & Krubitzer, L. (2017). Intracortical microstimulation maps of motor, somatosensory, and posterior parietal cortex in tree shrews (*tupaia belangeri*) reveal complex movement representations. *Cereb Cortex, 27*(2), 1439–1456.

Basso, M.A., & Wurtz, R.H. (1998). Modulation of neuronal activity in superior colliculus by changes in target probability. *J Neurosci, 18*, 7519–7534.

Baumann, M.A., Fluet, M.C., & Scherberger, H. (2009). Context-specific grasp movement representation in the macaque anterior intraparietal area. *Journal of Neuroscience, 29*, 6436–6448.

Bennur, S., & Gold, J.I. (2011). Distinct representations of a perceptual decision and the associated oculomotor plan in the monkey lateral intraparietal area. *J Neurosci, 31*, 913–921.

Brown, A.R., & Teskey, G.C. (2014). Motor cortex is functionally organized as a set of spatially distinct representations for complex movements. *J Neurosci, 34*, 13574–13585.

Brunet, T., & Arendt, D. (2016). From damage response to action potentials: early evolution of neural and contractile modules in stem eukaryotes. *Philos T R Soc B, 371*.

Butler, A.B., & Hodos, W. (2005). *Comparative vertebrate neuroanatomy: evolution and adaptation* (2nd ed.). Hoboken, NJ: Wiley-Interscience.

Calabresi, P., Picconi, B., Tozzi, A., Ghiglieri, V., & Di Filippo, M. (2014). Direct and indirect pathways of basal ganglia: a critical reappraisal. *Nat Neurosci, 17*, 1022–1030.

Caminiti, R., Ferraina, S., & Johnson, P.B. (1996). The sources of visual information to the primate frontal lobe: a novel role for the superior parietal lobule. *Cereb Cortex, 6*, 319–328.

Cazorla, M., de Carvalho, F.D., Chohan, M.O., Shegda, M., Chuhma, N., Rayport, S., Ahmari, S.E., Moore, H., & Kellendonk, C. (2014). Dopamine D2 receptors regulate the anatomical and functional balance of basal ganglia circuitry. *Neuron, 81*, 153–164.

Cisek, P. (2007). Cortical mechanisms of action selection: the affordance competition hypothesis. *Philos Trans R Soc Lond B Biol Sci, 362*, 1585–1599.

Cisek, P. (2012). Making decisions through a distributed consensus. *Curr Opin Neurobiol, 22*, 927–936.

Cisek, P., & Kalaska, J.F. (2005). Neural correlates of reaching decisions in dorsal premotor cortex: specification of multiple direction choices and final selection of action. *Neuron, 45*, 801–814.

Cisek, P., & Kalaska, J.F. (2010). Neural mechanisms for interacting with a world full of action choices. *Annu Rev Neurosci, 33*, 269–298.

Cisek, P., & Pastor-Bernier, A. (2014). On the challenges and mechanisms of embodied decisions. *Philos Trans R Soc Lond B Biol Sci, 369*.

Cisek, P., Puskas, G.A., & El-Murr, S. (2009). Decisions in changing conditions: the urgency-gating model. *J Neurosci, 29*, 11560–11571.

Coles, M.G.H., Gratton, G., Bashore, T.R., Eriksen, C.W., & Donchin, E. (1985). A psychophysiological investigation of the continuous-flow model of human information-processing. *J Exp Psychol Human, 11*, 529–533.

Comoli, E., Das Neves Favaro, P., Vautrelle, N., Leriche, M., Overton, P.G., & Redgrave, P. (2012). Segregated anatomical input to sub-regions of the rodent superior colliculus associated with approach and defense. *Front Neuroanat, 6*, 9.

Cooke, D.F., Padberg, J., Zahner, T., & Krubitzer, L. (2012). The functional organization and cortical connections of motor cortex in squirrels. *Cereb Cortex, 22*, 1959–1978.

Cox, S.M., Frank, M.J., Larcher, K., Fellows, L.K., Clark, C.A., Leyton, M., & Dagher, A. (2015). Striatal D1 and D2 signaling differentially predict learning from positive and negative outcomes. *Neuroimage, 109*, 95–101.

Cui, G., Jun, S.B., Jin, X., Pham, M.D., Vogel, S.S., Lovinger, D.M., & Costa, R.M. (2013). Concurrent activation of striatal direct and indirect pathways during action initiation. *Nature, 494*, 238–242.

Dayan, P., & Daw, N.D. (2008). Decision theory, reinforcement learning, and the brain. *Cogn Affect Behav Neurosci, 8*, 429–453.

Dekleva, B.M., Ramkumar, P., Wanda, P.A., Kording, K.P., & Miller, L.E. (2016). Uncertainty leads to persistent effects on reach representations in dorsal premotor cortex. *Elife, 5*.

DeLong, M.R. (1990). Primate models of movement disorders of basal ganglia origin. *Trends Neurosci, 13*, 281–285.

Desmurget, M., & Turner, R.S. (2008). Testing basal ganglia motor functions through reversible inactivations in the posterior internal globus pallidus. *J Neurophysiol, 99*, 1057–1076.

Desmurget, M., & Turner, R.S. (2010). Motor sequences and the basal ganglia: kinematics, not habits. *J Neurosci, 30*, 7685–7690.

Ding, L., & Gold, J.I. (2010). Caudate encodes multiple computations for perceptual decisions. *J Neurosci, 30*, 15747–15759.

Dorris, M.C., & Glimcher, P.W. (2004). Activity in posterior parietal cortex is correlated with the relative subjective desirability of action. *Neuron, 44*, 365–378.

Dudman, J.T., & Krakauer, J.W. (2016). The basal ganglia: from motor commands to the control of vigor. *Curr Opin Neurobiol, 37*, 158–166.

Erwin, D.H., Laflamme, M., Tweedt, S.M., Sperling, E.A., Pisani, D., & Peterson, K.J. (2011). The Cambrian conundrum: early divergence and later ecological success in the early history of animals. *Science, 334*, 1091–1097.

Filimon, F., Philiastides, M.G., Nelson, J.D., Kloosterman, N.A., & Heekeren, H.R. (2013). How embodied is perceptual decision making? Evidence for separate processing of perceptual and motor decisions. *Journal of Neuroscience, 33*, 2121–2136.

Frank, M.J. (2005). Dynamic dopamine modulation in the basal ganglia: a neurocomputational account of cognitive deficits in medicated and nonmedicated Parkinsonism. *J Cogn Neurosci, 17*, 51–72.

Frank, M.J. (2011). Computational models of motivated action selection in corticostriatal circuits. *Curr Opin Neurobiol, 21*, 381–386.

Freeze, B.S., Kravitz, A.V., Hammack, N., Berke, J.D., & Kreitzer, A.C. (2013). Control of basal ganglia output by direct and indirect pathway projection neurons. *J Neurosci, 33*, 18531–18539.

Friston, K. (2010). The free-energy principle: a unified brain theory? *Nat Rev Neurosci, 11*, 127–138.

Gallivan, J.P., Cavina-Pratesi, C., & Culham, J.C. (2009). Is that within reach? fMRI reveals that the human superior parieto-occipital cortex encodes objects reachable by the hand. *J Neurosci, 29*, 4381–4391.

Gess, R.W., Coates, M.I., & Rubidge, B.S. (2006). A lamprey from the Devonian period of South Africa. *Nature, 443*, 981–984.

Gharbawie, O.A., Stepniewska, I., & Kaas, J.H. (2011a). Cortical connections of functional zones in posterior parietal cortex and frontal cortex motor regions in new world monkeys. *Cerebral Cortex, 21*, 1981–2002.

Gharbawie, O.A., Stepniewska, I., Qi, H.X., & Kaas, J.H. (2011b). Multiple parietal-frontal pathways mediate grasping in macaque monkeys. *Journal of Neuroscience, 31*, 11660–11677.

Gibson, J.J. (1979). *The ecological approach to visual perception*. Boston: Houghton Mifflin.

Gold, J.I., & Shadlen, M.N. (2000). Representation of a perceptual decision in developing oculomotor commands. *Nature, 404*, 390–394.

Gold, J.I., & Shadlen, M.N. (2007). The neural basis of decision making. *Annu Rev Neurosci, 30*, 535–574.

Graziano, M.S. (2016). Ethological action maps: a paradigm shift for the motor cortex. *Trends Cogn Sci, 20*, 121–132.

Graziano, M.S., Taylor, C.S., & Moore, T. (2002). Complex movements evoked by microstimulation of precentral cortex. *Neuron, 34*, 841–851.

Grillner, S., & Robertson, B. (2015). The basal ganglia downstream control of brainstem motor centres – an evolutionarily conserved strategy. *Current Opinion in Neurobiology, 33*, 47–52.

Grillner, S., Robertson, B., & Stephenson-Jones, M. (2013). The evolutionary origin of the vertebrate basal ganglia and its role in action selection. *J Physiol-London, 591*, 5425–5431.

Gurney, K., Prescott, T.J., & Redgrave, P. (2001). A computational model of action selection in the basal ganglia. I. A new functional anatomy. *Biol Cybern, 84*, 401–410.

Horak, F.B., & Anderson, M.E. (1984a). Influence of globus pallidus on arm movements in monkeys. I. Effects of kainic acid-induced lesions. *J Neurophysiol, 52*, 290–304.

Horak, F.B., & Anderson, M.E. (1984b). Influence of globus pallidus on arm movements in monkeys. II. Effects of stimulation. *J Neurophysiol, 52*, 305–322.

Horwitz, G.D., Batista, A.P., & Newsome, W.T. (2004). Representation of an abstract perceptual decision in macaque superior colliculus. *J Neurophysiol, 91*, 2281–2296.

Janssen, P., & Shadlen, M.N. (2005). A representation of the hazard rate of elapsed time in macaque area LIP. *Nat Neurosci, 8*, 234–241.

Kaas, J.H., & Stepniewska, I. (2016). Evolution of posterior parietal cortex and parietal-frontal networks for specific actions in primates. *J Comp Neurol, 524*, 595–608.

Kennerley, S.W., Behrens, T.E., & Wallis, J.D. (2011). Double dissociation of value computations in orbitofrontal and anterior cingulate neurons. *Nat Neurosci, 14*, 1581–1589.

Kim, J.N., & Shadlen, M.N. (1999). Neural correlates of a decision in the dorsolateral prefrontal cortex of the macaque. *Nat Neurosci, 2*, 176–185.

Kitai, S.T., & Deniau, J.M. (1981). Cortical inputs to the subthalamus: intracellular analysis. *Brain Res, 214*, 411–415.

Klaes, C., Westendorff, S., Chakrabarti, S., & Gail, A. (2011). Choosing goals, not rules: deciding among rule-based action plans. *Neuron, 70*, 536–548.

Kravitz, A.V., Freeze, B.S., Parker, P.R., Kay, K., Thwin, M.T., Deisseroth, K., & Kreitzer, A.C. (2010). Regulation of Parkinsonian motor behaviours by optogenetic control of basal ganglia circuitry. *Nature, 466*, 622–626.

Lacalli, T.C. (1996). Frontal eye circuitry, rostral sensory pathways and brain organization in amphioxus larvae: evidence from 3D reconstructions. *Philos T Roy Soc B, 351*, 243–263.

Lacalli, T.C. (2008). Basic features of the ancestral chordate brain: a protochordate perspective. *Brain Res Bull, 75*, 319–323.

Leblois, A., Boraud, T., Meissner, W., Bergman, H., & Hansel, D. (2006). Competition between feedback loops underlies normal and pathological dynamics in the basal ganglia. *J Neurosci, 26*, 3567–3583.

Lepora, N.F., & Pezzulo, G. (2015). Embodied choice: how action influences perceptual decision making. *PLoS Comput Biol, 11*.

Mackie, G.O. (1970). Neuroid conduction and evolution of conducting tissues. *Q Rev Biol, 45*, 319–332.

Matelli, M., & Luppino, G. (2001). Parietofrontal circuits for action and space perception in the macaque monkey. *Neuroimage, 14*, S27–S32.

Mazzoni, P., Hristova, A., & Krakauer, J.W. (2007). Why don't we move faster? Parkinson's disease, movement vigor, and implicit motivation. *J Neurosci, 27*, 7105–7116.

McPeek, R.M., Han, J.H., & Keller, E.L. (2003). Competition between saccade goals in the superior colliculus produces saccade curvature. *J Neurophysiol, 89*, 2577–2590.

Miall, R.C. (1998). The cerebellum, predictive control and motor coordination. *Novartis Found Symp, 218*, 272–284; discussion 284–290.

Middleton, F.A., & Strick, P.L. (2000). Basal ganglia and cerebellar loops: motor and cognitive circuits. *Brain Res Brain Res Rev, 31*, 236–250.

Milner, A.D., & Goodale, M.A. (1995). *The visual brain in action*. Oxford; New York: Oxford University Press.

Mink, J.W. (1996). The basal ganglia: focused selection and inhibition of competing motor programs. *Prog Neurobiol, 50*, 381–425.

Mink, J.W., & Thach, W.T. (1991a). Basal ganglia motor control. II. Late pallidal timing relative to movement onset and inconsistent pallidal coding of movement parameters. *J Neurophysiol, 65*, 301–329.

Mink, J.W., & Thach, W.T. (1991b). Basal ganglia motor control. III. Pallidal ablation: normal reaction time, muscle cocontraction, and slow movement. *J Neurophysiol, 65*, 330–351.

Nakano, K. (2000). Neural circuits and topographic organization of the basal ganglia and related regions. *Brain Dev, 22*, Suppl 1, S5–16.

Nakayama, Y., Yamagata, T., & Hoshi, E. (2016). Rostrocaudal functional gradient among the pre-dorsal premotor cortex, dorsal premotor cortex and primary motor cortex in goal-directed motor behaviour. *Eur J Neurosci, 43*, 1569–1589.

Newell, A., & Simon, H.A. (1972). *Human problem solving.* Englewood Cliffs, NJ: Prentice-Hall.

Northcutt, R.G. (2012). Evolution of centralized nervous systems: two schools of evolutionary thought. *P Natl Acad Sci USA, 109*, 10626–10633.

O'Connell, R.G., Dockree, P.M., & Kelly, S.P. (2012). A supramodal accumulation-to-bound signal that determines perceptual decisions in humans. *Nature Neuroscience, 15*, 1729–1735.

O'Keefe, J., & Nadel, L. (1978). *The hippocampus as a cognitive map.* Oxford: Clarendon Press.

Ocana, F.M., Suryanarayana, S.M., Saitoh, K., Kardamakis, A.A., Capantini, L., Robertson, B., & Grillner, S. (2015). The lamprey pallium provides a blueprint of the mammalian motor projections from cortex. *Curr Biol, 25*, 413–423.

Padoa-Schioppa, C. (2011). Neurobiology of economic choice: a good-based model. *Annual Review of Neuroscience, 34*, 333–359.

Pasquereau, B., Nadjar, A., Arkadir, D., Bezard, E., Goillandeau, M., Bioulac, B., Gross, C.E., & Boraud, T. (2007). Shaping of motor responses by incentive values through the basal ganglia. *J Neurosci, 27*, 1176–1183.

Passingham, R.E., & Wise, S.P. (2012). *The neurobiology of the prefrontal cortex: anatomy, evolution, and the origin of insight.* Oxford: Oxford University Press.

Pastor-Bernier, A., & Cisek, P. (2011). Neural correlates of biased competition in premotor cortex. *J Neurosci, 31*, 7083–7088.

Pezzulo, G., & Cisek, P. (2016). Navigating the affordance landscape: feedback control as a process model of behavior and cognition. *Trends Cogn Sci, 20*, 414–424.

Piaget, J. (1963). *The origins of intelligence in children.* New York: W.W. Norton.

Platt, M.L., & Glimcher, P.W. (1997). Responses of intraparietal neurons to saccadic targets and visual distractors. *J Neurophysiol, 78*, 1574–1589.

Platt, M.L., & Glimcher, P.W. (1999). Neural correlates of decision variables in parietal cortex. *Nature, 400*, 233–238.

Puelles, L., Harrison, M., Paxinos, G., & Watson, C. (2013). A developmental ontology for the mammalian brain based on the prosomeric model. *Trends in Neurosciences, 36*, 570–578.

Pylyshyn, Z.W. (1984). *Computation and cognition: toward a foundation for cognitive science.* Cambridge, MA: MIT Press.

Ramkumar, P., Dekleva, B., Cooler, S., Miller, L., & Kording, K. (2016). Premotor and motor cortices encode reward. *PLoS One, 11*, e0160851.

Redgrave, P., Prescott, T.J., & Gurney, K. (1999). The basal ganglia: a vertebrate solution to the selection problem? *Neuroscience, 89*, 1009–1023.

Romanelli, P., Esposito, V., Schaal, D.W., & Heit, G. (2005). Somatotopy in the basal ganglia: experimental and clinical evidence for segregated sensorimotor channels. *Brain Res Brain Res Rev, 48*, 112–128.

Romo, R., Hernandez, A., & Zainos, A. (2004). Neuronal correlates of a perceptual decision in ventral premotor cortex. *Neuron, 41*, 165–173.

Sahibzada, N., Dean, P., & Redgrave, P. (1986) Movements resembling orientation or avoidance elicited by electrical-stimulation of the superior colliculus in rats. *Journal of Neuroscience, 6*, 723–733.

Saitoh, K., Menard, A., & Grillner, S. (2007). Tectal control of locomotion, steering, and eye movements in lamprey. *Journal of Neurophysiology, 97*, 3093–3108.

Samejima, K., Ueda, Y., Doya, K., & Kimura, M. (2005). Representation of action-specific reward values in the striatum. *Science, 310*, 1337–1340.

Schultz, W. (1997). Dopamine neurons and their role in reward mechanisms. *Curr Opin Neurobiol*, 7, 191–197.

Seo, M., Lee, E., & Averbeck, B.B. (2012). Action selection and action value in frontal-striatal circuits. *Neuron, 74*, 947–960.

Shadlen, M.N., & Newsome, W.T. (2001). Neural basis of a perceptual decision in the parietal cortex (area LIP) of the rhesus monkey. *J Neurophysiol, 86*, 1916–1936.

Shadlen, M.N., Kiani, R., Hanks, T.D., & Churchland, A.K. (2008). Neurobiology of decision making: an intentional framework. In C. Engel & W. Singer (Eds.), *Better than conscious? Decision making, the human mind, and implications for institutions*. Cambridge, MA: MIT Press.

Shimeld, S.M., & Holland, N.D. (2005). Amphioxus molecular biology: insights into vertebrate evolution and developmental mechanisms. *Can J Zool, 83*, 90–100.

Shu, D.G., Morris, S.C., Han, J., Zhang, Z.F., Yasui, K., Janvier, P., Chen, L., Zhang, X.L., Liu, J.N., Li, Y., & Liu, H.Q. (2003). Head and backbone of the Early Cambrian vertebrate Haikouichthys. *Nature, 421*, 526–529.

Stephenson-Jones, M., Samuelsson, E., Ericsson, J., Robertson, B., & Grillner, S. (2011). Evolutionary conservation of the basal ganglia as a common vertebrate mechanism for action selection. *Curr Biol, 21*, 1081–1091.

Stepniewska, I., Fang, P.C., & Kaas, J.H. (2005). Microstimulation reveals specialized subregions for different complex movements in posterior parietal cortex of prosimian galagos. *P Natl Acad Sci USA, 102*, 4878–4883.

Striedter, G.F. (2005). *Principles of brain evolution*. Sunderland, MA: Sinauer Associates, Inc.

Sugrue, L.P., Corrado, G.S., & Newsome, W.T. (2004). Matching behavior and the representation of value in the parietal cortex. *Science, 304*, 1782–1787.

Thura, D., & Cisek, P. (2014). Deliberation and commitment in the premotor and primary motor cortex during dynamic decision making. *Neuron, 81*, 1401–1416.

Thura, D., & Cisek, P. (2016). Modulation of premotor and primary motor cortical activity during volitional adjustments of speed-accuracy trade-offs. *J Neurosci, 36*, 938–956.

Thura, D., & Cisek, P. (2017). The basal ganglia do not select reach targets but control the urgency of commitment. *Neuron*, 95, 1160–1170 e1165.

Thura, D., Guberman, G., & Cisek, P. (2017). Trial-to-trial adjustments of speed-accuracy trade-offs in premotor and primary motor cortex. *J Neurophysiol, 117*, 665–683.

Thura, D., Beauregard-Racine, J., Fradet, C.W., & Cisek, P. (2012). Decision making by urgency gating: theory and experimental support. *J Neurophysiol, 108*, 2912–2930.

Thura, D., Cos, I., Trung, J., & Cisek, P. (2014). Context-dependent urgency influences speed-accuracy trade-offs in decision-making and movement execution. *J Neurosci, 34,* 16442–16454.

Tosches, M.A., & Arendt, D. (2013). The bilaterian forebrain: an evolutionary chimaera. *Current Opinion in Neurobiology, 23,* 1080–1089.

Tosoni, A., Galati, G., Romani, G.L., & Corbetta, M. (2008). Sensory-motor mechanisms in human parietal cortex underlie arbitrary visual decisions. *Nature Neuroscience, 11,* 1446–1453.

Turner, R.S., & Anderson, M.E. (1997). Pallidal discharge related to the kinematics of reaching movements in two dimensions. *J Neurophysiol, 77,* 1051–1074.

Turner, R.S., & Desmurget, M. (2010). Basal ganglia contributions to motor control: a vigorous tutor. *Curr Opin Neurobiol, 20,* 704–716.

Wallis, J.D. (2007). Orbitofrontal cortex and its contribution to decision-making. *Annu Rev Neurosci, 30,* 31–56.

Yttri, E.A., & Dudman, J.T. (2016). Opponent and bidirectional control of movement velocity in the basal ganglia. *Nature, 533,* 402–406.

6

HOW SEPARATE ARE REACHING AND GRASPING?

Adam G. Rouse, Kevin A. Mazurek, Zheng Liu, Gil Rivlis, and Marc H. Schieber

Introduction

Together, reaching and grasping constitute a behavior we all use numerous times each day. Picking up a coffee cup, opening a door, hammering a nail – all are actions that involve reaching out to grasp a physical target object that then is manipulated to achieve a goal. Reaching involves motion of the upper arm and forearm produced by rotation at the shoulder and elbow, and serves to transport the hand to the object. Grasping involves motion of the hand, i.e., the thumb and fingers, which first open wide enough and then close around the object, typically conforming to its shape upon physical contact. In the primary motor cortex (M1) the territory somatotopically representing the proximal upper extremity controls reaching while the territory representing the hand controls grasping. Visual information serving to guide reach-to-grasp movements likewise has been viewed as being processed along two separate cortico-cortical pathways. Information about the spatial location to which the arm must reach travels along a more dorsal pathway from the superior parietal cortex to the dorsal premotor cortex; information regarding the shape of the object the hand must grasp travels a more ventral pathway from the inferior parietal cortex to the ventral premotor cortex (Grafton, 2010; Jeannerod, Arbib, Rizzolatti, & Sakata, 1995; Karl & Whishaw, 2013; Rizzolatti & Luppino, 2001; Rizzolatti, Luppino, & Matelli, 1998). As separate processes, reaching and grasping can proceed at the same time: while the arm is accelerating and then decelerating to transport the hand, the fingers can be opening and then closing to grasp the object.

Thus conceptualized as two separate processes, reaching and grasping by and large each have been studied in experiments that isolate one or the other.

Numerous studies have examined reaching movements to various locations at which the same object (or no object) is grasped. And conversely, other studies have examined grasping a variety of different objects, all at the same location. While such reductionist constraints can be very useful experimentally, they also can constrain our understanding of how reaching and grasping behavior actually is accomplished in more natural situations. Such movements are not necessarily the simple sum of a reach plus a grasp, both performed at the same time. Furthermore, in natural situations grasping is rarely the final goal of a reach-to-grasp movement. Most often, the grasped object promptly is manipulated in some way, even if the manipulation is simply to carry the object from one place to another.

Here we review accumulating evidence that leads us to think differently about the separation between reaching and grasping, providing a new perspective. At multiple levels from the peripheral kinematics to the cerebral cortex, reaching and grasping each involve the entire upper extremity including both the arm and the hand. Rather than being separate processes that proceed concurrently, reaching and grasping can be viewed as two overlapping processes that proceed sequentially. The entire extremity first is projected toward the location of the target, and then shaped to grasp and manipulate the object. Beyond being merely semantic, this distinction becomes valuable for understanding how the nervous system controls natural reach-to-grasp movements.

Kinematics of Reach-to-Grasp Movements

Proximal Versus Distal

We tend to think of reaches as movements performed with the arm, involving rotations at the shoulder and elbow, whereas grasps are movements performed with the hand and involve rotations of the joints of the digits. Early studies indeed suggested that the kinematics of the arm in reaching and those of the hand in grasping were relatively independent of one another. The coordination of motion at the shoulder and elbow as human subjects reached out to grasp a rod in a constant location, for example, were found to be quite similar whether the hand began in an orientation that did not need to change or in an orientation that required the forearm to pronate or supinate during the reach (Lacquaniti & Soechting, 1982). Likewise, when subjects grasped an elongated object positioned at the same location on a tabletop but oriented at different angles, the transport velocity and acceleration of both the elbow and the wrist were unaffected by the object's orientation, even if the forearm had to be pronated to grasp the object (Stelmach, Castiello, & Jeannerod, 1994). Conversely, variation in the location to which the arm reached appeared to have little effect on the motion of joints distal to the elbow. When subjects reached to grasp the same object positioned at different distances from the body as illustrated in

Figure 6.1, for example, the peak transport velocity increased for larger distances, but the time course of grip size (i.e., the aperture measured between the tips of the thumb and index finger) remained constant (Jeannerod, 1984).

Subsequent studies, however, found the kinematics of reaching were not entirely independent of more distal aspects of the movement being performed, and vice versa distal kinematics were not independent of the location to which the arm reached. As human subjects reached to grasp an object positioned either

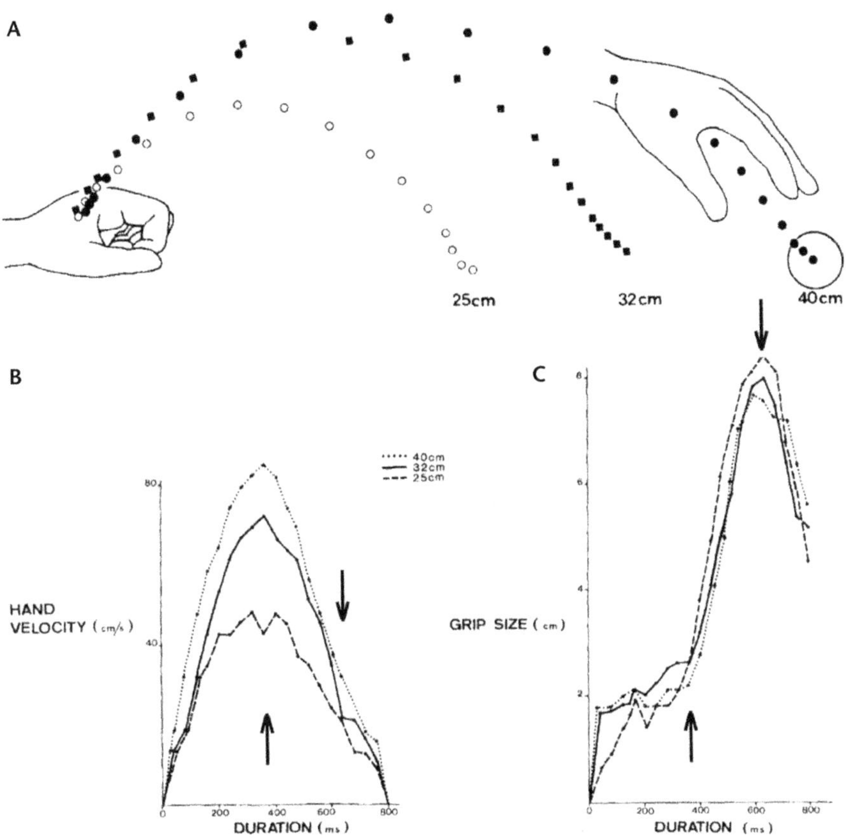

FIGURE 6.1 Reaching to grasp the same object at three different locations

Notes: A. A marker on the side of the hand was tracked in movie frames (50/s) as a human subject reached from the same starting location to grasp the same object placed at three different distances: 25, 32, or 40 cm. B. Hand velocity showed the typical bell-shaped time course. Peak velocity scaled with distance such that movement duration remained constant. C. Grip size (the distance between two markers on the tips of the thumb and index finger) increased and then decreased, following a similar time course for all three distances. Note that grip size (aperture) remained at about one-quarter of its peak value until the time of peak velocity (upward arrows) and then rapidly increased, attaining its peak value as the hand was decelerating, after about three-quarters of the movement duration (downward arrows). Modified from Jeannerod, 1984.

20 cm or 30 cm from the hand's starting position, reach velocity varied not only depending on distance, but also depending on whether a larger object was to be grasped with the whole hand or a small object was to be pinched between the thumb and index finger (Gentilucci et al., 1991), a finding which in part may reflect the general principle that movements are made more slowly when greater accuracy is required (Fitts, 1954). When human subjects matched the location and orientation of a rod held in the hand to the location and orientation of a target rod, orientation errors depended on target location and arm posture as well as on target orientation, suggesting that control of hand orientation was not independent of the arm posture produced by reaching (Soechting & Flanders, 1993). When subjects reached to various locations and actually grasped either a rod with the whole hand (Desmurget et al., 1996) or a small parallelepiped between the thumb and index finger (Gentilucci, Daprati, Gangitano, Saetti, & Toni, 1996), the kinematics of reaching depended not only on the location, but also on the orientation of the object. In humans, the reaching motion of the arm thus is adjusted depending on the orientation and grasp precision required of the hand.

Similar observations have been made in monkeys. In monkeys performing five different tasks – grasping a vertical rod, grasping an oblique rod, grasping a sphere, pressing a radial button, and pressing an oblique button – the posture of the arm was found to vary depending on the distal action being performed (Tillery, Ebner, & Soechting, 1995). In monkeys grasping cylinders of different size placed at the same location, reaching was faster when the object to be grasped was larger (Roy, Paulignan, Meunier, & Boussaoud, 2002), again reflecting the speed-accuracy tradeoff. And in monkeys that grasped and manipulated four different objects in up to eight different locations, rotations at the shoulder and elbow varied not only with location, but also depending on the object (Rouse & Schieber, 2015).

Conversely, grasping was found to vary to some extent depending on the location to which the arm reached. Maximum grip aperture in human subjects, for example, varied depending on reach direction, particularly when the object to be grasped was relatively small (Paulignan, Frak, Toni, & Jeannerod, 1997). And in monkeys, reaching to grasp the same cylinder to the left, center, or right elicited different kinematics not only of wrist velocity, but also of grip aperture (Roy et al., 2002). Moreover, as monkeys reached to grasp four objects in up to eight locations, angles at distal joints – including the carpometacarpal joint of the thumb and the metacarpophalangeal joints of the fingers – varied depending on the location to which the monkeys reached (Rouse & Schieber, 2015). Thus not only is the orientation and grasping motion of the hand dependent to some extent on the location to which the arm reaches, but the motion of the arm in reaching is affected by how the object is about to be grasped by the hand.

This interdependence of reaching and grasping actually should come as no surprise. The skeletal segments of the arm, forearm, hand, and digits all are linked mechanically by the elbow, wrist, and digital joints, none of which have all three rotational degrees of freedom. Any rotation at the shoulder or elbow therefore changes the angular orientation (attitude) of the fingers in earth-fixed space. Rotations at the wrist along with pronation/supination of the forearm provide some ability to maintain the attitude of the fingers as the shoulder and elbow rotate, but the extent of such compensation is limited mechanically. As such limits of compensation are approached, the motion of digits in grasping will vary depending on reach location. Conversely, the differences in object shapes and the manipulations to be performed elicit different grasps, which in turn require different final postures of the reaching arm and therefore different trajectories to arrive at the different postures.

Consider the action of reaching and grasping a jar to pick it up (Figure 6.2). The jar might be grasped from the side, taking advantage of the large contact area afforded between the skin and the glass surface (Figure 6.2A). Alternatively,

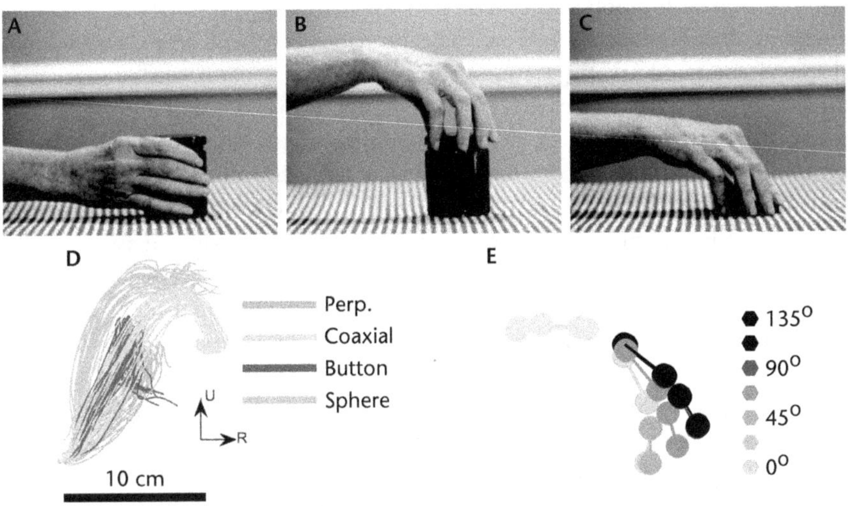

FIGURE 6.2 Interrelationships between reach location and grasp shape

Notes: A. A jar being grasped from the side. B. The same jar being grasped by the lid. C. The same lid being grasped when on the table by itself. Differences in the grasp used in A versus B are accompanied by differences in the location to which the arm reached. Differences in the location of the lid in B versus C are accompanied by more subtle differences in the angles of the wrist and digit joints. D and E illustrate the same interrelationships in multiple trials of reach-to-grasp movements performed by a monkey in a task that involved four different objects (perpendicular cylinder, coaxial cylinder, button, and sphere), each positioned at up to 8 different angular locations on a circle of 13 cm radius from the common center. D. Reach trajectories viewed from behind the monkey varied systematically depending on which of the four objects was positioned at the same location. E. Average angles between the palm (horizontal light gray circles at the top) and the index finger proximal and intermediate phalanges (each represented by a proximal and distal filled circle) varied depending on the angular location of the coaxial cylinder (color scale). D and E modified from Rouse and Schieber, 2015.

the jar might be grasped from above, taking advantage of the edge afforded by the lid (Figure 6.2B). These two grasp shapes certainly differ: the digits wrap around the cylindrical jar from the side in the first case, whereas the fingertips form a circle around the lid in the second. Beyond these differences in the hand shape used to form these two different grasps, the forearm must be more pronated when grasping from the side and more supinated when grasping from above. And the wrist also must be positioned lower when grasping from the side and higher when grasping from above. These differences in the position and orientation of the wrist mean that the final angles of the shoulder and elbow also must differ depending on which grasp is to be used. Hence the motion and shaping of the entire upper extremity must differ to some degree depending on the grasp to be applied.[1]

What if the lid was not on top of the jar, but instead was directly on the table (Figure 6.2C)? The circular grasp from above typically would be used to pick up the lid lying on the table. This lower location of the lid of course requires a lower location of the wrist which is achieved with somewhat different angles at the shoulder and elbow. The pronation of the forearm and overall shape of the hand are largely similar whether the lid is on top of the jar or directly on the table. Yet careful inspection reveals that the angles of the digital joints – metacarpophalangeal, proximal interphalangeal, and distal interphalangeal – all are slightly more extended when the lid is directly on the table instead of being on top of the jar. Hence the shaping of the hand as well as the arm differs slightly depending on the location at which the same grasp is to be applied.

These casual observations of a human subject are illustrated more quantitatively in data obtained from monkeys. Figure 6.2D shows how the reaching trajectory varied in multiple trials as a monkey reached from the same starting location to grasp and manipulate four different objects all positioned at the same location. Though the four objects all were at the same spatial location, the trajectory of the distal end of the forearm varied systematically depending on the object. The object-dependent differences in reach trajectory began shortly after movement onset and increased progressively during the movement. Conversely, Figure 6.2E shows the variation in the position of the proximal and middle phalanges of the index finger – and hence the metacarpophalangeal and proximal interphalangeal joint angles – shortly before contact with a single object when that same object was placed at different locations. These different locations necessitated slightly different angles of the index finger and the other digits as well.

Rather than thinking of reaches as movements performed with the arm and grasps as movements performed with the hand, further progress will require us to think of reaching and grasping each as movements involving the entire upper extremity. One aspect of these upper extremity movements depends on the location of the target object in peri-personal space; another depends on the size and shape of the object and how it is to be manipulated. Rather than being simply semantic, we will see below that this distinction becomes important for understanding how the nervous system controls natural reach-to-grasp movements.

Concurrent Versus Sequential

If reaching involves the arm while grasping involves the hand, then reaching and grasping can proceed concurrently. During reach-to-grasp movements joint rotations from proximal to distal certainly occur contemporaneously. Yet the motion of the extremity related to the location of the object and that related to the shape of the object do not necessarily proceed with parallel time courses.

Let us reexamine Figure 6.1, which reproduces seminal findings based on motion tracking of human reaching and grasping (Jeannerod, 1984), in more detail. Here, human subjects picked up the same object positioned at different distances from the shoulder. Frame-by-frame analysis of movie film (50 frames/sec) was used to track the motion of the distal forearm, thumb, and index finger, from which (i) the speed at which the hand was transported to the object's location (Hand Velocity), and (ii) the aperture between the thumb and index finger (Grip Size), each were calculated as a function of time. Hand velocity demonstrated the classic bell-shaped profile of constant duration, indicating that higher speeds were used to transport the hand longer distances in the same amount of time. Concurrently grip size increased and then decreased, showing a similar time course for all three distances to which the subject reached. Reaching and grasping proceeded at the same time.

Yet closer inspection of the grip size traces reveals additional, noteworthy features. Grip size, after an initial small increase (opening), increased only slightly more as the hand was accelerated toward the object. Midway through the movement, at the time of peak velocity (upward arrows), grip size had reached only about one-quarter of the aperture that eventually would be achieved in opening the hand. Not until the second half of the movements, as hand velocity was decreasing, did the grip open rapidly, reaching its peak size after about three-quarters of the overall movement time (downward arrows). Grip size then decreased – closing on the object – only in the final quarter of the movement. Most of grip opening and all of the closing thus occurred in the second half of the combined reach-to-grasp movement. A similar delay in the opening and closing of the grip aperture relative to the acceleration and deceleration of the arm endpoint is evident in other studies of the reach-to-grasp movements of both humans (Gentilucci et al., 1991) and monkeys (Roy et al., 2002).

More detailed findings were reported recently in monkeys performing a naturalistic reach-grasp-manipulate task (Rouse & Schieber, 2015). While the speed of the distal forearm showed the classic bell-shaped profile, analysis of individual metacarpophalangeal and proximal interphalangeal joint angles revealed that opening of the hand virtually halted in the midst of the reach, with full opening occurring in the second half of the movement and closing beginning shortly before contact with the object to be grasped. Moreover, analysis of variance over time revealed that many joint angles from the shoulder to the hand showed location-related variation being prominent in the first part of the movement prior to peak velocity and object-related variation predominating in the second part, as illustrated in Figure 6.3C. Such findings suggest that the entire extremity initially is projected toward the spatial location of the object, and later in the course of the movement the entire extremity is shaped to grasp and manipulate the object.

EMG Activity During Reach-to-Grasp Movements

Studies of electromyographic (EMG) activity by and large have focused either on proximal muscles during reaching or on distal muscles during grasping. The activity of proximal muscles has been studied by having human subjects reach to various locations without grasping different objects. During such movements, proximal EMG activity shows a tonic component that supports the limb against gravity and a phasic component that varies in amplitude and timing depending on the direction and speed of the reach (Buneo, Soechting, & Flanders, 1994; D'Avella, Fernandez, Portone, & Lacquaniti, 2008; Flanders, 1991; Flanders & Herrmann, 1992; Flanders, Pellegrini, & Geisler, 1996). Conversely, the distal muscle activity that drives the hand for grasping typically has been examined without reaching to different locations (Maier & Hepp-Reymond, 1995a, 1995b; Valero-Cuevas, 2000). Relating the activity of various muscles to the mechanics of the grasping hand is complex (Schaffelhofer, Sartori, Scherberger, & Farina, 2015b; Towles, Valero-Cuevas, & Hentz, 2013; Valero-Cuevas, 2005; Valero-Cuevas, Johanson, & Towles, 2003). Nevertheless, the specificity of different patterns of EMG activity across 8–12 forearm and intrinsic hand muscles for particular grasps has been demonstrated by decoding which of 6–8 objects differing in size and shape is about to be grasped (Brochier, Spinks, Umilta, & Lemon, 2004; Fligge, Urbanek, & van der Smagt, 2013). Decoding accuracy increases progressively during the movement, indicating that patterns of activation across the set of distal muscles become more object-specific as the movement evolves. Similarly, classification accuracy based on the joint angles of the four fingers has been shown to increase progressively during the movement (Santello & Soechting, 1998). Interestingly, decoding the object based on EMG activity approached 100% correct classification midway through the reach while classification based on finger joint angles did not peak until contact with the object, consistent with the progressive evolution of muscle activity causing that of joint angles.

Comparatively few studies have examined EMG during a variety of reach-to-grasp movements that involve different locations and different objects. In one

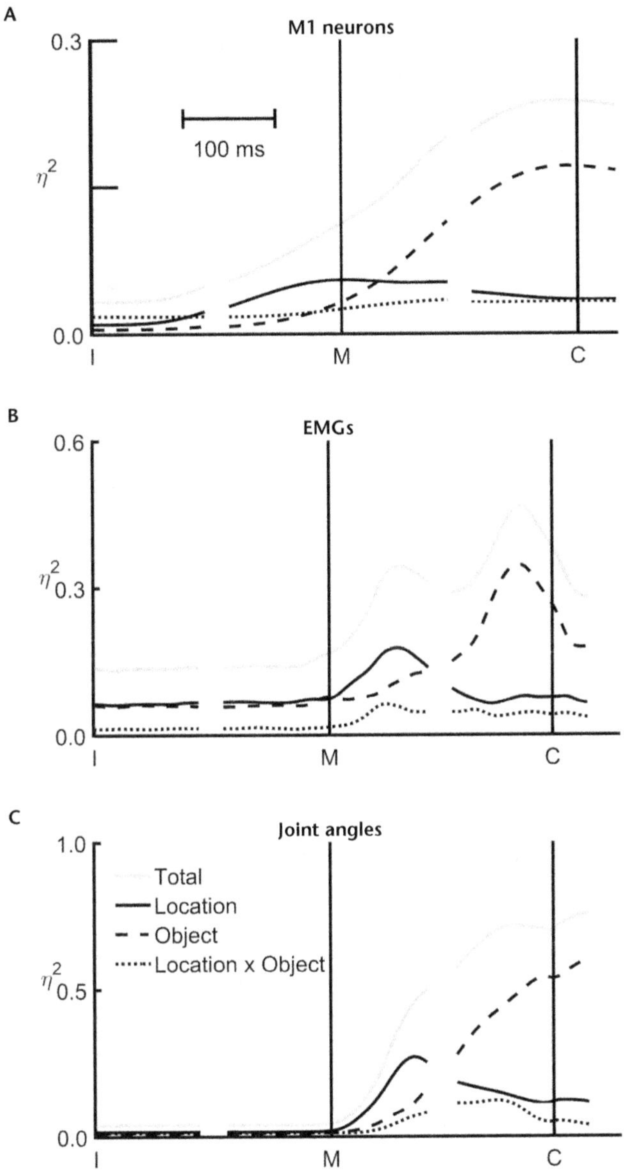

FIGURE 6.3 Location and object effect sizes as a function of time

Notes: In monkeys reaching to grasp four objects, each at up to eight locations, effect sizes (η^2) are shown as a function of time after the instruction about which object to grasp (I), around movement onset (M), and around the time of contact with the object (C) averaged across multiple M1 neurons (A), across EMG activity from multiple muscles (B), and across multiple joint angles (C). At all three levels, the main effect of location (solid black lines) peaked relatively early while the main effect of object (dashed lines) peaked later. Location × object interaction effects (dotted lines) remained relatively small throughout. The total of all three effects is shown as well (gray).

such study in monkeys, the EMG activity of shoulder and elbow muscles peaked on average 125 ms after movement onset, while the activity of extrinsic digit muscles peaked later at 221 ms (Stark, Asher, & Abeles, 2007), suggesting sequential activity first in proximal and then in distal muscles. Yet when human subjects reached to grasp three different objects at three different locations, the EMG activity of proximal muscles showed distinct patterns based not only on location, but on object as well (Martelloni, Carpaneto, & Micera, 2009). Analysis of variance over time showed two phases of EMG activity in muscles from the shoulder to the hand as monkeys performed a reach-grasp-manipulate task that dissociated location and object (Figure 6.3B): in the first phase EMG activity varied largely in relation to location, in the later phase the variation depended predominantly on object (Rouse & Schieber, 2016b). Such observations of EMG activity during more naturalistic reach-to-grasp movements, like observations of kinematics, support the notion that the entire extremity first is projected toward the target, and later is shaped to grasp and manipulate the object.

Cortical Activity Involved in Reach-to-Grasp Movements

Primary Motor Cortex

The Traditional Proximal Versus Distal Approach

Classical studies using electrical stimulation of the cortical surface demonstrated that the primary motor cortex (M1) was organized somatotopically (Penfield & Boldrey, 1937; Penfield & Rasmussen 1950; Woolsey, Erickson, & Gilson, 1979; Woolsey et al., 1952). Within the upper extremity territory, the hand representation was located laterally and also posteriorly in the anterior bank of the central sulcus. Representation of the proximal upper extremity was more medial and largely on the crown of the precentral gyrus. With the advent of more invasive techniques using penetrating microelectrodes for intracortical microstimulation (ICMS), the details of this arrangement have been refined in macaque monkeys. The macaque M1 has been shown to have a central core of distal digit and wrist representation that lies largely in the anterior bank of the central sulcus but extends to some degree on to the crown of the precentral gyrus (Kwan, MacKay, Murphy, & Wong, 1978; Park, Belhaj-Saif, Gordon, & Cheney, 2001). The core of distal representation is surrounded medially, anteriorly, and to some degree laterally by a "horseshoe" of more proximal representation. Much of the horseshoe of proximal representation lies on the surface of the precentral gyrus, but the medial and lateral extensions run down into the anterior bank of the central sulcus. Representations of movements of various parts of the upper extremity or contractions of individual muscles overlap extensively, both in the core of distal representation and in the horseshoe of proximal representation. In addition, the proximal and distal representations in the horseshoe and core have a large intermediary zone of overlap.

Although different representations overlap extensively in M1, the concept of somatotopic segregation has permitted the activity of M1 neurons to be examined in relation to movements confined to the arm, to the wrist, or to the fingers. Numerous studies therefore have used reaching movements of the arm, for example, to examine the relationships between the firing rates of macaque M1 neurons and the direction and speed of movement, the forces exerted, and/or the underlying muscle activation (Ashe & Georgopoulos, 1994; Churchland & Shenoy, 2007; Georgopoulos, Ashe, Smyrnis, & Taira, 1992; Georgopoulos, Kalaska, Caminiti, & Massey, 1982; Georgopoulos, Schwartz, & Kettner, 1986; Hatsopoulos, Xu, & Amit, 2007; Holdefer & Miller, 2002; Kalaska, 2009; Kalaska, Cohen, Hyde, & Prud'homme, 1989; Paninski, Fellows, Hatsopoulos, & Donoghue, 2004; Sergio, Hamel-Paquet, & Kalaska, 2005). Other studies constrained the arm and used wrist rotations to examine similar relationships (Evarts, 1968; 1969; Kakei, Hoffman, & Strick, 1999; Riehle, MacKay, & Requin, 1994; Schieber & Thach, 1985; Thach, 1978). Still other studies of neuron activity in M1 have focused on the action of the hand either in grasping or in relatively independent movements of the fingers (Lemon, Mantel, & Muir, 1986; Muir & Lemon, 1983; Poliakov & Schieber, 1999; Schieber & Hibbard, 1993; Spinks, Kraskov, Brochier, Umilta, & Lemon, 2008; Umilta, Brochier, Spinks, & Lemon, 2007). In all these situations, neural activity in M1 has been related to specific parts of the upper extremity.[2]

Are reaching and grasping then somatotopically segregated in M1 during naturalistic reach-to-grasp movements? During such movements, the kinematics of joints from the shoulder to the hand have been found to be encoded in and decodable from neural activity recorded via microelectrode arrays implanted in M1 on the crown of the macaque precentral gyrus (Bansal, Truccolo, Vargas-Irwin, & Donoghue, 2012; Saleh, Takahashi, & Hatsopoulos, 2012; Vargas-Irwin et al., 2010). These observations suggest that this portion of M1 contains a relatively complete representation of reach-to-grasp movements, despite the greater proportion of this territory being devoted to proximal representation. In a monkey performing reach-to-grasp movements involving either a precision grip or a side grip of the same object at the same location followed by an isometric pull, a microelectrode array implanted rostrally on the crown of the precentral gyrus near the border between the dorsal premotor cortex (PMd) and M1 showed two phases of activity in both movement-related potentials (MRPs) and spiking single- and multi-unit activity (Riehle, Wirtssohn, Grun, & Brochier, 2013). The early phase – during which a first MRP peak (P1) appeared – occurred prior to movement onset and was most pronounced on the medial aspect of the array, where somatosensory examination showed responses predominantly to passive movement of the proximal upper extremity. The later phase – including a second MRP peak (P2) as well as the peak discharge of most spiking units – occurred after movement onset and was most pronounced laterally where somatosensory responses arose predominantly from the distal extremity. These results suggest a spatiotemporal progression of two phases of neural activity on the crown of the precentral gyrus from a medial, proximal representation to a more lateral, distal representation (cf. Best, Suminski, Takahashi, Brown, & Hatsopoulos, 2016).

The same is not necessarily the case for the portion of macaque M1 in the anterior bank of the central sulcus. In a study that dissociated reach location versus object grasped (Rouse & Schieber, 2016a), location effects and objects effects each were found to be distributed widely throughout the upper extremity representation in the anterior bank of the central sulcus (Figure 6.4). Indeed, the firing rate of most single-units varied significantly both with location and with object. ICMS demonstrated that the sampled M1 territory covered the central core of distal representation as well as any flanking zones of proximal representation. Rather than propagating from proximal to distal zones, however, both location and object effects spread rapidly throughout the sampled territory. Nevertheless, two sequential phases were found here as well: location effects peaked first, around the time of movement onset, after which object effects

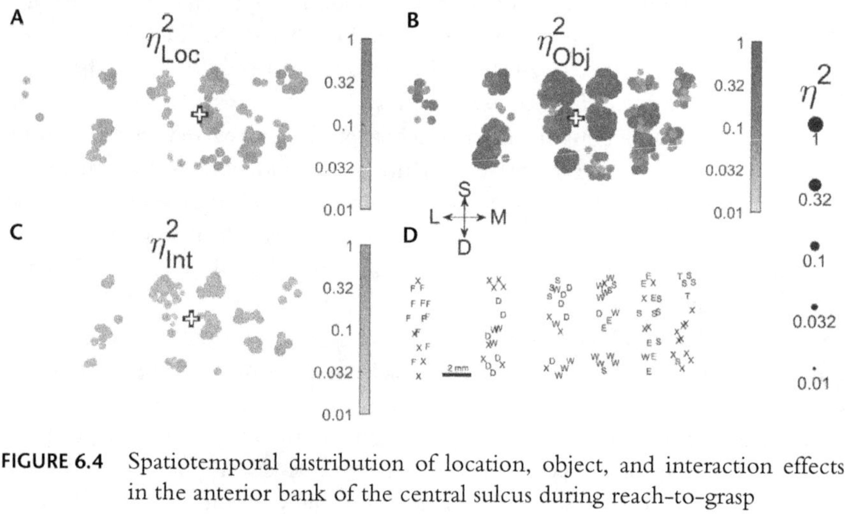

FIGURE 6.4 Spatiotemporal distribution of location, object, and interaction effects in the anterior bank of the central sulcus during reach-to-grasp

Notes: A–C. Circles in each panel represent location (A), object (B), and interaction effect sizes (C) for all single- and multi-units recorded through implanted microelectrode arrays in 14 sessions from one monkey. Both the grayscale and the diameter of each circle have been scaled proportional to the effect size (η^2) values from ANOVA at the moment of contact with an object. Each circle was plotted at the two-dimensional location of the electrode tip from which the unit was recorded. Black and white "+" signs mark the centroid location for each type of effect. This projection of the anterior bank of the left central sulcus is viewed as if from the posterior bank: Orientation crosshairs: S, superficial, up; D, deep, down; L, lateral, left; M, medial, right. D. The contralateral body part moved by threshold ICMS at each electrode tip: F – face, D – digits, W – wrist, E – elbow, S – shoulder, T – trunk, X – no response with currents up to 100 μA. Modified from Rouse and Schieber, 2016a.

became predominant, increasing progressively until just before contact (Figure 6.3A). Throughout the time course, the centroids for location effects and for object effects both were close to the center of the sampled territory, indicating that both location and object effects were spread similarly throughout (Figure 6.4). In the anterior bank of the central sulcus, the M1 activity projecting the extremity to a given location and the activity shaping the extremity to grasp a given object thus are not overtly segregated into proximal and distal zones. Rather, the activity projecting the extremity to the target location peaks early, while the activity shaping the extremity for grasping predominates later.

Old Versus New M1

Such differences in the spatiotemporal distribution of neural activity in M1 depending on whether recordings are made on the crown of the precentral gyrus versus the anterior bank of the central sulcus are consistent with a second type of anatomical organization in macaque M1 that has been identified recently. In M1, many neurons in layer V have corticofugal axons that project to interneurons in the spinal gray matter, either via the corticospinal tract or more indirectly through synaptic connections in the red nucleus or the pontomedullary reticular formation (Kuypers, 1987). A substantial fraction of M1 neurons, however, make monosynaptic connections directly on spinal α-motoneurons (Fetz & Cheney, 1980; Lemon et al., 1986; Shinoda, Yokota, & Futami, 1981). Interestingly, virtually all these "cortico-motoneuronal" (CM) cells are found in the caudal portion of M1 in the anterior bank of the central sulcus (Rathelot & Strick, 2009). Few if any CM cells are present in the more rostral portion of M1 on the crown of the precentral gyrus.[3] This distinction suggests a functional difference between the more caudal and phylogenetically "new" M1 with outputs that act monosynaptically on α-motoneurons versus the more rostral "old" M1 with corticospinal outputs that act through brainstem nuclei and spinal interneurons. New M1, with a larger proportion of distal than proximal representation, might play a greater role in more dexterous aspects of upper extremity movements, while old M1, with a larger proportion of proximal than distal representation, might play a greater role in more fundamental control of limb movement. Yet both old and new M1 participate in controlling naturalistic reach-to-grasp movements.

Further studies that explicitly compare old and new M1 will be needed to better understand the differences in their functional roles in controlling reach-to-grasp movements. A complication for such studies arises from the observations that the motion of the proximal arm in reaching varies depending on the object to be grasped, and that shape of the hand used in grasping will be adjusted depending on the object's location, as illustrated above in Figure 6.2. To disambiguate the role of the proximal versus distal upper extremity in reaching versus grasping, neural activity recorded during reach-to-grasp movements will need to be compared with input-output properties at a given recording site, defined with approaches such as spike- or stimulus-triggered averaging of EMG activity (Buys, Lemon, Mantel, & Muir, 1986; Cheney & Fetz, 1985; Fetz & Cheney, 1980; McKiernan, Marcario, Karrer, & Cheney, 1998; Park et al., 2001; Schieber & Rivlis, 2005), ICMS evoked movements (Donoghue, Leibovic, & Sanes, 1992; Gould, Cusick, Pons, & Kaas, 1986; Kwan et al., 1978), and/or responses to somatosensory stimulation (Fetz, Finocchio, Baker, & Soso, 1980; Lemon, 1981; Murphy, Kwan, MacKay, & Wong, 1978; Rosen & Asanuma, 1972; Soso & Fetz, 1980; Wong, Kwan, MacKay, & Murphy, 1978).

Premotor Cortex

Because of its strong projections to area 4, anteriorly adjacent Brodmann's area 6 on the lateral surface of the hemisphere has long been considered the "premotor" cortex (Wise, 1985). Anatomical studies have progressively subdivided area 6 into dorsal and ventral premotor areas (PMd and PMv), caudal and rostral subdivisions of each (F2 and F7 in PMd; F4 and F5 in PMv), with further subdivisions as well (e.g., F5a, F5b, F5c) (Belmalih et al., 2009; Rizzolatti & Luppino, 2001). Studies of neuronal activity in PMd (mostly in F2) by and large have used reaching movements to examine relationships between neural activity and a variety of movement parameters, especially during instructed delay periods during which the subject can plan movements before executing them (Cisek, Crammond, & Kalaska, 2003; Kalaska, 2009; Weinrich, Wise, & Mauritz, 1984). Studies of PMv have focused instead on grasping (Fluet, Baumann, & Scherberger, 2010; Raos, Umilta, Murata, Fogassi, & Gallese, 2006; Spinks et al., 2008), perhaps in large part because of the discovery of mirror neurons in PMv that discharge both as a monkey performs a particular grasp and as the monkey observes another individual perform a similar grasp (di Pellegrino, Fadiga, Fogassi, Gallese, & Rizzolatti, 1992; Gallese, Fadiga, Fogassi, & Rizzolatti, 1996). Decoding studies also have shown that reach direction can be decoded from PMd (Kemere et al., 2004), whereas grasp shape can be decoded from PMv (Carpaneto et al., 2011; Menz, Schaffelhofer, & Scherberger, 2015; Schaffelhofer, Agudelo-Toro, & Scherberger, 2015a; Townsend, Subasi, & Scherberger, 2011). These studies all have been consistent with the notion that PMd participates in control of reaching while PMv contributes to control of grasping.

Nevertheless, dichotomously assigning reaching to PMd and grasping to PMv may be an oversimplification. ICMS can evoke both proximal movements and distal movements in both PMd and PMv. In caudal PMd (F2), proximal movements are evoked medially and distal movement laterally (Raos, Franchi, Gallese, & Fogassi, 2003), whereas in PMv proximal movements are evoked caudally and distal movements rostrally (Gentilucci et al., 1988). Many neurons in F2, the caudal subdivision of PMd traditionally viewed as controlling reaching, have been shown to vary their discharge when monkeys grasp objects of different shapes, all positioned at the same reach location (Raos, Umilta, Gallese, & Fogassi, 2004). And conversely neurons in F4, the caudal subdivision of PMv, contribute to control of proximal movements (Gentilucci et al., 1988). Indeed, the kinematics of the entire arm and hand during reach-to-grasp movements can be decoded from neural activity recorded only from PMv (Bansal et al., 2012). A particularly relevant study examined single-unit activity as monkeys both reached to different locations and grasped different objects (Stark et al., 2007). Neurons throughout PMd and PMv were found to discharge both in relation to location and in relation to object (Figure 6.5), suggesting that the two major subdivisions of the premotor cortex each contribute both to reaching and to grasping. Furthermore, when ICMS-evoked movement and responses to passive movement were used to classify the sensorimotor properties of each recording site, location-related modulation and object-related modulation each were found both at proximal arm sites and at distal hand sites. Hence in spite of the trend for studies of PMd to emphasize reaching while studies of PMv emphasize grasping, PMd and PMv each participate in both aspects of naturalistic reach-to-grasp movements.

Moreover, although mirror neurons related to particular grasps have been studied most extensively in F5, where they appear concentrated in F5c and F5a (Belmalih et al., 2009; Nelissen, Luppino, Vanduffel, Rizzolatti, & Orban, 2005), neurons in PMd have been found to have mirror-like properties as well. In monkeys trained to perform reaching movements to acquire visual targets with a cursor, PMd neurons were active similarly during trials in which the monkey moved its arm to control the cursor and during other trials in which the motion of the cursor, though observed by the monkey, was controlled by

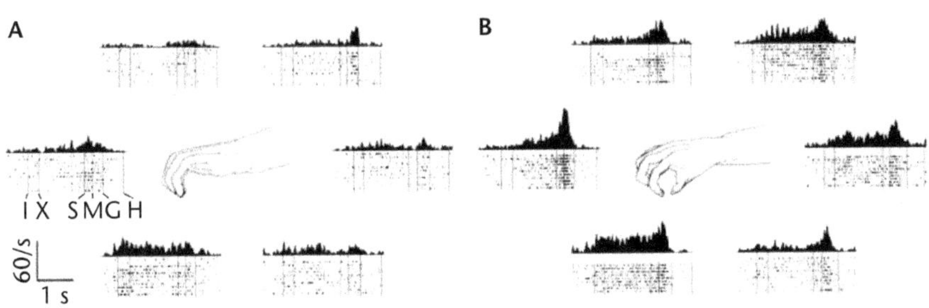

FIGURE 6.5 Firing rate variation with both location and object

Notes: Raster and histogram displays show the activity of a single unit recorded in PMd during an instructed delay task as a monkey performed either a power grip (A) or a precision grip (B) at each of six different locations. I – instruction on, X – instruction off, S – go signal, M – movement onset, G – grip achieved, H – hold period ends. For either the power grip or the precision grip, firing rate varied depending on location. For any location, firing rate varied depending on the grip. This unit discharged more in relation to precision grip than to power grip, yet threshold ICMS at the recording site evoked a proximal movement, elbow extension. Modified from Figure 6A of Stark et al., 2007.

an unseen individual instead of the monkey's own reaching movements (Cisek & Kalaska, 2004; Tkach, Reimer, & Hatsopoulos, 2007). Likewise in humans, observation of arm motion elicits fMRI activation in PMd (Casile et al., 2010). Studies have yet to examine whether similar neurons that mirror reaching movements of the arm are found in PMv, perhaps in subdivision F4, and whether neurons that mirror grasping movements are found in PMd.

Posterior Parietal Cortex

The premotor cortex receives considerable cortico-cortical input from the posterior parietal cortex. Though not entirely exclusive, dorsal regions of the posterior parietal cortex project predominantly to PMd, whereas ventral parietal regions project predominantly to PMv. Dorsal parietal regions are thought to receive visual information largely on the spatial location of targets, while the more ventral regions receive visual information on the shape of objects. Consistent with this dichotomy, neurons in the dorsally located parietal reach region (PRR) discharge strongly in relation to reaching movements made to different locations (Andersen, Snyder, Batista, Buneo, & Cohen, 1998; Batista & Andersen, 2001; Chang & Snyder, 2012), whereas those in the more ventral anterior intraparietal area (AIP) discharge in relation to the shape of objects about to be grasped (Baumann, Fluet, & Scherberger, 2009; Murata, Gallese, Luppino, Kaseda, & Sakata, 2000). Moreover, reaching movements can be decoded from PRR neuron populations (Mulliken, Musallam, & Andersen, 2008), and grasping from AIP (Menz et al., 2015; Schaffelhofer et al., 2015a).

But again, emerging studies indicate that this dorso-ventral separation of function in the parietal lobe is not as complete as previously thought. In the dorsomedial parieto-occipital junction, just posteromedial to PRR, neurons in area V6A have been found in separate studies to be related both to the location to which the arm reaches (Bosco, Breveglieri, Chinellato, Galletti, & Fattori, 2010; Fattori, Kutz, Breveglieri, Marzocchi, & Galletti, 2005; Galletti, Fattori, Kutz, & Battaglini, 1997) and to the object grasped (Fattori, Breveglieri, Bosco, Gamberini, & Galletti, 2015; Fattori, Breveglieri, Raos, Bosco, & Galletti, 2012; Fattori et al., 2010). Whether the more ventral AIP might contain neurons related to reach location is as yet unknown. Moreover, studies of the roles of the posterior parietal cortex have yet to examine more naturalistic movements in which the reach location and the object grasped are dissociated.

Conclusions

Historically, reaching and grasping have been approached as two separate processes that involve different portions of the upper extremity, different somatotopic regions within the primary motor cortex, and distinct regions within the premotor and posterior parietal cortices. Our understanding of the neural control of reaching and grasping therefore has been constrained largely to studies of reaching or of grasping. While much has been learned, here we have reviewed accumulating evidence that warrants reevaluation of this dichotomous approach. We suggest that reach-to-grasp movements might be viewed more accurately as movements of the entire upper extremity that first project the limb toward the target and then shape the entire upper extremity to an appropriate grasp for the impending manipulation. This revised view of reach-to-grasp movements can facilitate future progress in understanding the neural control of increasingly naturalistic movements and in developing neuroprosthetic devices that more closely emulate normal human movements.

Acknowledgements

This work was supported by grant R01 NS079664 from the National Institute of Neurological Disorders and Stroke. The authors thank Marsha Hayles for editorial comments.

Notes

1 This point is similar to the notion that in the special case of a precision grip, the entire trajectory will depend on the choice of points of contact for the tips of the thumb and index finger (Smeets & Brenner, 1999).

2 In retrospect, we can reconsider such studies that focused on a particular segment of the upper extremity from a more naturalistic point of view. During proximal reaching movements made with the arm, the hand may not have been entirely passive. The hand often held a joystick. Although the grasp may not have changed, the direction and magnitude of forces moving the joystick would have been exerted through the hand. Even when reaching to different locations on a touchscreen, the posture of the hand would have been maintained actively against interaction torques as the hand was accelerated and decelerated. During rotations at the wrist, the hand likewise either grasped a handle or pressed on a rotating paddle actively, whether isotonic or isometric. In all these instances, unobserved activation of hand and finger muscles may have been correlated to some extent with the motion of the arm or wrist. Consequently the M1 activity related to the parameters of the observed arm or wrist motion may not have been controlling only the arm or wrist, but the fingers as well. Similarly in studies where the arm reached from a standard starting location to grasp objects of different shape, the arm trajectory probably varied to some extent depending on the grasp, as illustrated above in Figure 6.2. Some of the M1 activity present during such grasps may have contributed to the necessary adjustment of shoulder, elbow, and wrist angles, even though the different objects all were positioned at the same peri-personal location. And in studies that restrained the arm during finger movements, the subjects nevertheless may have made relatively isometric contractions of proximal musculature that varied in relation to grasp shape or finger movements.

3 Human M1 is thought to have a larger proportion of CM cells than that found in non-human primates. Human M1 (Brodmann's area 4) also has been divided anatomically into posterior and anterior subdivisions (Geyer et al., 1996), but whether this distinction in humans also involves the location of the CM-cell population is as yet unknown.

References

Andersen, R.A., Snyder, L.H., Batista, A.P., Buneo, C.A., & Cohen, Y.E. (1998). Posterior parietal areas specialized for eye movements (LIP) and reach (PRR) using a common coordinate frame. *NovartisFoundSymp, 218*, 109–122.

Ashe, J., & Georgopoulos, A.P. (1994). Movement parameters and neural activity in motor cortex and area 5. *Cerebral Cortex, 4*, 590–600.

Bansal, A.K., Truccolo, W., Vargas-Irwin, C.E., & Donoghue, J.P. (2012). Decoding 3D reach and grasp from hybrid signals in motor and premotor cortices: spikes, multiunit activity, and local field potentials. *J Neurophysiol, 107*, 1337–1355.

Batista, A.P., & Andersen, R.A. (2001). The parietal reach region codes the next planned movement in a sequential reach task. *J Neurophysiol, 85*, 539–544.

Baumann, M.A., Fluet, M.C., & Scherberger, H. (2009). Context-specific grasp movement representation in the macaque anterior intraparietal area. *J Neurosci, 29*, 6436–6448.

Belmalih, A., Borra, E., Contini, M., Gerbella, M., Rozzi, S., & Luppino, G. (2009). Multimodal architectonic subdivision of the rostral part (area F5) of the macaque ventral premotor cortex. *J Comp Neurol, 512*, 183–217.

Best, M.D., Suminski, A.J., Takahashi, K., Brown, K.A., & Hatsopoulos, N.G. (2016). Spatio-temporal patterning in primary motor cortex at movement onset. *Cereb Cortex, 27*(2), 1491–1500.

Bosco, A., Breveglieri, R., Chinellato, E., Galletti, C., & Fattori, P. (2010). Reaching activity in the medial posterior parietal cortex of monkeys is modulated by visual feedback. *J Neurosci, 30*, 14773–14785.

Brochier, T., Spinks, R.L., Umilta, M.A., & Lemon, R.N. (2004). Patterns of muscle activity underlying object-specific grasp by the macaque monkey. *Journal of Neurophysiology, 92*, 1770–1782.

Buneo, C.A., Soechting, J.F., & Flanders, M. (1994). Muscle activation patterns for reaching: the representation of distance and time. *J Neurophysiol, 71*, 1546–1558.

Buys, E.J., Lemon, R.N., Mantel, G.W., & Muir, R.B. (1986). Selective facilitation of different hand muscles by single corticospinal neurones in the conscious monkey. *Journal of Physiology, 381*, 529–549.

Carpaneto, J., Umilta, M.A., Fogassi, L., Murata, A., Gallese, V., Micera, S., & Raos, V. (2011). Decoding the activity of grasping neurons recorded from the ventral premotor area F5 of the macaque monkey. *Neuroscience, 188*, 80–94.

Casile, A., Dayan, E., Caggiano, V., Hendler, T., Flash, T., & Giese, M.A. (2010). Neuronal encoding of human kinematic invariants during action observation. *Cereb Cortex, 20*, 1647–1655.

Chang, S.W., & Snyder, L.H. (2012). The representations of reach endpoints in posterior parietal cortex depend on which hand does the reaching. *J Neurophysiol, 107*, 2352–2365.

Cheney, P.D., & Fetz, E.E. (1985). Comparable patterns of muscle facilitation evoked by individual corticomotoneuronal (CM) cells and by single intracortical microstimuli in primates: evidence for functional groups of CM cells. *Journal of Neurophysiology, 53*, 786–804.

Churchland, M.M., & Shenoy, K.V. (2007). Temporal complexity and heterogeneity of single-neuron activity in premotor and motor cortex. *Journal of Neurophysiology, 97*, 4235–4257.

Cisek, P., Crammond, D.J., & Kalaska, J.F. (2003). Neural activity in primary motor and dorsal premotor cortex in reaching tasks with the contralateral versus ipsilateral arm. *J Neurophysiol, 89*, 922–942.

Cisek, P., & Kalaska, J.F. (2004). Neural correlates of mental rehearsal in dorsal premotor cortex. *Nature, 431*, 993–996.

D'Avella, A., Fernandez, L., Portone, A., & Lacquaniti, F. (2008). Modulation of phasic and tonic muscle synergies with reaching direction and speed. *Journal of Neurophysiology, 100*, 1433–1454.

Desmurget, M., Prablanc, C., Arzi, M., Rossetti, Y., Paulignan, Y., & Urquizar, C. (1996). Integrated control of hand transport and orientation during prehension movements. *Exp Brain Res, 110*, 265–278.

di Pellegrino, G., Fadiga, L., Fogassi, L., Gallese, V., & Rizzolatti, G. (1992). Understanding motor events: a neurophysiological study. *Experimental Brain Research, 91*, 176–180.

Donoghue, J.P., Leibovic, S., & Sanes, J.N. (1992). Organization of the forelimb area in squirrel monkey motor cortex: representation of digit, wrist, and elbow muscles. *Experimental Brain Research, 89*, 1–19.

Evarts, E.V. (1968). Relation of pyramidal tract activity to force exerted during voluntary movement. *Journal of Neurophysiology, 31*, 14–27.

Evarts, E.V. (1069). Activity of pyramidal tract neurons during postural fixation. *Journal of Neurophysiology, 32*, 375–385.

Fattori, P., Breveglieri, R., Bosco, A., Gamberini, M., & Galletti, C. (2015). Vision for prehension in the medial parietal cortex. *Cereb Cortex, 27*(2), 1149–1163.

Fattori, P., Breveglieri, R., Raos, V., Bosco, A., & Galletti, C. (2012). Vision for action in the macaque medial posterior parietal cortex. *J Neurosci, 32*, 3221–3234.

Fattori, P., Kutz, D.F., Breveglieri, R., Marzocchi, N., & Galletti, C. (2005). Spatial tuning of reaching activity in the medial parieto-occipital cortex (area V6A) of macaque monkey. *Eur J Neurosci, 22*, 956–972.

Fattori, P., Raos, V., Breveglieri, R., Bosco, A., Marzocchi, N., & Galletti, C. (2010). The dorsomedial pathway is not just for reaching: grasping neurons in the medial parieto-occipital cortex of the macaque monkey. *Journal of Neuroscience, 30*, 342–349.

Fetz, E.E., & Cheney, P.D. (1980). Postspike facilitation of forelimb muscle activity by primate corticomotoneuronal cells. *Journal of Neurophysiology, 44*, 751–772.

Fetz, E.E., Finocchio, D.V., Baker, M.A., & Soso, M.J. (1980). Sensory and motor responses of precentral cortex cells during comparable passive and active joint movements. *Journal of Neurophysiology, 43*, 1070–1089.

Fitts, P.M. (1954). The information capacity of the human motor system in controlling the amplitude of movement. *J Exp Psychol, 47*, 381–391.

Flanders, M. (1991). Temporal patterns of muscle activation for arm movements in three-dimensional space. *J Neurosci, 11*, 2680–2693.

Flanders, M., & Herrmann, U. (1992). Two components of muscle activation: scaling with the speed of arm movement. *J Neurophysiol, 67*, 931–943.

Flanders, M., Pellegrini, J.J., & Geisler, S.D. (1996). Basic features of phasic activation for reaching in vertical planes. *Experimental Brain Research, 110*, 67–79.

Fligge, N., Urbanek, H., & van der Smagt, P. (2013). Relation between object properties and EMG during reaching to grasp. *J Electromyogr Kinesiol, 23*, 402–410.

Fluet, M.C., Baumann, M.A., & Scherberger, H. (2010). Context-specific grasp movement representation in macaque ventral premotor cortex. *J Neurosci, 30*, 15175–15184.

Gallese, V., Fadiga, L., Fogassi, L., & Rizzolatti, G. (1996). Action recognition in the premotor cortex. *Brain, 119*, 593–609.

Galletti, C., Fattori, P., Kutz, D.F., & Battaglini, P.P. (1997). Arm movement-related neurons in the visual area V6A of the macaque superior parietal lobule. *European Journal of Neuroscience, 9*, 410–413.

Gentilucci, M., Castiello, U., Corradini, M.L., Scarpa, M., Umilta, C., & Rizzolatti, G. (1991). Influence of different types of grasping on the transport component of prehension movements. *Neuropsychologia, 29*, 361–378.

Gentilucci, M., Daprati, E., Gangitano, M., Saetti, M.C., & Toni, I. (1996). On orienting the hand to reach and grasp an object. *Neuroreport, 7*, 589–592.

Gentilucci, M., Fogassi, L., Luppino, G., Matelli, M., Camarda, R., & Rizzolatti, G. (1988). Functional organization of inferior area 6 in the macaque monkey. I. Somatotopy and the control of proximal movements. *ExpBrain Res, 71*, 475–490.

Georgopoulos, A.P., Ashe, J., Smyrnis, N., & Taira, M. (1992). The motor cortex and the coding of force. *Science, 256*, 1692–1695.

Georgopoulos, A.P., Kalaska, J.F., Caminiti, R., & Massey, J.T. (1982). On the relations between the direction of two-dimensional arm movements and cell discharge in primate motor cortex. *Journal of Neuroscience, 2*, 1527–1537.

Georgopoulos, A.P., Schwartz, A.B., & Kettner, R.E. (1986). Neuronal population coding of movement direction. *Science, 233*, 1416–1419.

Geyer, S., Ledberg, A., Schleicher, A., Kinomura, S., Schormann, T., Burgel, U., Klingberg, T., Larsson, J., Zilles, K., & Roland, P.E. (1996). Two different areas within the primary motor cortex of man. *Nature, 382*, 805–807.

Gould, H.J., Cusick, C.G., Pons, T.P., & Kaas, J.H. (1986). The relationship of corpus callosum connections to electrical stimulation maps of motor, supplementary motor, and the frontal eye fields in owl monkeys. *Journal of Comparative Neurology, 247*, 297–325.

Grafton, S.T. (2010). The cognitive neuroscience of prehension: recent developments. *Exp Brain Res, 204*, 475–491.

Hatsopoulos, N.G., Xu, Q., & Amit, Y. (2007). Encoding of movement fragments in the motor cortex. *Journal of Neuroscience, 27*, 5105–5114.

Holdefer, R.N., & Miller, L.E. (2002). Primary motor cortical neurons encode functional muscle synergies. *Experimental Brain Research, 146*, 233–243.

Jeannerod, M. (1984). The timing of natural prehension movements. *J Mot Behav, 16*, 235–254.

Jeannerod, M., Arbib, M.A., Rizzolatti, G., & Sakata, H. (1995). Grasping objects: the cortical mechanisms of visuomotor transformation. *Trends in Neurosciences, 18*, 314–320.

Kakei, S., Hoffman, D.S., & Strick, P.L. (1999). Muscle and movement representations in the primary motor cortex. *Science, 285*, 2136–2139.

Kalaska, J.F. (2009). From intention to action: motor cortex and the control of reaching movements. *Advances in Experimental Medicine and Biology, 629*, 139–178.

Kalaska, J.F., Cohen, D.A.D., Hyde, M.L., & Prud'homme, M. (1989). A comparison of movement direction-related versus load direction-related activity in primate motor cortex, using a two-dimensional reaching task. *Journal of Neuroscience, 9*, 2080–2102.

Karl, J.M., & Whishaw, I.Q. (2013). Different evolutionary origins for the reach and the grasp: an explanation for dual visuomotor channels in primate parietofrontal cortex. *Front Neurol, 4*, 208.

Kemere, C., Santhanam, G., Yu, B.M., Ryu, S., Meng, T., & Shenoy, K.V. (2004). Model-based decoding of reaching movements for prosthetic systems. *ConfProcIEEE Eng MedBiolSoc, 6*, 4524–4528.

Kuypers, H.G.J.M. (1987). Some aspects of the organization of the output of the motor cortex. In G. Bock, M. O'Connor, and J. Marsh (Eds.), *Motor areas of the cerebral cortex* (pp. 63–82). Chichester, Sussex: John Wiley & Sons.

Kwan, H.C., MacKay, W.A., Murphy, J.T., & Wong, Y.C. (1978). Spatial organization of precentral cortex in awake primates. II. Motor outputs. *Journal of Neurophysiology, 41*, 1120–1131.

Lacquaniti, F., & Soechting, J.F. (1982). Coordination of arm and wrist motion during a reaching task. *Journal of Neuroscience, 2*, 399–408.

Lemon, R.N. (1981). Functional properties of monkey motor cortex neurones receiving afferent input from the hand and fingers. *Journal of Physiology, 311*, 497–519.

Lemon, R.N., Mantel, G.W., & Muir, R.B. (1986). Corticospinal facilitation of hand muscles during voluntary movement in the conscious monkey. *Journal of Physiology, 381*, 497–527.

Maier, M.A., & Hepp-Reymond, M.C. (1995a). EMG activation patterns during force production in precision grip. I. Contribution of 15 finger muscles to isometric force. *Experimental Brain Research, 103*, 108–122.

Maier, M.A., & Hepp-Reymond, M.C. (1995b). EMG activation patterns during force production in precision grip. II. Muscular synergies in the spatial and temporal domain. *Experimental Brain Research, 103*, 123–136.

Martelloni, C., Carpaneto, J., & Micera, S. (2009). Characterization of EMG patterns from proximal arm muscles during object- and orientation-specific grasps. *IEEE Trans Biomed Eng, 56*, 2529–2536.

McKiernan, B.J., Marcario, J.K., Karrer, J.H., & Cheney, P.D. (1998). Corticomotoneuronal postspike effects in shoulder, elbow, wrist, digit, and intrinsic hand muscles during a reach and prehension task. *Journal of Neurophysiology, 80*, 1961–1980.

Menz, V.K., Schaffelhofer, S., & Scherberger, H. (2015). Representation of continuous hand and arm movements in macaque areas M1, F5, and AIP: a comparative decoding study. *J Neural Eng, 12*, 056016.

Muir, R.B., & Lemon, R.N. (1983). Corticospinal neurons with a special role in precision grip. *Brain Research, 261*, 312–316.

Mulliken, G.H., Musallam, S., & Andersen, R.A. (2008). Decoding trajectories from posterior parietal cortex ensembles. *J Neurosci, 28*, 12913–12926.

Murata, A., Gallese, V., Luppino, G., Kaseda, M., & Sakata, H. (2000). Selectivity for the shape, size, and orientation of objects for grasping in neurons of monkey parietal area AIP. *Journal of Neurophysiology, 83*, 2580–2601.

Murphy, J.T., Kwan, H.C., MacKay, W.A., & Wong, Y.C. (1978). Spatial organization of precentral cortex in awake primates. III. Input-output coupling. *Journal of Neurophysiology, 41*, 1132–1139.

Nelissen, K., Luppino, G., Vanduffel, W., Rizzolatti, G., & Orban, G.A. (2005). Observing others: multiple action representation in the frontal lobe. *Science, 310*, 332–336.

Paninski, L., Fellows, M.R., Hatsopoulos, N.G., & Donoghue, J.P. (2004). Spatiotemporal tuning of motor cortical neurons for hand position and velocity. *Journal of Neurophysiology, 91*, 515–532.

Park, M.C., Belhaj-Saif, A., Gordon, M., & Cheney, P.D. (2001). Consistent features in the forelimb representation of primary motor cortex in rhesus macaques. *Journal of Neuroscience, 21*, 2784–2792.

Paulignan, Y., Frak, V.G., Toni, I., & Jeannerod, M. (1997). Influence of object position and size on human prehension movements. *Experimental Brain Research, 114*, 226–234.

Penfield, W., & Boldrey, E. (1937). Somatic motor and sensory representation in the cerebral cortex of man as studied by electrical stimulation. *Brain, 37*, 389–443.

Penfield, W., & Rasmussen, T. (1950). *The cerebral cortex of man*. New York: Macmillan.

Poliakov, A.V., & Schieber, M.H. (1999). Limited functional grouping of neurons in the motor cortex hand area during individuated finger movements: a cluster analysis. *Journal of Neurophysiology, 82*, 3488–3505.

Raos, V., Franchi, G., Gallese, V., & Fogassi, L. (2003). Somatotopic organization of the lateral part of area F2 (dorsal premotor cortex) of the macaque monkey. *Journal of Neurophysiology, 89*, 1503–1518.

Raos, V., Umilta, M.A., Gallese, V., & Fogassi, L. (2004). Functional properties of grasping-related neurons in the dorsal premotor area F2 of the macaque monkey. *J Neurophysiol, 92*, 1990–2002.

Raos, V., Umilta, M.A., Murata, A., Fogassi, L., & Gallese, V. (2006). Functional properties of grasping-related neurons in the ventral premotor area F5 of the macaque monkey. *J Neurophysiol, 95*, 709–729.

Rathelot, J.A., & Strick, P.L. (2009). Subdivisions of primary motor cortex based on cortico-motoneuronal cells. *Proc Natl Acad Sci U S A, 106*, 918–923.

Riehle, A., MacKay, W.A., & Requin, J. (1994). Are extent and force independent movement parameters? Preparation- and movement-related neuronal activity in the monkey cortex. *Experimental Brain Research, 99*, 56–74.

Riehle, A., Wirtssohn, S., Grun, S., & Brochier, T. (2013). Mapping the spatio-temporal structure of motor cortical LFP and spiking activities during reach-to-grasp movements. *Front Neural Circuits, 7*, 48.

Rizzolatti, G., & Luppino, G. (2001). The cortical motor system. *Neuron, 31*, 889–901.

Rizzolatti, G., Luppino, G., & Matelli, M. (1998). The organization of the cortical motor system: new concepts. *Electroencephalography and Clinical Neurophysiology, 106*, 283–296.

Rosen, I., & Asanuma, H. (1972). Peripheral afferent inputs to the forelimb area of the monkey motor cortex: input-output relations. *Experimental Brain Research, 14*, 257–273.

Rouse, A.G., & Schieber, M.H. (2015). Spatiotemporal distribution of location and object effects in reach-to-grasp kinematics. *J Neurophysiol, 114*, 3268–3282.

Rouse, A.G., & Schieber, M.H. (2016a). Spatiotemporal distribution of location and object effects in primary motor cortex neurons during reach-to-grasp. *J Neurosci, 36*, 10640–10653.

Rouse, A.G., & Schieber, M.H. (2016b). Spatiotemporal distribution of location and object effects in the electromyographic activity of upper extremity muscles during reach-to-grasp. *J Neurophysiol, 115*, 3238–3248.

Roy, A.C., Paulignan, Y., Meunier, M., & Boussaoud, D. (2002). Prehension movements in the macaque monkey: effects of object size and location. *J Neurophysiol, 88*, 1491–1499.

Saleh, M., Takahashi, K., & Hatsopoulos, N.G. (2012). Encoding of coordinated reach and grasp trajectories in primary motor cortex. *J Neurosci, 32*, 1220–1232.

Santello, M., & Soechting, J.F. (1998). Gradual molding of the hand to object contours. *Journal of Neurophysiology, 79*, 1307–1320.

Schaffelhofer, S., Agudelo-Toro, A., & Scherberger, H. (2015a). Decoding a wide range of hand configurations from macaque motor, premotor, and parietal cortices. *J Neurosci, 35*, 1068–1081.

Schaffelhofer, S., Sartori, M., Scherberger, H., & Farina, D. (2015b). Musculoskeletal representation of a large repertoire of hand grasping actions in primates. *IEEE Trans Neural Syst Rehabil Eng, 23*, 210–220.

Schieber, M.H., & Hibbard, L.S. (1993). How somatotopic is the motor cortex hand area? *Science, 261*, 489–492.

Schieber, M.H., & Rivlis, G. (2005). A spectrum from pure post-spike effects to synchrony effects in spike-triggered averages of electromyographic activity during skilled finger movements. *Journal of Neurophysiology, 94*, 3325–3341.

Schieber, M.H., & Thach, W.T. (1985). Trained slow tracking. II. Bidirectional discharge patterns of cerebellar nuclear, motor cortex, and spindle afferent neurons. *Journal of Neurophysiology, 54*, 1228–1270.

Sergio, L.E., Hamel-Paquet, C., & Kalaska, J.F. (2005). Motor cortex neural correlates of output kinematics and kinetics during isometric-force and arm-reaching tasks. *Journal of Neurophysiology, 94*, 2353–2378.

Shinoda, Y., Yokota, J., & Futami, T. (1981). Divergent projection of individual corticospinal axons to motoneurons of multiple muscles in the monkey. *Neuroscience Letters, 23*, 7–12.

Smeets, J.B., & Brenner, E. (1999). A new view on grasping. *Motor Control, 3*, 237–271.

Soechting, J.F., & Flanders, M. (1993). Parallel, interdependent channels for location and orientation in sensorimotor transformations for reaching and grasping. *Journal of Neurophysiology, 70*, 1137–1150.

Soso, M.J., & Fetz, E.E. (1980). Responses of identified cells in postcentral cortex of awake monkeys during comparable active and passive joint movements. *Journal of Neurophysiology, 43*, 1090–1110.

Spinks, R.L., Kraskov, A., Brochier, T., Umilta, M.A., & Lemon, R.N. (2008). Selectivity for grasp in local field potential and single neuron activity recorded simultaneously from M1 and F5 in the awake macaque monkey. *Journal of Neuroscience, 28*, 10961–10971.

Stark, E., Asher, I., & Abeles, M. (2007). Encoding of reach and grasp by single neurons in premotor cortex is independent of recording site. *Journal of Neurophysiology, 97*, 3351–3364.

Stelmach, G.E., Castiello, U., & Jeannerod, M. (1994). Orienting the finger opposition space during prehension movements. *J Mot Behav, 26*, 178–186.

Thach, W.T. (1978). Correlation of neural discharge with pattern and force of muscular activity, joint position, and direction of intended next movement in motor cortex and cerebellum. *Journal of Neurophysiology, 41*, 654–676.

Tillery, S.I., Ebner, T.J., & Soechting, J.F. (1995). Task dependence of primate arm postures. *Exp Brain Res, 104*, 1–11.

Tkach, D., Reimer, J., & Hatsopoulos, N.G. (2007). Congruent activity during action and action observation in motor cortex. *Journal of Neuroscience, 27*, 13241–13250.

Towles, J.D., Valero-Cuevas, F.J., & Hentz, V.R. (2013). Capacity of small groups of muscles to accomplish precision grasping tasks. *Conf Proc IEEE Eng Med Biol Soc, 2013*, 6583–6586.

Townsend, B.R., Subasi, E., & Scherberger, H. (2011). Grasp movement decoding from premotor and parietal cortex. *Journal of Neuroscience*, 31, 14386–14398.

Umilta, M.A., Brochier, T., Spinks, R.L., & Lemon, R.N. (2007). Simultaneous recording of macaque premotor and primary motor cortex neuronal populations reveals different functional contributions to visuomotor grasp. *Journal of Neurophysiology, 98*, 488–501.

Valero-Cuevas, F.J. (2000). Predictive modulation of muscle coordination pattern magnitude scales fingertip force magnitude over the voluntary range. *Journal of Neurophysiology, 83*, 1469–1479.

Valero-Cuevas, F.J. (2005). An integrative approach to the biomechanical function and neuromuscular control of the fingers. *Journal of Biomechanics, 38*, 673–684.

Valero-Cuevas, F.J., Johanson, M.E., & Towles, J.D. (2003). Towards a realistic biomechanical model of the thumb: the choice of kinematic description may be more critical than the solution method or the variability/uncertainty of musculoskeletal parameters. *Journal of Biomechanics, 36*, 1019–1030.

Vargas-Irwin, C.E., Shakhnarovich, G., Yadollahpour, P., Mislow, J.M., Black, M.J., & Donoghue, J.P. (2010). Decoding complete reach and grasp actions from local primary motor cortex populations. *Journal of Neuroscience, 30*, 9659–9669.

Weinrich, M., Wise, S.P., & Mauritz, K.H. (1984). A neurophysiological study of the premotor cortex in the rhesus monkey. *Brain, 107*, 385–414.

Wise, S.P. (1985). The primate premotor cortex: past, present, and preparatory. *Annual Review of Neuroscience, 8*, 1–19.

Wong, Y.C., Kwan, H.C., MacKay, W.A., & Murphy, J.T. (1978). Spatial organization of precentral cortex in awake primates. I. Somatosensory inputs. *Journal of Neurophysiology, 41*, 1107–1119.

Woolsey, C.N., Erickson, T.C., & Gilson, W.E. (1979). Localization in somatic sensory and motor areas of human cerebral cortex as determined by direct recording of evoked potentials and electrical stimulation. *Journal of Neurosurgery, 51*, 476–506.

Woolsey, C.N., Settlage, P.H., Meyer, D.R., Sencer, W., Hamuy, T.P., & Travis, A.M. (1952). Patterns of localization in precentral and "supplementary" motor areas and their relation to the concept of a premotor area. *Res Pub Assoc Res Nerv Ment Dis, 30*, 238–264.

7

REPRESENTING VISUAL INFORMATION IN MOTOR TERMS FOR GRASP CONTROL

Kenneth F. Valyear

List of Abbreviations

AIP	monkey anterior intraparietal area
AIPC	human anterior intraparietal cortex
dPMC	human dorsal premotor cortex
F1	monkey primary motor hand area
F2	monkey dorsal premotor area
F5	monkey ventral premotor area
fMRI	functional magnetic resonance imaging
M1	human primary motor cortex
MIP	monkey medial intraparietal area
MVPA	multi-voxel pattern analysis
PPC	posterior parietal cortex
RS	repetition suppression
RSA	representational similarity analysis
S1	human primary sensory cortex
SPOC	human superior parieto-occipital cortex
TMS	transcranial magnetic stimulation
V6A	monkey medial posterior parietal area
vPMC	human ventral premotor cortex

One of the most important questions to be addressed in the field of movement science is how the brain translates sensory information about the world to a form that is understood by the motor system, and used to plan and control purposeful actions. This chapter reviews what is currently understood about the neural mechanisms responsible for the visual-to-motor transformations that are necessary for the control of object grasping. Similar principles are likely to apply to other sensory-to-motor transformations.

In the macaque monkey, the neural response mechanisms underlying the transformation of visual information to motor programmes for grasping have been studied extensively. I use these data as anchor points to evaluate evidence for and against the presence of similar mechanisms in the human brain, with an emphasis on the review of data derived from studies involving functional MRI (fMRI).

I then briefly review other evidence that extend beyond visual-to-motor transformations based on physical object properties. I focus on the role of task context and learned object features. Finally, I discuss future opportunities, where information is limited, and results are less certain.

Before proceeding, a few cautionary remarks on the comparative neurobiological approach are worth communicating. First, comparisons between species often span great distances in evolutionary timescales. Humans and macaque monkeys shared a common ancestor approximately 30 million years ago (Kay, Ross, & Williams, 1997). This is a long time, even in evolutionary terms. Second, the data I compare here are derived from very different methods. While monkey electrophysiological recordings directly reflect cellular responses, fMRI signals are some steps removed from neural activity (Logothetis & Wandell, 2004). Also, the spatial and temporal scales are widely dissimilar. Neural recordings reflect data from single cells, and with millisecond timing, while fMRI captures information about the responses of millions of cells, and with relatively course temporal precision.

Despite these challenges, the comparative approach is invaluable. The field of neuroscience benefits from additional constraints, not fewer (see Cisek & Pastor-Bernier, 2014, p. 3). It is of tremendous value to constrain hypotheses, and data interpretation, regarding human brain function on the basis of what is known from animal physiology. It is also worth appreciating that many of the limitations have specific, one-sided complications for results interpretation – they make conclusions about interspecies differences challenging, but interpretation of similarities is more straightforward. Indeed, as reviewed here, there are compelling similarities in the functional brain organization responsible for the visual control of grasping in humans and monkeys.

7.1 The Primate Cortical Visual System

7.1.1 Two Visual Pathways

Despite our conscious experience of the visual world as unitary, a wealth of evidence reveals that the neural organization of the primate visual system is modular. In their seminal paper, Ungerleider and Mishkin (1982) suggest that the primate cortical visual system is functionally divided: A pathway from occipital to inferotemporal cortex mediates 'object vision', while a separate pathway from occipital to posterior parietal cortex (PPC) mediates 'spatial vision' (see also Mishkin, 1972; Mishkin, Ungerleider, & Macko, 1983).

According to this framework, the two pathways contribute to our conscious visual experience of the world, but their functional units are tuned to different features of the visual array: the ventral 'what' pathway is tuned to intrinsic object features – like size and shape – and governs object recognition, whereas the dorsal 'where' pathway is tuned to extrinsic object features, like the spatial relations between objects, and governs their localization in space.

The 'what' vs 'where' model is supported by evidence from monkey lesion work (Pohl, 1973), neural recording (Gross & Mishkin, 1977; Gross, Rocha-Miranda, & Bender, 1972; Hyvarinen & Poranen, 1974; Robinson, Goldberg, & Stanton, 1978), and human neuropsychology (Balint, 1909; Hecaen and De Ajuriaguerra, 1954; Kimura, 1963; Ratcliff and Davies-Jones, 1972; Warrington, 1982; Warrington & James, 1967).

Goodale and Milner later re-characterize the roles of the dorsal and ventral visual pathways (Goodale & Milner, 1992; Goodale, Milner, Jakobson, & Carey, 1991; Milner & Goodale, 1995). They suggest that while the ventral stream mediates our conscious visual experience of the world, essential for the visual recognition of objects (vision-for-perception), the dorsal stream is dedicated to the visual guidance and control of actions, responsible for the (unconscious) transformation of visual information to appropriate motor outputs (vision-for-action).

This chapter focuses on the neural mechanisms underpinning visual-to-motor transformations for grasp control, for which brain areas in the dorsal visual pathway are known to play an essential role.

7.2 Monkey Neurophysiology

In a series of seminal experiments, Sakata and colleagues identify an area located at the anterior extent of the intraparietal sulcus, area AIP, and systematically characterize its visual and motor response properties (Murata, Gallese, Luppino, Kaseda, & Sakata, 2000; Sakata, Taira, Murata, & Mine, 1995; Taira, Mine, Georgopoulos, Murata, & Sakata, 1990). Recording activity from cells in area AIP during object grasping with and without vision available, and during object fixation alone, a group of cells are discovered that demonstrate matched visual (object) and motor (grasp) response selectivity. These cells respond maximally when viewing objects of specific size, shape, or orientation (visual-object selectivity), and, critically, also when grasping objects that share those features, even when grasping without vision (motor-grasp selectivity).

Cells with the same response properties are later identified in ventral premotor area F5 in the arcuate sulcus (Murata et al., 1997; Raos, Umilta, Murata, Fogassi, & Gallese, 2006; Rizzolatti et al., 1988), an area that shares dense reciprocal connections with AIP (Borra et al., 2008; Godschalk, Lemon, Kuypers, & Ronday, 1984; Luppino, Murata, Govoni, & Matelli, 1999). It was suggested that visual responses to objects in F5 "are neither visual nor intentional but represent the description of the presented object in motor terms" (Murata et al., 1997, p. 2229). This original interpretation remains the predominate hypothesis.

The mechanisms identified by these seminal discoveries constitute the most likely candidates for how the brain translates visual information to motor commands for grasp control. Matched visual-object and motor-grasp response selectivity is exactly the kind of neural response property expected from brain areas responsible for transforming visual information about the features of objects – size, shape, orientation – to corresponding motor parameters necessary for controlling the hand to interact with them (Fagg & Arbib, 1998; Jeannerod, Arbib, Rizzolatti, & Sakata, 1995; Luppino et al., 1999, p. 181).

In addition to areas AIP and F5, cells with visual-motor response congruency are present in dorsal premotor cortex, area F2, and the medial posterior parietal area V6A, located within the anterior bank of the parieto-occipital sulcus (Figure 7.1A). These areas have unique anatomical connections that constitute distinct pathways (Rizzolatti & Matelli, 2003). AIP and F5 are part of a dorsolateral pathway, while V6A and F2 are part of a dorsomedial pathway.

7.2.1 The Dorsolateral Pathway: AIP and F5

Areas AIP and F5 constitute the dorsolateral pathway for grasp control in the macaque monkey (Figure 7.1A). They are necessary for grasping. Temporary inactivation of either AIP (Gallese, Murata, Kaseda, Niki, & Sakata, 1994) or F5 (Fogassi et al., 2001) impairs the anticipatory shaping of the hand during grasping; grasping is made clumsy, and sometimes altogether unsuccessful.

Evidence supports the following model of the functions of the dorsolateral pathway. AIP represents visual information about the features of objects. Together with F5, AIP 'flags' the object features that are most relevant for intended actions, and the two areas collaborate to transform this information to corresponding sensory and motor parameters necessary for controlling the hand. F5 represents these parameters, and signals this information to M1, and ultimately to the spinal cord machinery for the control of hand and digit muscles.

AIP has fewer motor and more visual cells than F5, and unlike F5, AIP has cell types that only respond when vision is available – 'visual-dominant cells' (Murata et al., 1997, 2000; Raos et al., 2006). Further, multidimensional scaling methods reveal that while object-selectivity in AIP best reflects structural object features (Murata et al., 2000), object-selectivity in F5 is better explained by the hand configurations that are associated with grasping those objects (Raos et al., 2006).

Newer findings complement and extend these data. Schaffelhofer and Scherberger (2016) record concurrently from areas AIP, F5, and F1 while also

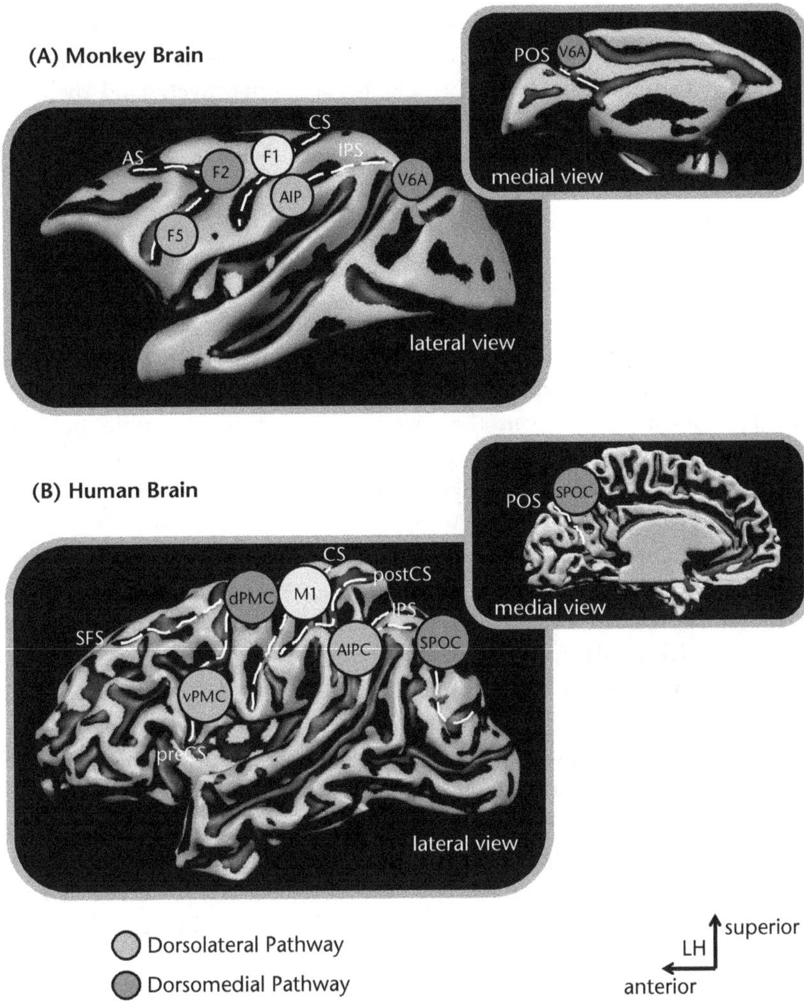

FIGURE 7.1 Grasp areas of the dorsolateral and dorsomedial pathways

Notes: Schematic representation of areas implicated in the visual-to-motor transformations necessary for the control of the hand for object grasping shown on the cortical surfaces of (A) the macaque monkey and (B) human brain. The cortical surfaces are defined at the grey–white matter boundary and partially inflated to reveal regions within the sulci. The left hemisphere (LH) is shown from lateral and medial views. Dashed lines indicate sulci: IPS, intraparietal sulcus; CS, central sulcus; AS, arcuate sulcus; POS, parieto-occipital sulcus; postCS, post-central sulcus; preCS, precentral sulcus; SFS, superior frontal sulcus. Monkey areas: AIP, anterior intraparietal area; F5, ventral premotor cortex; V6A, parieto-occipital cortex; F2, dorsal premotor cortex; F1, primary motor cortex. Human areas: AIPC, anterior intraparietal cortex; vPMC, ventral premotor cortex; SPOC, superior parieto-occipital cortex; dPMC, dorsal premotor cortex; M1, primary motor cortex. Stefan Everling generously provided the monkey MRI data used here for the cortical reconstruction.

measuring limb kinematics as monkeys view and grasp an extensive range of objects of different shapes and sizes. Neural responses and hand kinematics are transformed to multidimensional N- and J-space, respectively, and then ultimately to a common space for comparison.

The results demonstrate predominately visual-object coding in AIP and motor-grasp coding in areas F5 and M1. AIP shows strong modulation during object viewing, and the neural coding clearly reflects object shape. Conversely, area F5 responds only weakly during the view period, transiently representing a common encoding scheme with AIP, and then its activity strongly reflects hand kinematics during grasping, consistent with a motor-grasp encoding scheme.

Particularly informative, during conditions where objects are visually distinct but grasped similarly, AIP distinguishes between objects during the view phase, while F5 does not. However, the data also show that over time and prior to movement onset, AIP represents these same conditions more similarly, and also responds distinctly to visually identical objects when they are grasped differently.

Altogether, the data are consistent with an interpretation whereby over time AIP visually specifies a narrowing range of action possibilities, honing in on the object features that are essential for the upcoming grasp (for additional comments, see Valyear, 2016). This explains why responses to different objects that are grasped similarly show increasing representational similarity during planning, and why the same object that is grasped differently elicits distinct responses.

During the object view phase, AIP and F5 show common encoding. This suggests that during object viewing information is briefly shared between these areas. Although speculative, it is intriguing to consider the possibility that this sharing of information helps to shape responses in AIP, to narrow the number of action possibilities to a single set of visually defined grasp points on the target object. Notably, this interpretation can be considered consistent with the original accounts proposed by Sakata et al. (1995, p. 437; see also Murata et al., 2000, p. 2599).

Finally, the new results of Schaffelhofer and Scherberger (2016) also provide compelling evidence for the role of F5 in driving the activity of the primary motor hand area, F1. Over time, response encoding is increasingly similar between F5 and F1, with F5 modulation showing earlier onsets. These data complement and extend prior results derived from concurrent F5-F1 recordings (Spinks, Kraskov, Brochier, Umilta, & Lemon, 2008; Umilta, Brochier, Spinks, & Lemon, 2007).

7.2.2 The Dorsomedial Pathway: V6A and F2

The dorsomedial pathway includes areas MIP in the medial bank of the IPS, V6A in the anterior bank of the parieto-occipital sulcus, and dorsal premotor cortex, area F2 (Figure 7.1A). This pathway is important for the proximal control of the arm for reaching (Andersen, Andersen, Hwang, & Hauschild, 2014; Andersen & Cui, 2009; Caminiti, Ferraina, & Johnson, 1996; Crawford, Henriques, & Medendorp, 2011; Galletti, Fattori, Kutz, & Battaglini, 1997; Jeannerod, 1997; Wise, Boussaoud, Johnson, & Caminiti, 1997). Cells in V6A (Fattori, Kutz, Breveglieri, Marzocchi, & Galletti, 2005) and F2 (Cisek & Kalaska, 2005; Hoshi & Tanji, 2004, 2006) show arm-movement-direction selectivity.

Newer evidence indicates that areas V6A and F2 are also important for the distal control of the hand for grasping. Both V6A (Fattori, Breveglieri, Amoroso, & Galletti, 2004; Fattori et al., 2010) and F2 (Hendrix, Mason, & Ebner, 2009; Raos, Umilta, Gallese, & Fogassi, 2004; Stark, Asher, & Abeles, 2007) have cells that are tuned to object properties and the configuration of the hand for grasping, and some cells show visual-motor response congruency similar to the properties of AIP and F5 cells, discussed above. V6A cells are highly sensitive to wrist orientation (Fattori et al., 2009), and experimental lesioning of this area impairs both proximal and distal elements of grasp control (Battaglini et al., 2002).

It has been suggested that area V6A plays a special role in coordinating the proximal and distal elements of upper limb actions, and perhaps also in monitoring and correcting for errors in the spatiotemporal features of the moving arm and hand during reaching and grasping (Fattori et al., 2010; Fattori, Breveglieri, Bosco, Gamberini, & Galletti, 2015). F2 may also be important for using vision to monitor movements of the hand when grasping. When vision is removed during grasping, F2 cells no longer exhibit motor-grasp or visual-object selectivity (Raos et al., 2004).

7.3 Human Brain Areas

For each of the monkey brain areas reviewed above, there is evidence for a putative functional counterpart in the human brain (Figure 7.1B).

7.3.1 Anterior Intraparietal Cortex

Anterior intraparietal cortex (AIPC) may be a functional complement of monkey area AIP. In fMRI work, AIPC is reliably preferentially activated for grasping (Culham & Valyear, 2006; Grafton, 2010; see Valyear, Fitzpatrick, & McManus, 2017 for the results of a recent meta-analysis). Patients with damage to AIPC show disorganized anticipatory shaping of the hand to grasp (Binkofski et al., 1998), and transcranial magnetic stimulation (TMS) applied to AIPC impairs hand preshaping (Davare, Andres, Clerget, Thonnard, & Olivier, 2007; Rice, Tunik, Cross, & Grafton, 2007; Rice, Tunik & Grafton, 2006) and the ability to rapidly correct for sudden changes in the properties – size/orientation – of target objects during grasping (Tunik, Frey, & Grafton, 2005).

Newer work shows that temporarily reducing the excitability of AIPC disrupts grasp-specific interactions between ventral premotor cortex (putative functional complement of monkey area F5; see section 7.3.2) and the hand area of primary motor cortex (M1) (Davare, Rothwell, & Lemon, 2010).

7.3.2 Ventral Premotor Cortex

The suggested functional counterpart to monkey area F5 in humans is known as ventral premotor cortex (vPMC), located within the precentral sulcus, overlapping with the inferior frontal and precentral gyri. Reports of vPMC as selectively active for grasping in fMRI studies are much less common than AIPC (although see Binkofski et al., 1998; Cavina-Pratesi et al., 2010). The reasons for this are not clear, although may relate to a focus in the field on group-averaged data; when individual-level results are systematically characterized, some individuals reliably demonstrate grasp-selective activity in vPMC (Valyear, Mattos, Philip, Kaufman, & Frey, 2017). Task complexity is likely another critical factor. Studies involving more complex hand actions, like using tools, consistently identify vPMC (Brandi, Wohlschlager, Sorg, & Hermsdorfer, 2014; Valyear, Gallivan, McLean, & Culham, 2012).

Data from TMS work clearly indicate an essential role for vPMC in the control of grasp. Paired-pulse stimulation methods demonstrate the functional influence of vPMC on M1 and the control of hand muscles during object grasping (Davare, Lemon, & Olivier, 2008; Davare, Montague, Olivier, Rothwell, & Lemon, 2009), and repetitive TMS to vPMC impairs digit coordination during grasp control (Davare, Andres, Cosnard, Thonnard, & Olivier, 2006).

7.3.3 Dorsal Premotor Cortex

Dorsal premotor cortex (dPMC) is the putative functional complement of monkey area F2. Like F2, human dPMC has been implicated in both reach (arm transport) and grasp (hand preshaping) control. With fMRI, dPMC is the second most consistently reported brain area that is preferentially activated during object grasping, next only to AIPC (Valyear, Fitzpatrick, & McManus, 2017, meta-analysis), and TMS to dPMC disrupts the temporal coupling between grip and lift components of reach-to-grasp actions (Davare et al., 2006). Together, this evidence suggests that dPMC is important for the control of both hand movements and manipulative forces – i.e. kinetics – during object grasping.

A wealth of fMRI data indicate that dPMC is involved in the control of the arm for reaching (Astafiev et al., 2003; Bernier & Grafton, 2010; Beurze, de Lange, Toni, & Medendorp, 2007; Medendorp, Goltz, Crawford, & Vilis, 2005; Prado et al., 2005; Tosoni, Galati, Romani, & Corbetta, 2008). These studies also commonly report superior parieto-occipital cortex (putative functional complement of monkey area V6A; see Section 7.3.4) and posterior intraparietal cortex (putative functional complement of monkey area MIP) as preferentially active for reaching.

7.3.4 Superior Parieto-Occipital Cortex

Mounting evidence suggests that an area known as superior parieto-occipital cortex (SPOC), located within the superior parietal lobule anterior to the parietal occipital sulcus along the midline of the cortex, may be functionally similar to monkey area V6A (Pitzalis, Fattori, & Galletti, 2015; Vesia & Crawford, 2012).

TMS to SPOC disrupts reaching performance (Vesia, Prime, Yan, Sergio, & Crawford, 2010), and damage here can result in optic ataxia, characterized by difficulties using visual information to guide and control the arm/hand for actions (Andersen et al., 2014; Balint, 1909; Jakobson, Archibald, Carey, & Goodale, 1991; Karnath & Perenin, 2005; Perenin & Vighetto, 1988). As noted above, fMRI data reliably reveal SPOC (and other areas within posterior intraparietal cortex) as preferentially responsive during tasks that require reaching-to-contact targets (Astafiev et al., 2003; Bernier & Grafton, 2010; Beurze et al., 2007; Connolly, Andersen, & Goodale, 2003; Medendorp et al., 2005; Prado et al., 2005; Tosoni et al., 2008; Tosoni et al., 2015). This evidence is consistent with the role of V6A in arm control.

Newer data suggest that SPOC is also important for grasp control. Some fMRI studies report SPOC as preferentially active for grasping (Culham et al., 2003; Monaco et al., 2011; Monaco et al., 2014), and suggest tuning to visual object properties and hand configurations for grasping, reviewed below.

7.3.5 Measuring Response Properties with fMRI

Do the human brain areas introduced above demonstrate response properties similar to those described in monkeys? I believe that addressing this question is fundamental to understanding how the human brain mediates the functional use of the hand.

It can be challenging to reveal insights about the response properties of brain areas using fMRI. For example, let's say we want to test the hypothesis that AIPC represents hand configurations for grasping, and we set up two conditions – A and B – differing only in the configuration of the hand used to grasp. If distinct but similar proportions of cells in AIPC represent either hand configuration, directly contrasting the two conditions will not identify AIPC.[1] The activity per condition will subtract away. This illustrates how conventional fMRI contrast methods may yield inconclusive results, or even motivate false conclusions about the response properties of brain areas.

Thankfully, other fMRI design and analyses methods have been developed that can, for certain conditions, do better. Here, I discuss two such methods.

One method is known as fMRI repetition suppression (fMRI-RS), or fMRI-adaptation, and the other is known as multi-voxel pattern analysis (MVPA). I use our example above to illustrate the basic principles of each approach.

7.3.5.1 fMRI-RS

Generally, repeated events – stimuli/actions – result in reduced fMRI activity levels. The thinking here is that when two or more successive events activate the same or similar neural representations, the strengths of the associated fMRI responses are reduced. The fMRI-RS approach takes advantage of these repetition-related reductions in fMRI signals to probe the response properties of human brain areas.

The logic is as follows. If a brain area is important for processing a specific stimulus or response feature, when that feature is repeated, reduced fMRI activity levels – fMRI-RS – are predicted. The repeated feature successively activates overlapping neural representations, and as a consequence the responses associated with those representations are reduced. Conversely, a change in this feature is predicted to activate non-overlapping, non-suppressed neural representations. As such, greater fMRI activity levels are predicted when compared to a condition involving feature repetition. This relative increase is sometimes referred to as a rebound, or release from RS, and this is interpreted as evidence that this brain area is important for processing that particular stimulus or response feature.

Let's revisit our example above. If hand configuration A is repeated, those cells that represent this hand configuration – i.e. that specify the corresponding sensorimotor parameters necessary to shape the hand in this way – are predicted to respond in ways that translate to reduced fMRI activity levels. At the cellular level, this may reflect various mechanisms, including decreased neural firing rates, shortened firing durations, and/or fewer numbers of responsive cells (Grill-Spector, Henson, & Martin, 2006; James & Gauthier, 2006; Krekelberg, Boynton, & van Wezel, 2006; Wiggs & Martin, 1998). If we compare this to a condition where hand configuration A is non-repeated (for e.g., preceded by hand configuration B), maximal (non-supressed) fMRI response levels are predicted.

Using this approach, AIPC is predicted to show greater fMRI activity for non-repeated vs repeated conditions: B-A/A-B > A-A/B-B. This result would suggest that AIPC is involved in specifying hand configurations for grasping.

7.3.5.2 MVPA

Different from conventional fMRI contrast methods, MVPA tests predictions about the spatial patterns of fMRI activity associated with different conditions. A common variant of this method uses classifier algorithms (e.g. support vector machines) to test whether the spatial configuration of voxels within brain areas

– their 'representational geometry' – can be used to reliably (above chance) discriminate between experimental conditions (Kamitani & Tong, 2005).

Consider again our example above. While hand configurations A and B may lead to similar overall activity in AIPC, indistinguishable with direct contrast methods, the relative spatial patterns of voxel-wise activity in AIPC that are associated with hand configurations A and B may be distinct. If this is true, MVPA classification methods can be used to reliably discriminate these conditions based on the spatial patterns of associated responses in AIPC. As such, we would conclude that AIPC distinctly represents these hand configurations for grasping.

Another variant of MVPA known as representational similarity analysis (RSA) tests for possible relationships between the spatial patterns of fMRI activity within brain areas and different features of stimuli or behaviours in multidimensional space (Kriegeskorte & Kievit, 2013). RSA can not only reveal whether brain areas represent particular experimental conditions, but also the organizational principles that define their response properties.

For a recent review of MVPA methods applied to the study of the control of actions, generally, see Gallivan and Culham (2015).

7.4 Physical Object Properties

> The shapes and sizes of objects, in fact, are perceived in relation to the hands, as graspable or not graspable, in terms of their affordances for manipulation.
>
> *(Gibson, 1986, p. 224)*

A number of studies have used fMRI-RS and MVPA to investigate the response properties of brain areas important for representing physical object properties for the purpose of grasping. Here I will briefly review these data bearing in mind the basic monkey neurophysiological response properties of areas important for grasping, reviewed above.

Specifically, according to the monkey data, human AIPC, SPOC, v-, and dPMC are predicted to show visual response selectivity for object properties – shape, size, and orientation – and motor response selectivity for different hand configurations used for grasping. We can think of these predictions as consistent with *visual-object* and *motor-grasp* coding, respectively.

All four brain areas – AIPC, SPOC, v-, and dPMC – should show fMRI-RS for repeated visual-object and motor-grasp features, and MVPA methods are predicted to reveal distinct spatial patterns of activity associated with visual-object and motor-grasp features in all four areas.

7.4.1 Shape

Using an RS design, Kroliczak, McAdam, Quinlan, and Culham (2008) independently vary target object shape and hand configuration required to grasp (Figure 7.2). AIPC shows RS when either the object shape or grasp is repeated, consistent with visual-object and motor-grasp coding, respectively. This is precisely in line with the neural recording data from monkey area AIP.

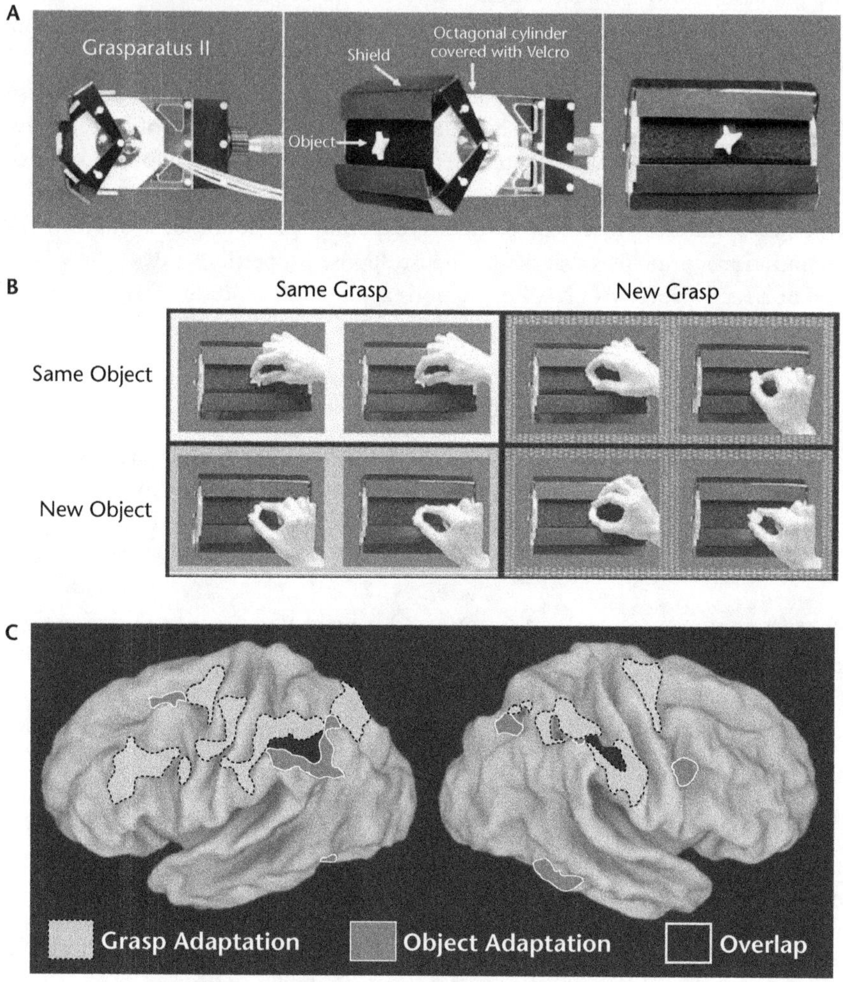

FIGURE 7.2 fMRI-RS for repeated grasp and object features

Notes: Participants grasp real three-dimensional objects in the MRI scanner (A). Grasp (motor) and object (visual) features are either repeated or changed, independently (B). Group results are shown on a three-dimension cortical representation (C). Anterior intraparietal cortex shows evidence for both visual-object and motor-grasp coding, while dorsal and ventral premotor cortices show evidence for motor-grasp coding. Reproduced from Kroliczak et al. (2008).

Their results also reveal evidence for motor-grasp coding in v- and dPMC. These areas show a rebound from RS when grasp is new. This evidence is also consistent with the monkey data from areas F5 and F2, respectively.

However, only vPMC shows evidence for visual-object coding, and unexpectedly, these effects are evident only in the right hemisphere, ipsilateral to the hand used for grasping. These results are inconsistent with predictions from monkey neurophysiology showing that both F5 and F2 are visually tuned to object properties for grasping.

As one possible explanation, perhaps v- and dPMC more strongly represent the final committed action, and as a result show insensitivity to object repeats when grasp is new. This interpretation is consistent with other monkey neural recording data (e.g. Schaffelhofer & Scherberger, 2016).

It is unclear why SPOC does not show RS; monkey data from V6A predicts both visual-object and motor-grasp coding. Objects vary in shape and orientation, and different grasps involve different wrist orientations.

Findings from a recent study by Fabbri and colleagues (2016) complement and extend these data. These authors use RSA to determine which of several different visual-object and motor-grasp factors best explain the spatial patterns of regional brain activity during object viewing and grasping tasks. They include a large set of objects of different shapes and sizes, and participants grasp these objects under conditions of high versus low precision demands (finger-thumb opposition versus whole-hand grasping), using two or five digits. Participants also passively view these same objects.

The results are somewhat surprising. With respect to object features, object elongation is the factor that best explains the response patterns in areas AIPC, SPOC, and dPMC, outperforming the models that specify only object size or shape. Object elongation also best explains responses in the lateral occipitotemporal cortex (a key part of the ventral visual pathway; see Section 7.1.1). The activity in vPMC is not well explained by any object model.

Motorically, the digit model – how many digits were used to grasp – demonstrates the highest explanatory power for sensorimotor areas, including AIPC and d- and vPMC, as well as primary sensory and motor cortices, S1/M1. This result can (at least in part) be interpreted as response sensitivity to different hand configurations for grasping, and thus, is largely consistent with other data from fMRI-RS reviewed here. Notably, a region the authors refer to as posterior-SPOC shows sensitivity to precision demands.

7.4.2 Size

Other fMRI-RS designs target object size. Monaco and colleagues (2014) vary the size of target objects independently from the size of their grasp-relevant dimension. Specifically, participants grasp objects along their vertical axes, and the overall size of the object is varied independently of this 'grasp axis'. For example, the same rectangular block may be shown – repeated size – but rotated such that the width of the grasp-relevant dimension is changed. Wrist orientation is held constant, so that a change in grasp specifically translates to a change in the extent to which the hand needs to open – the grasp aperture. They also include a view-only condition, to investigate whether RS depends on grasping or can arise purely visually.

The results for AIPC are complex. A release from RS is observed only when both size and the grasp-axis are changed, and only for the grasping condition. The explanation for this pattern is unclear. If AIPC specifies grasp possibilities, a release from RS is expected whenever an object change occurs. Alternatively, if AIPC only represents grasp-relevant dimensions, and is otherwise insensitive to other object features, changing the grasp-relevant dimension should invariably lead to a release from RS. Instead, changes to both features – size and grasp axis – are necessary.

The results for SPOC are more straightforward to interpret. Bilaterally, SPOC shows sensitivity to changes in the grasp axis, independent of the task. This suggests visual-object and motor-grasp coding, consistent with various monkey physiological data from area V6A, reviewed above.

Similar results are shown for the lateral occipitotemporal cortex, while other visual areas in medial occipital cortex show sensitivity to object size. Finally, dPMC is also sensitive to the grasp-relevant dimension of objects, but only during grasping, consistent with a strictly motor-grasp encoding scheme.

In a newer RS study (Monaco, Sedda, Cavina-Pratesi, & Culham, 2015), object size and location are varied independently, and participants grasp or view objects. The data suggest that AIPC represents object size independent of location, and specific to when participants are required to grasp. Here, changing size changes the grasp-relevant properties, and thus these results are consistent with the prior findings just discussed (Monaco et al., 2014).

Additionally, the newer data reveal selectivity to both object size and position within SPOC and dPMC. This is consistent with a role in both distal and proximal control, as suggested by the monkey data from areas V6A and F2. Size- and position-selectivity in SPOC and dPMC is specific to the grasp task, and the RS effects are not additive, consistent with the possibility that these features are represented separately in these areas, coded in distinct neural units. Why size encoding in SPOC is specific to the grasping condition in this study but generalizes to both the grasp and view-only conditions in the prior work (Monaco et al., 2014) is yet unclear.

7.4.3 Orientation

Using fMRI-RS, Monaco et al. (2011) investigate response selectivity for object orientation in three conditions: Grasp, Reach (to touch, without shaping the hand to grasp), and View. The results demonstrate RS for repeated orientation in SPOC and dPMC, specific to the Grasp condition.

These data may reflect motor response coding necessary for configuring the wrist and forearm to accommodate the object orientations used, as predicted by monkey neurophysiological results from V6A (Fattori et al., 2009; Fattori et al., 2010) and F2 (Raos et al., 2004). Another area within posterior IPS showed similar effects.

Conversely, AIPC does not show orientation sensitivity in this study. The authors suggest that this may reflect comparatively fewer numbers of cells tuned to object orientation in monkey area AIP, consistent with data from Murata et al. (2000). However, newer neural data reveal high proportions of cells in monkey AIP that are tuned to object orientation (Baumann et al., 2009), so this explanation seems unlikely. Alternatively, perhaps greater variation in the number of target object orientations is required to reliably detect orientation selectivity in AIPC. This study used one object presented at two orientations.

With 2D pictures of familiar graspable objects, we also fail to detect RS for object orientation in AIPC (Rice, Valyear, Goodale, Milner, & Culham, 2007), and instead find sensitivity to object orientation within a more posterior region of the IPS, at the lateral junction of the parietal and occipital cortices (Valyear, Culham, Sharif, Westwood, & Goodale, 2006; Rice, Valyear, et al., 2007). We view this result as consistent with our other data suggesting that responses to images of familiar tools within the IPS reflects higher-level cognitive motor representations related to their conventional use, and not implicit grasp plans (Valyear, Cavina-Pratesi, Stiglick, & Culham, 2007; see Section 7.5.2).

The results of a study by Fabbri, Caramazza, and Lingnau (2010) provide additional insights about the motoric coding of forearm and wrist posture for grasping. Participants grasp or reach-to-touch spherical objects presented at different locations. An adaptation design is used whereby 1–8 successive actions directed to the same target in space (the adapted movement direction) are followed by a 'test' response to the same or different target location. Reaching to different locations tended to require not only different forearm movements, but also wrist orientations for object interaction, and the angular difference between adapted and test responses was parametrically varied.

The results reveal direction selectivity in a number of areas including AIPC (and medial IPS) dPMC, SPOC, and M1. SPOC shows the highest degree of direction selectivity, and right-lateralized SPOC shows invariance as to which hand is used to perform actions – effector independence. AIPC is also sensitive to the type of action performed, showing a greater rebound from adaptation when both the target position and action-type are new.

Sensitivity to target locations in AIPC may (primarily) reflect changing wrist orientation. This account accommodates other results that demonstrate motor-grasp coding (Kroliczak et al., 2008) but relative insensitivity to arm transport (Cavina-Pratesi et al., 2010) and object position (Monaco et al., 2015). Conversely, SPOC and dPMC are known to play an important role in both arm transport and wrist posture. Strong direction selectivity in SPOC is entirely consistent with the data from V6A showing that wrist posture and arm direction are both strongly represented (Fattori et al., 2009; Fattori et al., 2010). Taken together, these results provide compelling evidence for complementary functionality between these areas in humans and monkeys.

With the same target-object arrangement, Fabbri, Strnad, Caramazza, and Lingnau (2014) use MVPA to investigate motor-grasp response selectivity when participants are instructed to use either precision or whole-hand grasps. AIPC, SPOC, and both the v- and dPMC show both grasp-type and arm-direction selectivity. Again, different object locations (reach directions) require different wrist orientations for object interaction, and this likely contributed to their findings.

7.4.4 Summary

7.4.4.1 Motor-Grasp Coding

Overall, the findings from fMRI-RS and MVPA studies are highly consistent with predictions from monkey neurophysiology, and suggest that all four human brain areas – AIPC, v- and dPMC, and SPOC – play an important role in configuring the hand for grasping according to object size, shape, and orientation. AIPC, dPMC, and SPOC are most consistently implicated, while vPMC is implicated, but not as consistently.

7.4.4.2 Visual-Object Coding

The data regarding the visual encoding of object features in AIPC, v/dPMC, and SPOC are less clear. When participants view objects without grasping (or planning to grasp), the majority of studies fail to identify visual-object coding in these brain areas. Consequently, response specificity for object properties during grasping may be explained solely on the basis of motor programming (i.e. may reflect motor-grasp coding related to configuring the hand accordingly).

Alternatively, these results may reflect sensitivity to task context. Action intention may 'gate' visual response selectivity such that when grasping is not required or planned, the visual-object coding properties of these brain areas are 'masked' from detection. Other work indicates that attention and fMRI-RS interact during visual object perception (Murray & Wojciulik, 2004). Notably, this interpretation can also accommodate the Kroliczak et al. (2008) results – in this study participants are always required to grasp and visual-object coding (in the form of object-specific RS) is detected.

7.5 Context and Learned Object Properties

7.5.1 Context

Size, shape, and orientation are not the only factors that matter for how we shape our hands to grasp objects; what we intend to do with those objects is essential, too. Various behavioural studies clearly demonstrate this. When participants grasp the same objects for different purposes, the anticipatory shaping of the hand differs, and the early steps of actions reflect the particular mechanics – movement and force requirements – of later steps (Ansuini, Giosa, Turella, Altoe, & Castiello, 2008; Ansuini, Santello, Massaccesi, & Castiello, 2006; Armbrüster & Spijkers, 2006; Cohen & Rosenbaum, 2011; Marteniuk, MacKenzie, Jeannerod, Athenes, & Dugas, 1987; Rosenbaum & Jorgensen, 1992).

Our own behavioural work has shown this, too. Using a collection of kitchen tools that have the same physical handles, we show that during grasping the hand moves differently for different tools when the task is to demonstrate their use, but not when the task is to move those same tools to a new location (Valyear, Chapman, Gallivan, Mark, & Culham, 2011). By examining the relative spatial configuration of the thumb and fingers just before tools are picked up, we see that these task differences reflect the anticipated mechanical features of the next steps of actions.

Other evidence suggests that task context changes the way object properties are specified by the visual system. The properties of non-target objects in a search task have different effects on behaviour depending on the type of actions that are planned (Bekkering & Neggers, 2002; Pavese & Buxbaum, 2002; see also Humphreys & Riddoch, 2001, for related patient data). Similarly, we show that visuomotor priming – decreases in response times to initiate actions as a consequence of preceding visual events – depends on task context (Valyear et al., 2011), consistent with an interaction between the visual specification of action possibilities and task-defined cognitive set (Bub, Masson, & Bukach, 2003). Also, in patients who demonstrate 'utilization behaviour' (Lhermitte, 1983; Shallice, Burgess, Schon, & Baxter, 1989), task context plays a driving role in triggering impulsive grasp responses (Humphreys & Riddoch, 2000; Riddoch, Edwards, Humphreys, West, & Heafield, 1998; Riddoch, Humphreys, & Edwards, 2000a; , Riddoch, Humphreys, & Edwards, 2000b).

Above (Section 7.5), I mention the possibility that task context may determine whether visual-object coding is identified in AIPC, v- and dPMC, and SPOC. The majority of fMRI-RS findings indicate insensitivity to changes in object size, shape, and/or orientation when objects are merely passively viewed. Future work specifically designed to test the hypothesis that action context changes the visual specification of affordances will be of value. A major challenge will be to disentangle the visual from motor responses. Paradigms that both *combine and separate* vision and action, to test for within- and across-modality RS, may be particularly useful (see Section 7.6.1).

Notably, the response properties of cells in monkey areas AIP (Baumann et al., 2009) and F5 (Fluet, Baumann, & Scherberger, 2010) show sensitivity to task context. When monkeys are trained to grasp a target object with either a precision or whole-hand grasp depending on a learned (otherwise arbitrary) fixation cue, cells in AIP and F5 show visual-object coding during the premovement phase of a delayed grasping task, but only after the task cue is presented (see also Schaffelhofer & Scherberger, 2016). In other words, the visual-object coding properties of cells are only detected after task goals are specified. These data demonstrate that task context modulates how visual information is represented in motor terms in monkey areas AIP and F5.

7.5.2 Learned Use

> Tools, as well as musical instruments or sport materials, are objects which cannot be characterized merely by their geometrical properties like size, shape or orientation. They have additional properties that cannot be detected unless one knows what the object is for and how to use it.
>
> *(Jeannerod & Jacob, 2005, p. 306)*

It has been suggested that viewing tools involves the specification of action possibilities related to their learned use. Responses to initiate grasping actions are influenced by visual information about the functional properties of tools (Bub et al., 2003; Bub, Masson, & Cree, 2008), and these effects reflect access to predicted action outcomes according to their conventional use (Masson, Bub, & Breuer, 2011). Consistent with these data, intraparietal and premotor areas are preferentially active when viewing pictures of tools (Chao & Martin, 2000; Creem-Regehr & Lee, 2005), and our own work suggests that this activity is not merely attributable to perceived graspability (Valyear et al., 2007). Specifically, we show that viewing hand-held tools relative to other familiar graspable but non-tool objects (e.g. a book; a tomato) preferentially activates an area within the intraparietal sulcus, near grasp-related AIPC.[2]

On the basis of these data, we hypothesize that experience is capable of reshaping the neural mechanisms that underpin the specification of visual information as action possibilities. Over time, and with sufficient experience the same neural mechanisms that mediate the visual-to-motor specification of physical object properties come to visually specify learned object properties in motor terms.

Using a novel fMRI-RS design, we provide critical support for this model (Valyear et al., 2012). Participants grasp and pantomime the use of small plastic replicas of common kitchen tools, all with the same shaped handles (Figure 7.3). Tool use actions are preceded by a visual preview (prime) of either the same or different tools.

The results reveal fMRI-RS within brain areas previously implicated as important for grasping and using tools, overlapping with AIPC and vPMC, and nearby but lateral to dPMC (Figure 7.4).

Our design controls for motor and somatosensory factors. The contrasts that identify fMRI-RS comprise conditions involving the same executed actions. The results are not attributable to potential differences in the extent, duration,

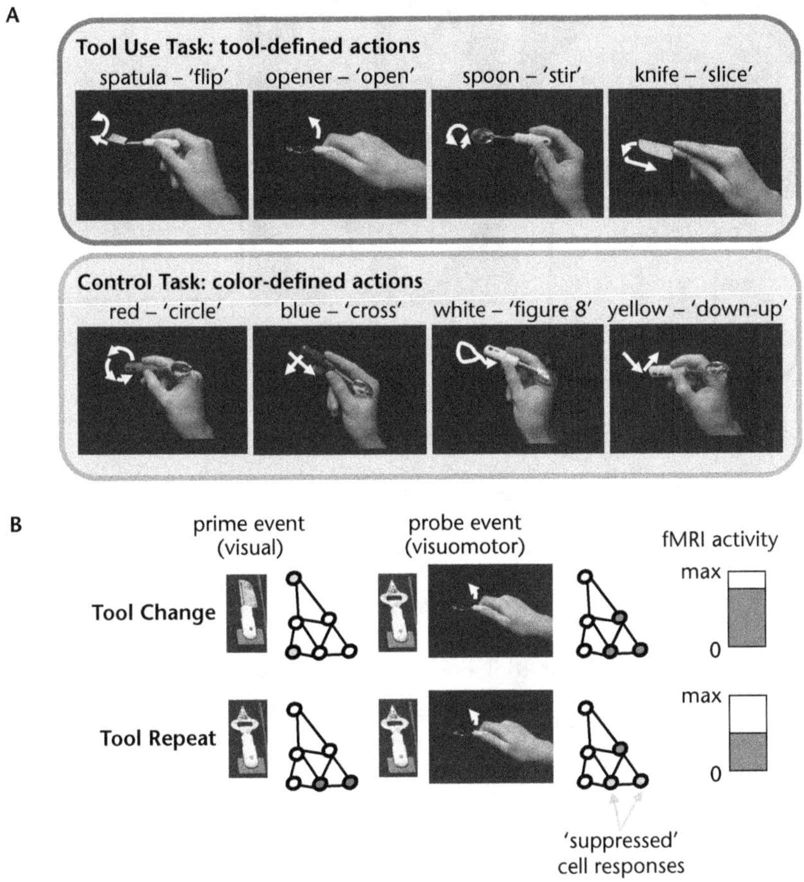

FIGURE 7.3 fMRI-RS for tool use: methods

Notes: Participants grasp and demonstrate the use of familiar tools in the scanner. Movements are either consistent with learned actions (Tool Use Task) or depend on arbitrary associations (Control Task) (A). Repetition suppression is predicted to result in decreased fMRI activity for the Tool Repeat condition, which may reflect reduced responses from 'suppressed' cell populations (B). Reproduced from Valyear et al. (2012).

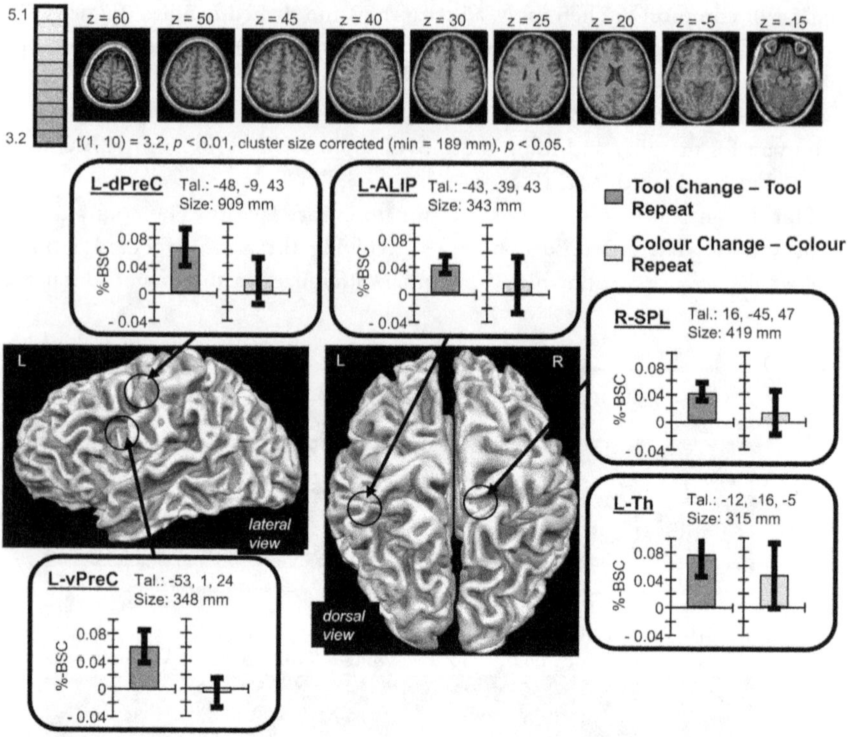

FIGURE 7.4 fMRI-RS for tool use: results

Notes: Group results reveal significant repetition suppression for the Tool Use Task with left anterior intraparietal (L-AIP), left dorsal precentral cortex (L-dPreC), left ventral precentral cortex (L-vPreC), left thalamus (L-Th), and the right superior parietal lobule (R-SPL). Bar plots reflect group mean differences per Tool Use and Control Tasks, with 95% confidence intervals. Reproduced from Valyear et al. (2012).

or complexity of different tool use actions. Instead, they must reflect visual responses to tools during prime events that precede those actions.

We also use a control task to rule out the possibility that these effects reflect correct vs incorrect action cuing, and to critically evaluate a key aspect of our hypothesis – that the strengths of learned visual-to-motor associations are essential. Participants grasp the same tools as used in our tool use task, but perform meaningless actions with them according to newly learned arbitrary rules based on the colours of their handles. The results indicate no fMRI-RS for the control task (Figure 7.4). This is inconsistent with the idea that fMRI-RS for the tool use task reflects correct vs incorrect action cuing.

Altogether, our data support the hypothesis that familiar tools are visually represented in motor terms related to their conventional use, and that the mechanisms underlying visual-to-motor transformations are shaped by experience.

7.6 Future Directions

7.6.1 Visual-Motor Response Congruency

Despite the relatively compelling evidence in favour of functionally similar response properties between monkey and human brain areas that are important for grasp control, a clear demonstration of visual-motor response congruency in the human brain is missing. Do human brain areas show evidence for matched visual-object and motor-grasp coding?

Future work may borrow from methods developed by Dinstein and colleagues designed to test for evidence of mirror-neuron-like responses in human brain areas (Dinstein, Gardner, Jazayeri, & Heeger, 2008; Dinstein, Hasson, Rubin, & Heeger, 2007). Mirror neurons respond during both action performance and when actions are observed, and critically, for some mirror cells there is close correspondence between their motor and visual response specificity – i.e. the same/similar actions activate these cells maximally during both performance and observation (di Pellegrino, Fadiga, Fogassi, Gallese, & Rizzolatti, 1992; Gallese, Fadiga, Fogassi, & Rizzolatti, 1996; Rizzolatti, Fadiga, Gallese, & Fogassi, 1996; for review see Rizzolatti & Craighero, 2004). In other words, these cells show visual-motor response congruency analogous to the response properties described for viewing and grasping objects reviewed above (Section 7.2).

Using a novel fMRI-RS design, Dinstein et al. (2007) test for RS related to (1) repeated observed hand actions, (2) repeated performed hand actions, and critically, (3) repeated observed-performed, and (4) performed-observed actions. The first two conditions test for *within-modality* RS (visual-visual, motor-motor, respectively), while the second two conditions test for *across-modality* RS (visual-motor, motor-visual, respectively). As a characteristic response property of mirror neurons, visual-motor response congruency predicts both within- and across-modality RS.

Their results show overlapping within-modality RS for both repeated visual-visual and motor-motor conditions within AIPC and vPMC, but critically, no evidence for across-modality RS.

Using similar logic but with MVPA, Dinstein et al. (2008) test for above chance within- and across-modality decoding, again with respect to observing and performing hand actions. That is, in this study Dinstein and colleagues test if the spatial patterns of activity associated with observing a particular hand action accurately classify the patterns of activity associated with observing (within-modality) or performing (across-modality) that same hand action.

Consistent with their previous fMRI-RS data, their results show accurate within- but not across-modality decoding in AIPC.

Similar methods can be applied to test for visual-motor response congruency for viewing and grasping objects. Based on the monkey data, both within- (view-view, grasp-grasp) and across-modality (view-grasp, grasp-view) RS, and above chance MVPA decoding, are predicted for human areas AIPC, SPOC, v- and dPMC. Such results would provide compelling evidence for visual-motor response congruency in humans, similar to those response properties described in monkey areas AIP, V6A, F5, and F2.

7.6.2 Linking Brain Function and Behaviour

Systematic investigations of the potential links between fMRI-RS for hand actions and behavioural motor repetition ('priming') results are lacking, and this represents exciting opportunities for future research.

As reviewed above, there is plenty of evidence showing that brain areas implicated in the control of the hand for grasping – AIPC, SPOC, v- and dPMC – show fMRI-RS when specific motor features of grasping actions – like hand shape, wrist orientation, and grip aperture – are repeated. At the same time, a host of other behavioural data indicate that recent movement history influences the performance of future hands actions, including effects of grasp repetition on the way the hand is shaped to grasp (Cohen & Rosenbaum, 2004, 2011; Dixon, McAnsh, & Read, 2012; Kelso, Buchanan, & Murata, 1994; Rosenbaum & Jorgensen, 1992; Schutz, Weigelt, Odekerken, Klein-Soetebier, & Schack, 2011; Short & Cauraugh, 1997).

However, data that connect these sets of findings are sparse. Future work designed to investigate the putative links between fMRI-RS and recent motor history has the potential to greatly advance our conceptual understanding of the underlying causes and behavioural significance.

We recently made new steps in this direction. First, we demonstrate decreased response times to initiate successive grasping actions when the same hand is repeated, even when the movement features of those actions are new (Valyear & Frey, 2014). Next, we replicate these behavioural results in the MR scanner, and, critically, show concurrent fMRI-RS within areas of PPC, overlapping with AIPC and more posterior-medial aspects of the intraparietal cortex (Valyear & Frey, 2015).

These data provide new support for a theory of recent motor history effects we call the Planning Efficiency Hypothesis. According to this hypothesis, behavioural effects of recent motor history reflect increased neural processing efficiency due to the reuse of motor plans (Rosenbaum, Chapman, Weigelt, Weiss, & van der Wel, 2012). Our results are consistent with this model. Reductions in response times to initiate actions are accompanied by fMRI-RS within areas important for action planning. Both effects are consistent with more efficient action planning when features of recent actions are reused to plan new actions.

Our data, however, are merely suggestive about the potential links between behavioural effects of recent motor history and fMRI-RS. To establish casual relationships, more direct evidence is required. For example, if TMS to areas showing fMRI-RS was found to abolish or attenuate behavioural effects of recent motor history, this would provide evidence for their causal links.

In the area of conceptual/semantic processing, clear progress has been made in connecting fMRI-RS to behavioural repetition priming (Schacter, Wig, & Stevens, 2007). Here, both TMS (Thiel et al., 2005; Wig, Grafton, Demos, & Kelley, 2005) and pharmacological manipulations (Thiel, Henson, Morris, Friston, & Dolan, 2001) have proven valuable. Future investigations of the putative relationships between fMRI-RS for repeated actions and behavioural effects of recent motor history stand to benefit from these examples.

7.7 Conclusions

This chapter reviews what is understood about the brain mechanisms that govern the visual control of grasping. Much of what is known stems from direct neural recordings in monkeys. Several key brain areas have been discovered, and their response properties have been meticulously characterized. Some cells represent visual information in motor terms. They specify object properties – shape, size, orientation – in terms that are used by the motor system to shape the hand during grasping.

Comparative evidence generally supports the hypothesis that functionally similar brain areas are present in humans. This evidence includes data from neuroimaging, neuropsychology, and brain stimulation. Specific brain areas within posterior parietal and premotor cortex are implicated, and their macro-level functional organization is consistent with predictions made from monkey data.

Determining the response properties of human brain areas is particularly challenging, however. I discussed two fMRI methods that are useful – fMRI-RS and MVPA. Altogether, the evidence from this work is promising, yet not always consistent, nor decisive. While motor response properties largely match predictions from monkey physiology, results concerning visual response properties are complex, and definitive tests for congruency between visual and motor responses have not been investigated.

There are ample opportunities for meaningful advancements. Links between fMRI and behavioural data are lacking. Future work that combines fMRI with brain stimulation is likely to break new ground, and continuing efforts to integrate data across species and methods are essential.

Understanding how the brain uses vision to control the hand for grasping has far-reaching basic science and clinical implications. The principles learned will inform our understanding of other sensorimotor transformations, and guide and constrain new neurobiological models of more complex manual behaviours. A better understanding of the neural control of grasping may also improve brain-machine-interface solutions for patients with paralysis (Aflalo et al., 2015), and rehabilitation interventions for stroke survivors (van Vliet, Pelton, Hollands, Carey, & Wing, 2013), and hand transplant recipients (Valyear, Mattos, Philip, Kaufman, & Frey, 2017).

Notes

1 Of course, if the proportions of cells devoted to either hand configuration differ significantly, then a direct contrast between conditions will identify AIPC.
2 It is worth acknowledging that the non-tool objects we use in this study also have learned visuomotor associations beyond their graspable properties. For example, it can be argued that a book is associated with a characteristic set of hand actions – to hold, open, and turn its pages. As such, we interpret our results as (at least partly) related to differences in the *strengths* of learned visuomotor associations between tools and non-tool objects.

References

Aflalo, T., Kellis, S., Klaes, C., Lee, B., Shi, Y., Pejsa, K., Shanfield, K., Hayes-Jackson, S., Aisen, M., Heck, C., Liu, C., & Andersen, R.A. (2015). Neurophysiology. Decoding motor imagery from the posterior parietal cortex of a tetraplegic human. *Science, 348*, 906–910.

Andersen, R.A., Andersen, K.N., Hwang, E.J., & Hauschild, M. (2014). Optic ataxia: from Balint's syndrome to the parietal reach region. *Neuron, 81*, 967–983.

Andersen, R.A., & Cui, H. (2009). Intention, action planning, and decision making in parietal-frontal circuits. *Neuron, 63*, 568–583.

Ansuini, C., Giosa, L., Turella, L., Altoe, G., & Castiello, U. (2008). An object for an action, the same object for other actions: effects on hand shaping. *Exp Brain Res, 185*, 111–119.

Ansuini, C., Santello, M., Massaccesi, S., & Castiello, U. (2006). Effects of end-goal on hand shaping. *J Neurophysiol, 95*, 2456–2465.

Armbrüster, C., & Spijkers, W. (2006). Movement planning in prehension: do intended actions influence the initial reach and grasp movement? *Motor Control, 10*, 311–329.

Astafiev, S.V., Shulman, G.L., Stanley, C.M., Snyder, A.Z., Van, Essen, D.C., Corbetta, M. (2003). Functional organization of human intraparietal and frontal cortex for attending, looking, and pointing. *J Neurosci, 23*, 4689–4699.

Balint, R. (1909). Seelenhämmung des 'Schauens', optische Ataxie, räumliche Störungen des Aufmersamkeit. *Monastchrift für Psychiatrie und Neurologie, 25*, 51–81. English translation: Harvey (1995).

Battaglini, P.P., Muzur, A., Galletti, C., Skrap, M., Brovelli, A., & Fattori, P. (2002). Effects of lesions to area V6A in monkeys. *Exp Brain Res, 144*, 419–422.

Baumann, M.A., Fluet, M.C., & Scherberger, H. (2009). Context-specific grasp movement representation in the macaque anterior intraparietal area. *J Neurosci, 29*, 6436–6448.

Bekkering, H., & Neggers, S.F. (2002). Visual search is modulated by action intentions. *Psychol Sci, 13*, 370–374.

Bernier, P.M., & Grafton, S.T. (2010). Human posterior parietal cortex flexibly determines reference frames for reaching based on sensory context. *Neuron, 68*, 776–788.

Beurze, S.M., de Lange, F.P., Toni, I., & Medendorp, W.P. (2007). Integration of target and effector information in the human brain during reach planning. *J Neurophysiol, 97*, 188–199.

Binkofski, F., Dohle, C., Posse, S., Stephan, K.M., Hefter, H., Seitz, R.J., & Freund, H.J. (1998). Human anterior intraparietal area subserves prehension: a combined lesion and functional MRI activation study. *Neurology, 50*, 1253–1259.

Borra, E., Belmalih, A., Calzavara, R., Gerbella, M., Murata, A., Rozzi, S., & Luppino, G. (2008). Cortical connections of the macaque anterior intraparietal (AIP) area. *Cereb Cortex, 18*, 1094–1111.

Brandi, M.L., Wohlschlager, A., Sorg, C., & Hermsdorfer, J. (2014). The neural correlates of planning and executing actual tool use. *J Neurosci, 34*, 13183–13194.

Bub, D.N., Masson, M.E., & Bukach, C.M. (2003). Gesturing and naming: the use of functional knowledge in object identification. *Psychol Sci, 14*, 467–472.

Bub, D.N., Masson, M.E., & Cree, G.S. (2008). Evocation of functional and volumetric gestural knowledge by objects and words. *Cognition, 106*, 27–58.

Caminiti, R., Ferraina, S., & Johnson, P.B. (1996). The sources of visual information to the primate frontal lobe: a novel role for the superior parietal lobule. *Cereb Cortex, 6*, 319–328.

Cavina-Pratesi, C., Monaco, S., Fattori, P., Galletti, C., McAdam, T.D., Quinlan, D.J., Goodale, M.A., & Culham, J.C. (2010). Functional magnetic resonance imaging reveals the neural substrates of arm transport and grip formation in reach-to-grasp actions in humans. *J Neurosci, 30*, 10306–10323.

Chao, L.L., & Martin, A. (2000). Representation of manipulable man-made objects in the dorsal stream. *Neuroimage, 12*, 478–484.

Cisek, P., & Kalaska, J.F. (2005). Neural correlates of reaching decisions in dorsal premotor cortex: specification of multiple direction choices and final selection of action. *Neuron, 45*, 801–814.

Cisek, P., & Pastor-Bernier, A. (2014). On the challenges and mechanisms of embodied decisions. *Philos Trans R Soc Lond B Biol Sci, 369*.

Cohen, R.G., & Rosenbaum, D.A. (2004). Where grasps are made reveals how grasps are planned: generation and recall of motor plans. *Exp Brain Res, 157*, 486–495.

Cohen, R.G., & Rosenbaum, D.A. (2011). Prospective and retrospective effects in human motor control: planning grasps for object rotation and translation. *Psychol Res, 75*, 341–349.

Connolly, J.D., Andersen, R.A., & Goodale, M.A. (2003). FMRI evidence for a 'parietal reach region' in the human brain. *Exp Brain Res, 153*, 140–145.

Crawford, J.D., Henriques, D.Y., & Medendorp, W.P. (2011). Three-dimensional transformations for goal-directed action. *Annu Rev Neurosci, 34*, 309–331.

Creem-Regehr, S.H., & Lee, J.N. (2005). Neural representations of graspable objects: are tools special? *Brain Res Cogn Brain Res, 22*, 457–469.

Culham, J.C., Danckert, S.L., DeSouza, J.F., Gati, J.S., Menon, R.S., & Goodale, M.A. (2003). Visually guided grasping produces fMRI activation in dorsal but not ventral stream brain areas. *Exp Brain Res, 153*, 180–189.

Culham, J.C., & Valyear, K.F. (2006). Human parietal cortex in action. *Curr Opin Neurobiol, 16*, 205–212.

Davare, M., Andres, M., Clerget, E., Thonnard, J.L., & Olivier, E. (2007). Temporal dissociation between hand shaping and grip force scaling in the anterior intraparietal area. *J Neurosci, 27*, 3974–3980.

Davare, M., Andres, M., Cosnard, G., Thonnard, J.L., & Olivier, E. (2006). Dissociating the role of ventral and dorsal premotor cortex in precision grasping. *J Neurosci, 26*, 2260–2268.

Davare, M., Lemon, R., & Olivier, E. (2008). Selective modulation of interactions between ventral premotor cortex and primary motor cortex during precision grasping in humans. *J Physiol, 586*, 2735–2742.

Davare, M., Rothwell, J.C., & Lemon, R.N. (2010). Causal connectivity between the human anterior intraparietal area and premotor cortex during grasp. *Curr Biol, 20*, 176–181.

Davare, M., Montague, K., Olivier, E., Rothwell, J.C., & Lemon, R.N. (2009). Ventral premotor to primary motor cortical interactions during object-driven grasp in humans. *Cortex, 45*, 1050–1057.

di Pellegrino, G., Fadiga, L., Fogassi, L., Gallese, V., & Rizzolatti, G. (1992). Understanding motor events: a neurophysiological study. *Exp Brain Res, 91*, 176–180.

Dinstein, I., Gardner, J.L., Jazayeri, M., & Heeger, D.J. (2008). Executed and observed movements have different distributed representations in human aIPS. *J Neurosci, 28*, 11231–11239.

Dinstein, I., Hasson, U., Rubin, N., & Heeger, D.J. (2007). Brain areas selective for both observed and executed movements. *J Neurophysiol, 98*, 1415–1427.

Dixon, P., McAnsh, S., & Read, L. (2012). Repetition effects in grasping. *Can J Exp Psychol, 66*, 1–17.

Fabbri, S., Caramazza, A., & Lingnau, A. (2010). Tuning curves for movement direction in the human visuomotor system. *J Neurosci, 30*, 13488–13498.

Fabbri, S., Strnad, L., Caramazza, A., & Lingnau, A. (2014). Overlapping representations for grip type and reach direction. *Neuroimage, 94*, 138–146.

Fabbri, S., Stubbs, K.M., Cusack, R., & Culham, J.C. (2016). Disentangling representations of object and grasp properties in the human brain. *J Neurosci, 36*, 7648–7662.

Fagg, A.H., & Arbib, M.A. (1998). Modeling parietal-premotor interactions in primate control of grasping. *Neural Netw, 11*, 1277–1303.

Fattori, P., Breveglieri, R., Amoroso, K., & Galletti, C. (2004). Evidence for both reaching and grasping activity in the medial parieto-occipital cortex of the macaque. *Eur J Neurosci, 20*, 2457–2466.

Fattori, P., Breveglieri, R., Bosco, A., Gamberini, M., & Galletti, C. (2015). Vision for prehension in the medial parietal cortex. *Cereb Cortex, 7*(2), 1149–1163.

Fattori, P., Breveglieri, R., Marzocchi, N., Filippini, D., Bosco, A., & Galletti, C. (2009). Hand orientation during reach-to-grasp movements modulates neuronal activity in the medial posterior parietal area V6A. *J Neurosci, 29*, 1928–1936.

Fattori, P., Kutz, D.F., Breveglieri, R., Marzocchi, N., & Galletti, C. (2005). Spatial tuning of reaching activity in the medial parieto-occipital cortex (area V6A) of macaque monkey. *Eur J Neurosci, 22*, 956–972.

Fattori, P., Raos, V., Breveglieri, R., Bosco, A., Marzocchi, N., & Galletti, C. (2010). The dorsomedial pathway is not just for reaching: grasping neurons in the medial parieto-occipital cortex of the macaque monkey. *J Neurosci, 30*, 342–349.

Fluet, M.C., Baumann, M.A., & Scherberger, H. (2010). Context-specific grasp movement representation in macaque ventral premotor cortex. *J Neurosci, 30*, 15175–15184.

Fogassi, L., Gallese, V., Buccino, G., Craighero, L., Fadiga, L., & Rizzolatti, G. (2001). Cortical mechanism for the visual guidance of hand grasping movements in the monkey: a reversible inactivation study. *Brain, 124*, 571–586.

Gallese, V., Fadiga, L., Fogassi, L., & Rizzolatti, G. (1996). Action recognition in the premotor cortex. *Brain, 119*(Pt 2), 593–609.

Gallese, V., Murata, A., Kaseda, M., Niki, N., & Sakata, H. (1994). Deficit of hand preshaping after muscimol injection in monkey parietal cortex. *Neuroreport, 5*, 1525–1529.

Galletti, C., Fattori, P., Kutz, D.F., & Battaglini, P.P. (1997). Arm movement-related neurons in the visual area V6A of the macaque superior parietal lobule. *Eur J Neurosci, 9*, 410–413.

Gallivan, J.P., & Culham, J.C. (2015). Neural coding within human brain areas involved in actions. *Curr Opin Neurobiol, 33*, 141–149.

Gibson, J.J. (1986). *The ecological approach to visual perception*. New York: Psychology Press.

Godschalk, M., Lemon, R.N., Kuypers, H.G., & Ronday, H.K. (1984). Cortical afferents and efferents of monkey postarcuate area: an anatomical and electrophysiological study. *Exp Brain Res, 56*, 410–424.

Goodale, M.A., & Milner, A.D. (1992). Separate visual pathways for perception and action. *Trends Neurosci, 15*, 20–25.

Goodale, M.A., Milner, A.D., Jakobson, L.S., & Carey, D.P. (1991). A neurological dissociation between perceiving objects and grasping them. *Nature, 349*, 154–156.

Grafton, S.T. (2010). The cognitive neuroscience of prehension: recent developments. *Exp Brain Res, 204*, 475–491.

Grill-Spector, K., Henson, R., & Martin, A. (2006). Repetition and the brain: neural models of stimulus-specific effects. *Trends Cogn Sci, 10*, 14–23.

Gross, C.G., & Mishkin, M. (1977). The neural basis of stimulus equivalence across retinal translation. In S. Harnard, R. Doty, J. Jaynes, L. Goldstein, & G. Krauthamer (Eds), *Lateralization of the nervous system* (pp. 109–122). New York: Academic Press.

Gross, C.G., Rocha-Miranda, C.E., & Bender, D.B. (1972). Visual properties of neurons in inferotemporal cortex of the macaque. *J Neurophysiol, 35*, 96–111.

Harvey, M. (1995). Translation of 'Psychic paralysis of gaze, optic ataxia, and spatial disorder of attention' by Rudolph Bálint. *Cognitive Neuropsychology, 12*, 265–281.

Hecaen, H., & De Ajuriaguerra, J. (1954). Balint's syndrome (psychic paralysis of visual fixation) and its minor forms. *Brain, 77*, 373–400.

Hendrix, C.M., Mason, C.R., & Ebner, T.J. (2009). Signaling of grasp dimension and grasp force in dorsal premotor cortex and primary motor cortex neurons during reach to grasp in the monkey. *J Neurophysiol, 102*, 132–145.

Hoshi, E., & Tanji, J. (2004). Functional specialization in dorsal and ventral premotor areas. *Prog Brain Res, 143*, 507–511.

Hoshi, E., & Tanji, J. (2006). Differential involvement of neurons in the dorsal and ventral premotor cortex during processing of visual signals for action planning. *J Neurophysiol, 95*, 3596–3616.

Humphreys, G.W., & Riddoch, M.J. (2000). One more cup of coffee for the road: object-action assemblies, response blocking and response capture after frontal lobe damage. *Exp Brain Res, 133*, 81–93.

Humphreys, G.W., & Riddoch, M.J. (2001). Detection by action: neuropsychological evidence for action-defined templates in search. *Nat Neurosci, 4*, 84–88.

Hyvarinen, J., & Poranen, A. (1974). Function of the parietal associative area 7 as revealed from cellular discharges in alert monkeys. *Brain, 97*, 673–692.

Jakobson, L.S., Archibald, Y.M., Carey, D.P., & Goodale, M.A. (1991). A kinematic analysis of reaching and grasping movements in a patient recovering from optic ataxia. *Neuropsychologia, 29*, 803–809.

James, T.W., & Gauthier, I. (2006) Repetition-induced changes in BOLD response reflect accumulation of neural activity. *Hum Brain Mapp, 27*, 37–46.

Jeannerod, M. (1997). *The cognitive neuroscience of action*. Oxford; Cambridge, MA: Blackwell.

Jeannerod, M., Arbib, M.A., Rizzolatti, G., & Sakata, H. (1995). Grasping objects: the cortical mechanisms of visuomotor transformation. *Trends Neurosci, 18*, 314–320.

Jeannerod, M., & Jacob, P. (2005). Visual cognition: a new look at the two-visual systems model. *Neuropsychologia, 43*, 301–312.

Kamitani, Y., & Tong, F. (2005). Decoding the visual and subjective contents of the human brain. *Nat Neurosci, 8*, 679–685.

Karnath, H.O., & Perenin, M.T. (2005). Cortical control of visually guided reaching: evidence from patients with optic ataxia. *Cereb Cortex, 15*, 1561–1569.

Kay, R.F., Ross, C., & Williams, B.A. (1997). Anthropoid origins. *Science, 275*, 797–804.

Kelso, J.A.S., Buchanan, J.J., & Murata, T. (1994). Multifunctionality and switching in the coordination dynamics of reaching and grasping. *Human Movement Science, 13*, 63–94.

Kimura, D. (1963). Right temporal-lobe damage: perception of unfamiliar stimuli after damage. *Arch Neurol, 8*, 264–271.

Krekelberg, B., Boynton, G.M., & van Wezel, R.J. (2006). Adaptation: from single cells to BOLD signals. *Trends Neurosci, 29*, 250–256.

Kriegeskorte, N., & Kievit, R.A. (2013). Representational geometry: integrating cognition, computation, and the brain. *Trends Cogn Sci, 17*, 401–412.

Kroliczak, G., McAdam, T.D., Quinlan, D.J., & Culham, J.C. (2008). The human dorsal stream adapts to real actions and 3D shape processing: a functional magnetic resonance imaging study. *J Neurophysiol, 100*, 2627–2639.

Lhermitte, F. (1983). 'Utilization behaviour' and its relation to lesions of the frontal lobes. *Brain, 106*(Pt 2), 237–255.

Logothetis, N.K., & Wandell, B.A. (2004). Interpreting the BOLD signal. *Annu Rev Physiol, 66*, 735–769.

Luppino, G., Murata, A., Govoni, P., & Matelli, M. (1999). Largely segregated parietofrontal connections linking rostral intraparietal cortex (areas AIP and VIP) and the ventral premotor cortex (areas F5 and F4). *Exp Brain Res, 128*, 181–187.

Marteniuk, R.G., MacKenzie, C.L., Jeannerod, M., Athenes, S., & Dugas, C. (1987). Constraints on human arm movement trajectories. *Can J Psychol, 41*, 365–378.

Masson, M.E., Bub, D.N., & Breuer, A.T. (2011). Priming of reach and grasp actions by handled objects. *J Exp Psychol Hum Percept Perform, 27*(5), 1470–1484.

Medendorp, W.P., Goltz, H.C., Crawford, J.D., & Vilis, T. (2005). Integration of target and effector information in human posterior parietal cortex for the planning of action. *J Neurophysiol, 93*, 954–962.

Milner, A.D., & Goodale, M.A. (1995). *The visual brain in action*. Oxford, New York: Oxford University Press.

Mishkin, M (1972). Cortical visual areas and their interactions. In A.G. Karczmar & J.C. Eccles (Eds), *Brain and human behavior* (pp. 187–208). Berlin: Springer.

Mishkin, M., Ungerleider, L.G., & Macko, K.A. (1983). Object vision and spatial vision: two cortical pathways. *Trends in Neurosciences, 6*, 414–417.

Monaco, S., Cavina-Pratesi, C., Sedda, A., Fattori, P., Galletti, C., & Culham, J.C. (2011). Functional magnetic resonance adaptation reveals the involvement of the dorsomedial stream in hand orientation for grasping. *J Neurophysiol, 106*, 2248–2263.

Monaco, S., Chen, Y., Medendorp, W.P., Crawford, J.D., Fiehler, K., & Henriques, D.Y. (2014). Functional magnetic resonance imaging adaptation reveals the cortical networks for processing grasp-relevant object properties. *Cereb Cortex, 24*, 1540–1554.

Monaco, S., Sedda, A., Cavina-Pratesi, C., & Culham, J.C. (2015). Neural correlates of object size and object location during grasping actions. *Eur J Neurosci, 41*, 454–465.

Murata, A., Fadiga, L., Fogassi, L., Gallese, V., Raos, V., & Rizzolatti, G. (1997). Object representation in the ventral premotor cortex (area F5) of the monkey. *J Neurophysiol, 78*, 2226–2230.

Murata, A., Gallese, V., Luppino, G., Kaseda, M., & Sakata, H. (2000). Selectivity for the shape, size, and orientation of objects for grasping in neurons of monkey parietal area AIP. *J Neurophysiol, 83*, 2580–2601.

Murray, S.O., & Wojciulik, E. (2004). Attention increases neural selectivity in the human lateral occipital complex. *Nat Neurosci, 7*, 70–74.

Pavese, A., & Buxbaum, L.J. (2002). Action matters: The role of action plans and affordances in selection for action. *Visual Cognition, 9*, 559–590.

Perenin, M.T., & Vighetto, A. (1988). Optic ataxia: a specific disruption in visuomotor mechanisms. I. Different aspects of the deficit in reaching for objects. *Brain, 111*(Pt 3), 643–674.

Pitzalis, S., Fattori, P., & Galletti, C. (2015). The human cortical areas V6 and V6A. *Vis Neurosci, 32*, E007.

Pohl, W. (1973). Dissociation of spatial discrimination deficits following frontal and parietal lesions in monkeys. *J Comp Physiol Psychol, 82*, 227–239.

Prado, J., Clavagnier, S., Otzenberger, H., Scheiber, C., Kennedy, H., & Perenin, M.T. (2005). Two cortical systems for reaching in central and peripheral vision. *Neuron, 48*, 849–858.

Raos, V., Umilta, M.A., Gallese, V., & Fogassi, L. (2004). Functional properties of grasping-related neurons in the dorsal premotor area F2 of the macaque monkey. *J Neurophysiol, 92*, 1990–2002.

Raos, V., Umilta, M.A., Murata, A., Fogassi, L., & Gallese, V. (2006). Functional properties of grasping-related neurons in the ventral premotor area F5 of the macaque monkey. *J Neurophysiol, 95*, 709–729.

Ratcliff, G., & Davies-Jones, G.A. (1972). Defective visual localization in focal brain wounds. *Brain, 95*, 49–60.

Rice, N.J., Tunik, E., Cross, E.S., & Grafton, S.T. (2007). On-line grasp control is mediated by the contralateral hemisphere. *Brain Res, 1175*, 76–84.

Rice, N.J., Tunik, E., & Grafton, S.T. (2006). The anterior intraparietal sulcus mediates grasp execution, independent of requirement to update: new insights from transcranial magnetic stimulation. *J Neurosci, 26*, 8176–8182.

Rice, N.J., Valyear, K.F., Goodale, M.A., Milner, A.D., & Culham, J.C. (2007). Orientation sensitivity to graspable objects: an fMRI adaptation study. *Neuroimage, 36*, Suppl 2, T87–93.

Riddoch, M.J., Edwards, M.G., Humphreys, G.W., West, R., & Heafield, T. (1998). Visual affordances direct action: neuropsychological evidence from manual interference. *Cognitive Neuropsychology, 15*, 645–683.

Riddoch, M.J., Humphreys, G.W., & Edwards, M.G. (2000a). Visual affordances and object selection. In S. Monsell & J. Driver (Eds), *Attention and performance XVIII* (pp. 603–626). Cambridge, MA: MIT Press.

Riddoch, M.J., Humphreys, G.W., & Edwards, M.G. (2000b). Neuropsychological evidence distinguishing object selection from action (effector) selection. *Cogn Neuropsychol, 17*, 547–562.

Rizzolatti, G., Camarda, R., Fogassi, L., Gentilucci, M., Luppino, G., & Matelli, M. (1988). Functional organization of inferior area 6 in the macaque monkey. II. Area F5 and the control of distal movements. *Exp Brain Res, 71*, 491–507.

Rizzolatti, G., & Craighero, L. (2004). The mirror-neuron system. *Annu Rev Neurosci, 27*, 169–192.

Rizzolatti, G., Fadiga, L., Gallese, V., & Fogassi, L. (1996). Premotor cortex and the recognition of motor actions. *Brain Res Cogn Brain Res, 3*, 131–141.

Rizzolatti, G., & Matelli, M. (2003). Two different streams form the dorsal visual system: anatomy and functions. *Exp Brain Res, 153*, 146–157.

Robinson, D.L., Goldberg, M.E., & Stanton, G.B. (1978). Parietal association cortex in the primate: sensory mechanisms and behavioral modulations. *J Neurophysiol, 41*, 910–932.

Rosenbaum, D.A., Chapman, K.M., Weigelt, M., Weiss, D.J., & van der Wel, R. (2012). Cognition, action, and object manipulation. *Psychol Bull, 138*, 924–946.

Rosenbaum, D.A., & Jorgensen, M.J. (1992). Planning macroscopic aspects of manual control. *Human Movement Science, 11*, 61–69.

Sakata, H., Taira, M., Murata, A., & Mine, S. (1995). Neural mechanisms of visual guidance of hand action in the parietal cortex of the monkey. *Cereb Cortex, 5*, 429–438.

Schacter, D.L., Wig, G.S., & Stevens, W.D. (2007). Reductions in cortical activity during priming. *Curr Opin Neurobiol, 17*, 171–176.

Schaffelhofer, S., & Scherberger, H. (2016). Object vision to hand action in macaque parietal, premotor, and motor cortices. *Elife, 5*, e15278.

Schutz, C., Weigelt, M., Odekerken, D., Klein-Soetebier, T., & Schack, T. (2011). Motor control strategies in a continuous task space. *Motor Control, 15*, 321–341.

Shallice, T., Burgess, P.W., Schon, F., & Baxter, D.M. (1989). The origins of utilization behaviour. *Brain, 112*(Pt 6), 1587–1598.

Short, M.W., & Cauraugh, J.H. (1997). Planning macroscopic aspects of manual control: end-state comfort and point-of-change effects. *Acta Psychol (Amst), 96*, 133–147.

Spinks, R.L., Kraskov, A., Brochier, T., Umilta, M.A., & Lemon, R.N. (2008). Selectivity for grasp in local field potential and single neuron activity recorded simultaneously from M1 and F5 in the awake macaque monkey. *J Neurosci, 28*, 10961–10971.

Stark, E., Asher, I., & Abeles, M. (2007). Encoding of reach and grasp by single neurons in premotor cortex is independent of recording site. *J Neurophysiol, 97*, 3351–3364.

Taira, M., Mine, S., Georgopoulos, A.P., Murata, A., & Sakata, H. (1990). Parietal cortex neurons of the monkey related to the visual guidance of hand movement. *Exp Brain Res, 83*, 29–36.

Thiel, A., Haupt, W.F., Habedank, B., Winhuisen, L., Herholz, K., Kessler, J., Markowitsch, H.J., & Heiss, W.D. (2005). Neuroimaging-guided rTMS of the left inferior frontal gyrus interferes with repetition priming. *Neuroimage, 25*, 815–823.

Thiel, C.M., Henson, R.N., Morris, J.S., Friston, K.J., & Dolan, R.J. (2001). Pharmacological modulation of behavioral and neuronal correlates of repetition priming. *J Neurosci, 21*, 6846–6852.

Tosoni, A., Galati, G., Romani, G.L., & Corbetta, M. (2008). Sensory-motor mechanisms in human parietal cortex underlie arbitrary visual decisions. *Nat Neurosci, 11*, 1446–1453.

Tosoni, A., Pitzalis, S., Committeri, G., Fattori, P., Galletti, C., & Galati, G. (2015). Resting-state connectivity and functional specialization in human medial parieto-occipital cortex. *Brain Struct Funct, 220*, 3307–3321.

Tunik, E., Frey, S.H., & Grafton, S.T. (2005). Virtual lesions of the anterior intraparietal area disrupt goal-dependent on-line adjustments of grasp. *Nat Neurosci, 8*, 505–511.

Umilta, M.A., Brochier, T., Spinks, R.L., & Lemon, R.N. (2007). Simultaneous recording of macaque premotor and primary motor cortex neuronal populations reveals different functional contributions to visuomotor grasp. *J Neurophysiol, 98*, 488–501.

Ungerleider, L.G., & Mishkin, M. (1982). Two cortical visual systems. In D.J. Ingle, M.A. Goodale, & R.J.W. Masfield (Eds), *Analysis of visual behavior* (pp. 549–586). Cambridge, MA: MIT Press.

Valyear, K.F. (2016). The hand that 'sees' to grasp. *Elife, 5*.

Valyear, K.F., Cavina-Pratesi, C., Stiglick, A.J., & Culham, J.C. (2007). Does tool-related fMRI activity within the intraparietal sulcus reflect the plan to grasp? *Neuroimage, 36*, Suppl 2, T94–T108.

Valyear, K.F., Chapman, C.S., Gallivan, J.P., Mark, R.S., & Culham, J.C. (2011). To use or to move: goal-set modulates priming when grasping real tools. *Exp Brain Res, 212*, 125–142.

Valyear, K.F., Culham, J.C., Sharif, N., Westwood, D., & Goodale, M.A. (2006). A double dissociation between sensitivity to changes in object identity and object orientation in the ventral and dorsal visual streams: a human fMRI study. *Neuropsychologia, 44*, 218–228.

Valyear, K.F., Fitzpatrick, A.M., & McManus, E.F. (2017). The neuroscience of human tool use. In J. Kaas (Ed.), Evolution of nervous systems (pp. 341–353, 2nd ed.). Oxford: Elsevier.

Valyear, K.F., & Frey, S.H. (2014). Hand selection for object grasping is influenced by recent motor history. *Psychon Bull Rev, 21*, 566–573.

Valyear, K.F., & Frey, S.H. (2015). Human posterior parietal cortex mediates hand-specific planning. *Neuroimage, 114*, 226–238.

Valyear, K.F., Gallivan, J.P., McLean, D.A., & Culham, J.C. (2012). fMRI repetition suppression for familiar but not arbitrary actions with tools. *J Neurosci, 32*, 4247–4259.

Valyear, K.F., Mattos, D., Philip, B., Kaufman, C., & Frey, S.H. (2017). Grasping with a new hand: improved performance and normalized grasp-selective brain responses despite persistent functional changes in primary motor cortex and low-level sensory and motor impairments. *NeuroImage*. doi: 10.1016/j.neuroimage.2017.09.052.

Vesia, M., & Crawford, J.D. (2012). Specialization of reach function in human posterior parietal cortex. *Exp Brain Res, 221*, 1–18.

Vesia, M., Prime, S.L., Yan, X., Sergio, L.E., & Crawford, J.D. (2010). Specificity of human parietal saccade and reach regions during transcranial magnetic stimulation. *J Neurosci, 30*, 13053–13065.

van Vliet, P., Pelton, T.A., Hollands, K.L., Carey, L., & Wing, A.M. (2013). Neuroscience findings on coordination of reaching to grasp an object: implications for research. *Neurorehabil Neural Repair, 27*, 622–635.

Warrington, E.K. (1982). Neuropsychological studies of object recognition. *Philos Trans R Soc Lond B Biol Sci, 298*, 15–33.

Warrington, E.K., & James, M. (1967). Disorders of visual perception in patients with localized cerebral lesions. *Neuropsychologia, 5*, 253–266.

Wig, G.S., Grafton, S.T., Demos, K.E., & Kelley, W.M. (2005). Reductions in neural activity underlie behavioral components of repetition priming. *Nat Neurosci, 8*, 1228–1233.

Wiggs, C.L., & Martin, A. (1998). Properties and mechanisms of perceptual priming. *Curr Opin Neurobiol, 8*, 227–233.

Wise, S.P., Boussaoud, D., Johnson, P.B., & Caminiti, R. (1997). Premotor and parietal cortex: corticocortical connectivity and combinatorial computations. *Annu Rev Neurosci, 20*, 25–42.

PART III

On the Planning and Control of Reach-to-Grasp Behavior in Adults

8

THE CONTROL OF THE REACH-TO-GRASP MOVEMENT

Jeroen B. J. Smeets and Eli Brenner

8.1 Kinematics of Reaching to Grasp: Movements of the Digits in Space

The description of how we move when reaching to grasp an object started with the work of Marc Jeannerod (Jeannerod 1981, 1984, 1986). He and many others describe grasping in terms of coordinating the movement of the hand towards the object (i.e., movement of the hand relative to the object) with the pre-shaping and closing of the hand (i.e., movement of the digits relative to each other). In this chapter, we choose a different approach. We focus on the trajectories of the digits relative to the object, and relate them to the sensory information that is used to generate these trajectories. We have discussed at several places why we think this is both a very efficient way of describing grasping and one that is closely related to the way grasping is controlled (Smeets and Brenner 1999, 2016).

In this chapter, we divide the problem of making a reach-to-grasp movement into two phases. First, grasping points on the object are selected. These are the positions at which the digits will make contact with the object's surface. Subsequently, the movement trajectories towards the grasping points are shaped.

We limit ourselves to describing grasping with the finger and thumb, often referred to as a "precision grip." As our description is in terms of individual fingers, one can easily extend it to more than two digits. However, for the selection of grasping points, one has to realize that the thumb often opposes all the other fingers, because the anatomy of the hand often suggests such a configuration. When grasping with more than two digits, one can take this into account and simplify the selection problem by combining the fingers into a single "virtual finger." One can subsequently choose positions for the individual fingers that comply with this virtual finger position (Baud-Bovy and Soechting 2001; Iberall et al. 1986). The digits are then each guided to their chosen positions.

8.2 Selection of Grasping Points

8.2.1 Object Shape and Orientation

When choosing grasping points for the digits, one has to make sure that the object will not slip when one starts to exert grip and lift forces. For this, there are two requirements: the sum of the moments of these vectors should be zero, and each force vector should be within the cone of friction. All combinations of positions of the digits with associated force vectors that fulfill these conditions will lead to a stable grasp. In order to use this grasp to lift the object without it tipping, the line connecting the digits (opposition axis) should pass through or above the object's center of mass. The area on the object that can be used for this is sometimes referred to as opposition space (Iberall et al. 1986; Jeannerod 1999; Stelmach et al. 1994).

If the opposition axis does not pass through or above the center of mass, one can exert torques to prevent the object from rotating. Therefore, subjects usually choose a grasping axis that passes near the center of mass (Lederman and Wing 2003). However, this choice is systematically biased in several ways. For instance, when grasping a 10 cm bar with the right hand, subjects grasp about 5 mm to the right of the center of mass, and when grasping with the left hand, there is a similar bias to the left (Paulun et al. 2014). Visual illusions that

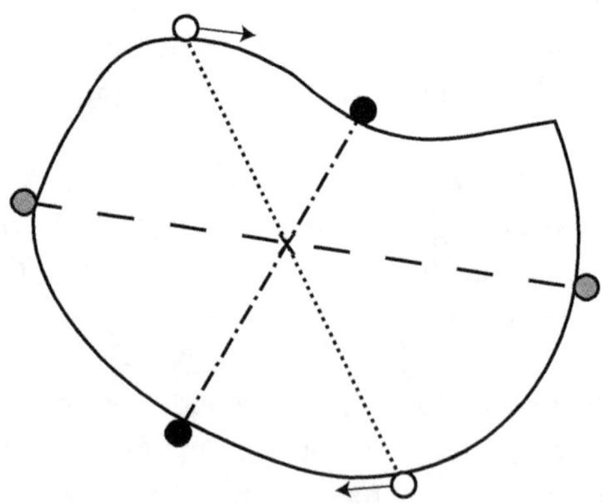

FIGURE 8.1 An object with two possible opposition axes (dashed and dashed-dotted lines), and one set of grasping points (circles connected by dotted line) that will slip in the direction of the arrows if a grip force is applied and the friction is low

influence the perceived location of the center of mass add a bias that corresponds to the effect of the illusion (Ellis et al. 1999).

For objects with a circular shape, there are an infinite number of opposition axes that are equivalent in terms of opposition space. Nevertheless, subjects have a very clear preference for a certain grip orientation (Rosenbaum et al. 2001; Schot et al. 2010). Such preference has been interpreted in terms of striving for comfort (Cruse et al. 1993). The preferred orientation is presumably based on physical experience, and is therefore based on the anatomy of the arm and hand. When choosing an orientation, one does not actually compare the comfort of the two configurations, but must associate visual information on the opposition axes with memory of earlier experiences using the corresponding postures (Rosenbaum et al. 2001). Although one might think that it is useful to be able to see a grasping point before moving to it, the visibility of grasping points does not affect the choice between possible opposition axes, so apparently the posture is more important (Voudouris et al. 2012a).

When an object is not circular-symmetric, there are generally several possible opposition axes. The various options might differ in their stability. A stable opposition axis allows for more variations in finger placement while leaving the ability to lift the object intact (e.g., the dotted line in Figure 8.1 illustrates a more stable opposition axis than does the continuous line). For an elliptical object, there are two stable opposition axes: the principle axes. Grasping along the short axes is more stable than grasping along the long axis. When allowed to choose, subjects do indeed have a preference for the short one. However, this preference is not absolute, as it decreases if the short axis' orientation differs considerably from the anatomically preferred orientation (Cuijpers et al. 2004).

Is this choice of grasping point really made before the movement starts, or does it evolve during the execution of the grasp? One of the findings on grasping elliptical objects is that subjects do not grasp exactly along one of the principle axes of the ellipse, but have a systematic bias. One might interpret this as an indication that the contact points are not fully planned, but that subjects adjust their plan based on the comfort they experience during execution. However, this interpretation is incorrect, because the deviation is already clear early in the movement, so the planned grip is not restricted to one of the principle axes (Cuijpers et al. 2006). One might argue that such a choice must be due to a misperception of spatial properties (i.e., object shape). In a study that explicitly addressed the question of whether movements are planned in advance, subjects had to grasp circular and elliptical objects (Hesse et al. 2008). As mentioned in the previous paragraph, a circular object is normally grasped with the digits in a preferred orientation. Nevertheless, when subjects had viewed an elliptical object more than 2 seconds before viewing the circular target, their grasp angle was clearly influenced by the orientation of the elliptical object, showing that the grip orientation is indeed planned well in advance (Hesse et al. 2008).

When grasping an object along a given opposition axis, there are still two possibilities: the thumb can be positioned at either of the two ends. In anatomical terms, it can be grasped with the lower arm pronated or supinated. Subjects generally have a clear preference for one of the two, depending on the object's orientation, with only a small range of orientations for which both options are equally likely (Stelmach et al. 1994). If the visual information is biased (due to a visual illusion) it is the illusory orientation that determines the choice (Crajé et al. 2008). This suggests that choosing grasping points is based on the same processing of visual information as perceptual judgments. However, two tasks being based on the same processing of visual information does not imply that one should be able to perform both tasks, as they may differ in other aspects. Indeed, there are neurological patients that can perceptually discriminate object shapes without being able to choose adequate grasping points and vice versa (Goodale et al. 1994b).

Thus, the choice of contact points is determined in advance, considering the stability of the grip. However, people do not always choose the most stable grip orientation, because the comfort of the required posture is also considered, as well as other task demands. This is possible because the digits are soft and flexible, so their placement is often not extremely critical, making it possible to trade off grip stability and required grip force against using a comfortable posture and the extent to which dexterity is required.

8.2.2 Influence of the Trajectory

In the previous section, we provided ample evidence that the choice of grasping points is based on visual judgments about the object combined with experience with the comfort of grasping postures. This description of the choice process neglects any influences or constraints on the movement of the hand before it reaches the target. This is clearly an over-simplification. For instance, when grasping a 4 cm diameter cylinder from various starting positions, a small (<20 percent) but systematic dependence of the grasp orientation on the initial orientation of the hand has been found (Hesse and Deubel 2009): the more clockwise the initial hand orientation, the more clockwise the grip when the cylinder was grasped. In addition to the initial orientation of the hand, its initial position also has a small effect on grip orientation (Schot et al. 2010). In these experiments, subjects were shown the target object while their hand was already at the starting position, so those postural parameters were available when selecting grasping points.

Do subjects also take the planned trajectory into account? Voudouris and colleagues devoted several studies to this question. In one study (Voudouris et al. 2012b), obstacles were placed near the trajectories towards the digits' anticipated grasping points. The effect of these obstacles was surprisingly small. A consistent effect on the choice was only found for obstacles that were placed near the thumb. Interestingly, similar small effects on the contact points can be induced when the "obstacles" are actually only images in the background. When grasping a target disk that was placed on a screen displaying the images of four disks surrounding the target (roughly resembling the Ebbinghaus illusion, Figure 8.2), the placement of the digits on the target depended systematically on the locations of the images of the disks (de Grave et al. 2005). It was completely evident that the disks were just images, so the effect of obstacles on grasping might not be guided by an evaluation of the risks of colliding with the obstacles (which was zero for the images), but might rather be based on an automatic visuomotor association between visible structures and path selection.

The movement of the hand on its way to the object does not seem to contribute to the grip orientation when grasping. The grip orientation was unaffected both when the movement of the hand was perturbed during the movement (Grea et al. 2000), and when the digits had to deviate from the planned movement due to obstacles (Voudouris et al. 2010). On the other hand, the planning does take future comfort into account: when picking up a bar in order to place it in a specific orientation elsewhere, the comfort at the moment of placement is more relevant for choosing a grip than the comfort when picking it up (Rosenbaum et al. 1992). A possible reason for this "end-state comfort" effect is that comfort corresponds to being able to perform precisely, and the motor constraints for successfully placing a bar are more severe than for successful grasping (Hughes et al. 2012).

Occasionally, the situation that was considered when selecting the grasping points will have changed by the time the grasping movement starts. If one's task

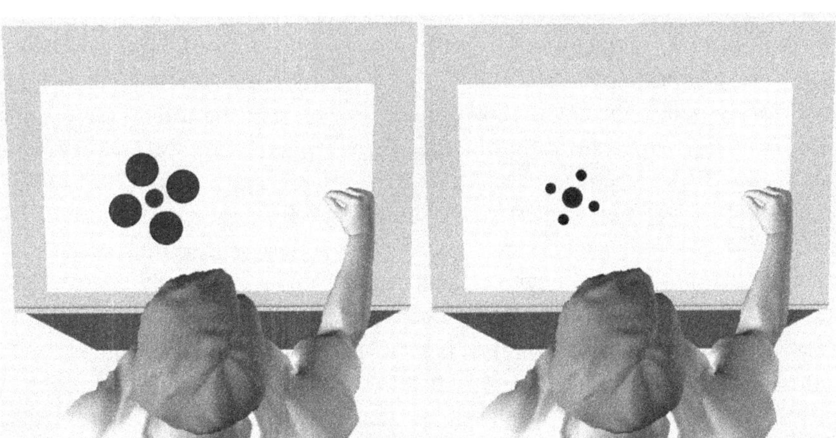

FIGURE 8.2 Grasping the central disk in a configuration resembling the Ebbinghaus illusion (de Grave et al. 2005). The images surrounding the central disk not only make the disk on the right appear larger than that on the left, but also influences the choice of grasping points

is to grasp the lit object, and suddenly a different object is lit, it is clear that one has to reselect grasping points (Paulignan et al. 1991b). What happens if the changes are more subtle? To investigate this, subjects were asked to grasp a cube that was oriented in a way that made people tend to grasp it along one of the axes, but close to the border with orientations for which they would select the other axis. Once subjects started their reach-to-grasp movement, the cube rotated slightly so that it was now at an orientation for which the other grasping axis would normally be preferred. Although it was possible to stick to the chosen grasping axis, on many trials subjects switched grasping axes at a very short latency: 160 ms (Voudouris et al. 2013). This is in line with the suggestion that the selection of grasping points is an automatic process.

Thus, the choice of grasping points can be changed in response to occurrences during the movement, and sometimes depends to a modest extent on the surrounding and on the path towards the object, but the choice appears primarily to depend on the object and the posture of the arm when making contact with it.

8.2.3 Eye Movements

Knowing the eye movements that are made during a task can provide insight into what information is deemed the most interesting at a certain time. As information about the target of a movement is needed before making the movement, knowing where people are looking might reveal how they plan their grasping movements. When grasping objects to place them elsewhere, subjects generally look at a position on the object near where their digit will make contact with it. At least, they do so if subjects can only see one grasping point. They direct their gaze near that point until just before making contact with the object (Johansson et al. 2001).

If both grasping points are visible, many studies report a tendency to look near the grasping point for the index finger rather than the thumb for various types of objects and viewing geometries (Brouwer et al. 2009; Cavina-Pratesi and Hesse 2013; de Grave et al. 2008; Desanghere and Marotta 2011; Voudouris et al. 2016). This tendency is in some way related to grasping, as subjects (initially) direct their gaze to the center of the object during free viewing or when required to make perceptual judgments (Brouwer et al. 2009; Desanghere and Marotta 2011). However, what aspect of grasping causes this tendency is not clear: many hypotheses could be rejected (Voudouris et al. 2016).

If a position on an object is relevant for grasping, and therefore generally fixated, occluding that part of the object does not keep subjects from fixating that location (de Grave et al. 2008). Apparently, the aim of these fixations is not to obtain detailed visual information about the contact position, but rather to obtain extra-retinal information on its location, a conclusion that is in line with the lack of effect of direct visibility on the choice of grasping points that we discussed in section 8.2.4.

8.3 Shaping the Movements

8.3.1 Basic Shape

Once the grasping points are chosen, the digits move towards these points. They do not move along straight paths, but follow curved trajectories so that the digits first move away from the straight line to the grasping point to a position a few centimeters from the object's surface, and then curve towards their grasping points. The points where the trajectories are furthest from a straight line correspond to the maximum grip aperture.

A simple way to model the kinematics of these movements is by finding the smoothest possible movement trajectory that ends at the grasping point with zero velocity and a deceleration perpendicular to the orientation of the surface (Smeets and Brenner 1999). The location of the point at which this trajectory is furthest from a straight line follows directly from the model. When moving to position on a surface, the model predicts that the maximal deviation of the trajectory from a straight line scales with distance from the starting point from a plane through that surface (with a gain of 0.8). In other words: if a grasping point is moved by 1 cm perpendicular to the local surface orientation, the maximal deviation from a straight line will change by 0.8 cm. This is indeed how subjects behave (Smeets and Brenner 2001). When applied to both digits at the same time, the prediction is that maximum grip aperture scales with object size, with a slope of 0.8, again in line with experimentally observed behavior (reviewed by Smeets and Brenner 1999).

Some features of grasping movements are not captured in this simple description. A first one is that maximum grip aperture is not only affected by the distance between the grasping points, but also by other dimensions of the target object. For instance, when grasping an ellipse along its minor axis, the maximum grip aperture is larger than when grasping a circular object with the same size (Cuijpers et al. 2004). A second finding that is not captured by the simple description is that when starting with an open hand to grasp a large object, subjects start by closing the grip and subsequently reopening it (Hesse and Deubel 2009; Meulenbroek et al. 2001; Smeets and Brenner 2002).

By using a dynamics-based model instead of a kinematic one, these additional features of grasping behavior can also be incorporated. In this model, the basic trajectory formation is caused by each digit being attracted towards its grasping point and repelled by other parts of the object (and by other objects), and by assuming that the digits have a preferred distance from each other (Verheij et al. 2012). In this model, the combination of the digits being repelled by other parts of the object and attracted to the grasping points gives rise to curved movements of the digits such that they approach the grasping points more or less perpendicularly (see Figure 8.3). Note that the attracting and repelling forces in this model describe the visuomotor interaction, but they are not

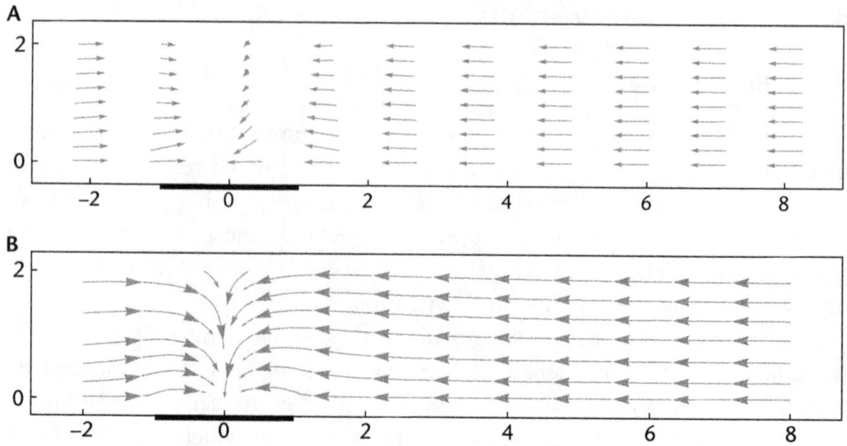

FIGURE 8.3 (A) Example of a force field that is the combination of a repelling force perpendicular to the surface (2 cm wide) and an attractive force towards the grasping point at its center. (B) Possible resulting movement trajectories. This is a caricature of the fields used by Verheij et al. (2012) to model grasping trajectories

directly measurable. This model can be used to make quantitative predictions for hitherto untested situations, such as for the effect of various object shapes on grip formation. Such experiments have been performed and confirmed the predictions to a large extent (Borchers et al. 2014; Verheij et al. 2014a).

Accuracy of grasping is limited by precision of visual localization of the grasping point and proprioceptive localisation of the digits (Smeets and Brenner 2008). However, as the movement is stopped by the grasping surface, the strategy to approach the surface closer to perpendicular can ensure a better precision on the grasping surface (Smeets and Brenner 1999). This is indeed what is observed. Moreover, an equivalent of Fitts' law (Fitts 1954) has also been observed in grasping behavior: the smaller the contact area for the digits, the slower the movement (Zaal and Bootsma 1993).

8.3.2 Individual Differences

Most of the papers discussed in this chapter discuss grasping behavior as a behavior that is common to all humans. A few studies, however, have focused on variations in behavior. For instance, Bongers and colleagues reported that in addition to the gradual opening and closing of the hand that is reported in most studies, sometimes participants open their hand directly to a plateau value, and other trials show a more bumpy behavior (Bongers et al. 2012). Also the timing of opening of the hand and moving of the hand differs: some subjects keep their digits together during the initial part of the movement, and open their hand only during the second half of the movement, whereas others start opening their hand directly at movement onset (Cuijpers et al. 2004).

These variations in movement patterns can be better understood when expressed as the movements of the individual digits through space. If one compares the movement trajectory of the tip of a digit when it is on its way to touch an object with the trajectory of that digit when reaching to grasp the object, they are remarkably similar (Smeets et al. 2010). If one combines a trajectory for the index-finger with one of the thumb, the timing of the pattern of opening and closing is governed by the trajectories of the individual digits. The idiosyncratic grasp formation corresponds to the idiosyncratic differences in the trajectories of the single-digit movements (Smeets and Brenner 2016).

8.3.3 On-Line Control

The stereotypical grasping pattern suggest that these movements are preprogrammed ballistic movements that are executed without peripheral feedback. This is definitely not the case: grasping movements are under visual control. The initial demonstration (Paulignan et al. 1991a, 1991b) showed responses to changing the illumination of objects that did not actually change size or position. The interpretation of these data is that the changes are not on-line adjustments of the movements, but switches to newly planned movements to the alternative target location (Smeets et al. 2002), and thus reselection of grasping points, as discussed in section 8.2.2.

In addition to being able to reprogram movements when the situation changes substantially, one would expect goal-directed movements of the digits in grasping to continuously be adjusted to accommodate improved location information (de Brouwer et al. 2014) or compensate for any movement errors. Consequently, they should follow any changes in the grasping locations that occur during the movement with a very short latency (~100 ms; Smeets et al. 2016), just as the digits do in other goal-directed movements. This is indeed what is found when an object rotates (Desmurget et al. 1996; Voudouris et al. 2013), changes size (van de Kamp et al. 2009; van de Kamp and Zaal 2007) or changes position (Grea et al. 2002). Similar behavior has also been reported for responses to changes in object size in bimanual grasping (Zaal and Bongers 2014).

The responses to a perturbation of the location of a target of a goal-directed hand movement depends on the timing: the later the perturbation during the movement, the more vigorous the response (Liu and Todorov 2007; Oostwoud Wijdenes et al. 2011). Is this also the case in grasping? This has not been investigated directly, but indirect evidence can be found in reported latencies of adjustments. When using a conservative detection threshold to determine a latency, one will find shorter latencies for more vigorous responses (Oostwoud Wijdenes et al. 2014). Responses to perturbations of object size in grasping movements have been reported to have shorter latencies when applied later in the movement (Hesse and Franz 2009; van de Kamp et al. 2009). We propose that this is because they are more vigorous. So the control of grasping is also similar to the control of goal-directed movements of the digits in this respect.

Goal-directed hand movements are not only adjusted when the target is perturbed, but also when feedback about the hand is perturbed (Brenner and Smeets 2003; Franklin and Wolpert 2008). Volcic and Domini (2016) performed a similar experiment in grasping. They changed the feedback of the hand subtly once grip aperture had reached the size of the object. When the hand opened further than the size of the object, the additional finger and thumb movement was either magnified or reduced. Subjects compensated for this manipulation, by moving the digits less far or further apart, respectively. This correction only happened when the viewing geometry allowed subjects to see the distance between their finger and the grasping surface. Thus, altogether, it is evident that grasping movements are under continuous visual control.

8.3.4 Obstacles

The repelling forces in the dynamic model are not only caused by other parts of the object that is grasped, but also by other objects near the grasping trajectory. The obstacles affect both the spatial and temporal aspects of the movement. The temporal effect of an obstacle near the grasping location of a digit is to slow down the grasping movement (Biegstraaten et al. 2003; Mon-Williams et al. 2001; Tresilian 1998; Voudouris et al. 2012b). As grasping with an obstacle near the grasping location requires a more accurate movement, this finding is in line with the well-known speed-accuracy trade-off. It can also be understood in terms of the dynamical model: obstacles correspond to a repelling force, so that the sum of the attracting and repelling force is smaller, leading to a slower movement. The second effect of an obstacle is that it changes the curvature of the digit's path, leading to change in maximum grip aperture. This effect is easy to understand if we consider an obstacle as a repeller, although other model interpretations have also been proposed (Vaughan et al. 2001).

We already mentioned in section 8.2.2 that pictorial elements that could be interpreted as obstacles influence the choice of grasping points (de Grave et al. 2005). Such images that can be interpreted as obstacles also affect the trajectories of the digits towards grasping points: moving images of flankers in the Ebbinghaus illusion closer to the target results in a larger maximum grip aperture (Haffenden et al. 2001). In interpreting this experiment, it is difficult to separate the obstacle effect from the effect of the illusion itself (Kopiske et al. 2016), as will be discussed in section 8.3.6.

The biggest obstacle for our hands in our daily office routines is the table. Does the table indeed influence our movements as an obstacle? One might think so, as our hand movements that start and end on a table curve upwards, away from the table. However, this interpretation is not valid. Verheij and colleagues removed (parts of) the table, and found virtually no effect, provided that the small horizontal surface near the start and the target object ensured an upward start and a target approach from above (Verheij et al. 2013b). An object in between these positions at table height did not affect the curvature. Only objects that were higher than the starting position affected the vertical curvature of the digits during a grasping movement (Verheij et al. 2014b). An additional (minor) cause of the vertical curvature of grasping movements is gravity: this was shown by rotating the participant together with starting position and target relative to gravity. The curvature of the digits paths in the grasping movement was only mildly affected by this manipulation, corresponding to a direct (downward) effect of gravity (Verheij et al. 2013a).

8.3.5 The Moment of Contact

In our description of trajectory formation, the geometry around the grasping point plays an essential role: the digits are not just closing towards the grasping points, but tend to approach the local surface perpendicularly (Smeets and Brenner 1999), which can be interpreted as the surface acting as a repeller (Verheij et al. 2012). A straightforward way to test this is to vary the orientation of the contact surface while keeping the grasping points constant. If one does so (and makes sure that the edges of the object do not act directly as obstacles), one finds that the trajectories of the digits vary with local surface orientation (Kleinholdermann et al. 2007).

One might expect that both digits will contact the object at the same moment. However, this is not the case. Several studies report that the index finger makes contact tens of millimeters before the thumb (Biegstraaten et al. 2006; Brouwer et al. 2009; Cavina-Pratesi and Hesse 2013; Reilmann et al. 2001; Voudouris et al. 2012a). These studies report that the early arrival of the index finger is associated with a late departure. Such a tendency to synchronize movements of the digits rather than the moment of contact has also been reported for bimanual pointing movements (Boessenkool et al. 1999).

It is well known that haptic (tactile) information is essential for object manipulation: haptic feedback ensures that contact forces are scaled to prevent slip and excessive forces (Johansson and Flanagan 2009). It is less well known that haptic feedback is essential to shape the digits movements in grasping. If one has to pantomime grasping, i.e., perform the action without any haptic feedback, the basic shape of the trajectories and properties such as the scaling with object size are different (Goodale et al. 1994a). If a real object is present, but variations in visual information on object shape are inconsistent with the haptically experienced shape, some subjects neglect the visual information (Cuijpers et al. 2008). The presence of veridical haptic feedback is also essential for a patient with a ventral lesion to be able to normally grasp objects (Schenk 2012).

After contact, the digits start applying forces to the surface in order to provide adequate grip and lift forces to lift the object without it slipping. These lifting forces are not perpendicular to the surface, but directed a bit upward. One might expect that the movement of the digits just before contact is already aligned with this direction, and that variations in the approach direction correspond to variations in initial force direction. This is not the case: the initial contact force is directed downwards (Flanagan et al. 1999) and the direction of approach is not clearly correlated with the initial build-up of force (Biegstraaten et al. 2006). This independence is in line with switching between neural control strategies when transitioning from motion to force production (Venkadesan and Valero-Cuevas 2008). The (relative) timing of the digits' movements and the forces exerted after contact are not accounted for by the above-mentioned models of grasping.

8.3.6 Illusions

Since 1995, the study of grasping objects in an illusory context has become extremely popular. In that year, a paper was published that showed that the Ebbinghaus illusion affected perceptual judgments more than it influenced peak grip aperture in grasping (Aglioti et al. 1995). This result was interpreted by the authors as support for the idea that perception and action are based on two parallel visual processing streams, a ventral one that is susceptible to contextual illusions that underlies perceptual judgments and a dorsal one that neglects the context to control goal-directed movements. Other researches questioned this interpretation (Franz and Gegenfurtner 2008; Smeets and Brenner 2006). A first reason for questioning this interpretation is that the original paper showed that there was an effect of the illusion, so the claim in the title of the paper that the illusion did not deceive the hand is exaggerated. Moreover, as mentioned in section 8.3.4, the discussion as to whether there is an effect of the illusion is confounded with possible interpretations of any differences that are found in terms of obstacles. The second reason is that when appropriately matching the perceptual task to the grasping task, the effects appeared to be equal for both tasks (Franz et al. 2000; Pavani et al. 1999). We won't go further into the specifics of this debate, but refer the reader to other reviews (Schenk et al. 2011; Westwood and Goodale 2011).

Before discussing the third reason, we will discuss an attempt to settle the discussion on whether illusions affect the trajectories in grasping. It has been argued that illusions might affect the planning, but not the execution of the grasping movement, which would then result in a gradual decrease of the effect of illusions such as the Ebbinghaus illusion during the course of the movement (Glover and Dixon 2002). Although the experimental results seem to support this claim, their interpretation is not without pitfalls. In line with the need to precisely equate the tasks that was mentioned in the previous paragraph, one can only draw a definitive conclusion from the time course of the effect of the illusion if one can predict how a constant illusion effect would influence the trajectories of the digits. Doing so predicts that a constant illusion effect results in an illusion effect on the digits' trajectories (and thus grip aperture) that gradually decreases as the movement unfolds (Smeets et al. 2003).

A year after Aglioti et al.'s original publication, their interpretation that actions are immune to illusions was questioned by us (Brenner and Smeets 1996). We showed that the Ponzo illusion (Figure 8.4) did not affect the movements of the digits towards the object, but did affect the forces used to lift the object. The third objection is thus that grip aperture might be immune to the illusion, but motor control is more than opening a hand. As grasping is moving digits to positions, there is no reason to expect an illusion that does not affect positions to influence the trajectories of the digits to these positions. On the other hand, to lift an object, a visual estimate of weight is required, and size can provide such an estimate (Gordon et al. 1991). The lack of effect of the Ponzo illusion on grip aperture despite a clear effect on lift forces has been replicated and extended to grip force by Jackson and Shaw (2000). Several later studies confirmed the lack of effect of the Ponzo illusion on the trajectory of the digits (Ganel et al. 2008), although one study reported an effect in the first few trials (Whitwell et al. 2016).

The Müller-Lyer illusion has been much more popular than the Ponzo illusion for studying grasping behavior. Most papers reported an effect of the illusion, but its effect on maximum grip aperture when grasping the central bar was much smaller that its perceptual effect. In their extensive review of this literature, Bruno and Franz (2009) argue that two factors might underlie this difference. The first is that one would not expect an equal effect, because (as discussed in section 8.3.1) a 1 cm real change in object size should only lead to a 0.8 cm larger grip aperture. The second is that in grasping you can correct your

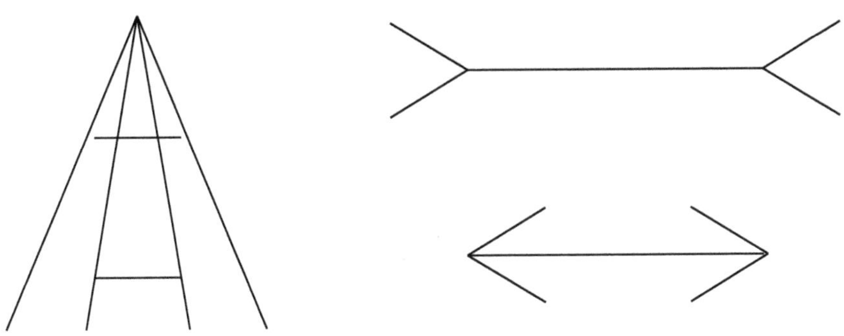

FIGURE 8.4 Two size illusions that have been used in grasping research. In the Ponzo illusion (left) and the Müler-Lyer illusion (right) the two horizontal lines are equally long, but the context lines make the upper line look larger than the lower line

movements based on visual feedback during the movement. On the other hand, others have argued that the effects might not be due to the illusory size at all, but due to the fins being considered as obstacles that change the precision requirements. An argument for this account is based on the analysis of the trajectories near the moment of contact. If a larger-than-normal peak grip aperture were due to misperception of the bar's size, one would expect a prolonged low velocity phase before contact, as contact would not have been obtained at the expected position. Such a prolonged deceleration phase was not observed (Biegstraaten et al. 2007).

Other illusions were tested to rule out the possibility that movement trajectories are only influenced by illusions because the latter are interpreted as obstacles. Neither the diagonal illusion (Stottinger et al. 2009; Stottinger and Perner 2006) nor the empty space illusion (Stottinger et al. 2012) affected peak grip aperture during grasping. To reconcile all these studies, one could argue that pure *size* illusions do not affect the movements of the digits in grasping (and thus peak grip aperture) because the digits move towards positions. The fact that some illusory figures affect grasping might be due to other effects, such as a misperception of the location of the grasping points or parts of the figures being treated as obstacles.

8.3.7 Moving Targets

Grasping moving targets or catching them has one additional factor which makes the task more complicated than grasping static target: there is no direct visual information about the location of the grasping points to guide your movement. In some way, predictive information has to be used. A seminal study is the experiment by Savelsbergh and colleagues, who studied the timing of the closing movements when catching an approaching ball (Savelsbergh et al. 1991). They manipulated one source of information (optical expansion) by sometimes letting the ball deflate as it approached, and found that subjects delayed their hand closure, as if they expected the ball to arrive later. Other sources of information about (motion in) depth such as binocular disparity and retinal size also play a role in this timing (van der Kamp et al. 1997).

Despite differences between the sources of information and between the way in which the distance between target and hand is reduced (actively in grasping, passively in catching), the timing of events is remarkably similar (van de Kamp et al. 2010). Not only the temporal, but also the spatial aspects of the digits' trajectories are very similar when grasping stationary and moving objects (Schot et al. 2011). This is not a coincidence, but related to the fact that the geometry of the task for the digits is very similar once the hand has started the reach-to-grasp movement: the hand and object are approaching each other, and the digits have to move in such a way that they will approach the object's surface more or less orthogonally.

8.4 Summary

In this chapter, we showed that a close examination of the task geometry can explain how our digits move when we reach out to grasp an object. The laws that govern this behavior are similar to the ones that describe single-digit goal-directed movements. As with any movement, the selection of endpoint depends on why one is making the movement, and the trajectory is under continuous visual guidance, influenced by constraints imposed by obstacles and by the need to be precise.

References

Aglioti S., DeSouza J. F. X., Goodale M. A., 1995 "Size-contrast illusions deceive the eye but not the hand." *Current Biology* **5** 679–685.

Baud-Bovy G., Soechting J. F., 2001 "Two virtual fingers in the control of the tripod grasp" *Journal of Neurophysiology* **86** 604–615.

Biegstraaten M., de Grave D. D. J., Brenner E., Smeets J. B. J., 2007 "Grasping the Muller-Lyer illusion: not a change in perceived length" *Experimental Brain Research* **176** 497–503.

Biegstraaten M., Smeets J. B. J., Brenner E., 2003 "The influence of obstacles on the speed of grasping" *Experimental Brain Research* **149** 530–534.

Biegstraaten M., Smeets J. B. J., Brenner E., 2006 "The relation between force and movement when grasping an object with a precision grip" *Experimental Brain Research* **171** 347–357.

Boessenkool J. J., Nijhof E. J., Erkelens C. J., 1999 "Variability and correlations in bi-manual pointing movements" *Human Movement Science* **18** 525–552.

Bongers R. M., Zaal F., Jeannerod M., 2012 "Hand aperture patterns in prehension" *Human Movement Science* **31** 487–501.

Borchers S., Verheij R., Smeets J. B. J., Himmelbach M., 2014 "The influence of object height on maximum grip aperture in empirical and modeled data" *Journal of Experimental Psychology-Human Perception and Performance* **40** 889–896.

Brenner E., Smeets J. B. J., 1996 "Size illusion influences how we lift but not how we grasp an object" *Experimental Brain Research* **111** 473–476.

Brenner E., Smeets J. B. J., 2003 "Fast corrections of movements with a computer mouse" *Spatial Vision* **16** 365–376.

Brouwer A. M., Franz V. H., Gegenfurtner K. R., 2009 "Differences in fixations between grasping and viewing objects" *Journal of Vision* **9.**

Bruno N., Franz V. H., 2009 "When is grasping affected by the Müller-Lyer illusion?: A quantitative review" *Neuropsychologia* **47** 1421–1433.

Cavina-Pratesi C., Hesse C., 2013 "Why do the eyes prefer the index finger? Simultaneous recording of eye and hand movements during precision grasping" *Journal of Vision* **13.**

Crajé C., van der Kamp J., Steenbergen B., 2008 "The effect of the "rod-and-frame" illusion on grip planning in a sequential object manipulation task" *Experimental Brain Research* **185** 53–62.

Cruse H., Brüwer M., Dean J., 1993 "Control of three- and four-joint arm movement: strategies for a manipulator with redundant degrees of freedom" *Journal of Motor Behavior* **25** 131–139.

Cuijpers R. H., Brenner E., Smeets J. B. J., 2006 "Grasping reveals visual misjudgements of shape" *Experimental Brain Research* **175** 32–44.

Cuijpers R. H., Brenner E., Smeets J. B. J., 2008 "Consistent haptic feedback is required but it is not enough for natural reaching to virtual cylinders" *Human Movement Science* **27** 857–872.

Cuijpers R. H., Smeets J. B. J., Brenner E., 2004 "On the relation between object shape and grasping kinematics" *Journal of Neurophysiology* **91** 2598–2606.

de Brouwer A. J., Brenner E., Medendorp W. P., Smeets J. B. J., 2014 "Time course of the effect of the Muller-Lyer illusion on saccades and perceptual judgments" *Journal of Vision* **14(1)**:4 1–11.

de Grave D. D. J., Biegstraaten M., Smeets J. B. J., Brenner E., 2005 "Effects of the Ebbinghaus figure on grasping are not only due to misjudged size" *Experimental Brain Research* **163** 58–64.

de Grave D. D. J., Hesse C., Brouwer A. M., Franz V. H., 2008 "Fixation locations when grasping partly occluded objects" *Journal of Vision* **8** 5:1–11.

Desanghere L., Marotta J. J., 2011 " "Graspability" of objects affects gaze patterns during perception and action tasks" *Experimental Brain Research* **212** 177–187.

Desmurget M., Prablanc C., Arzi M., Rossetti Y., Paulignan Y., Urquizar C., 1996 "Integrated control of hand transport and orientation during prehension movements" *Experimental Brain Research* **110** 265–278.

Ellis R. R., Flanagan J. R., Lederman S. J., 1999 "The influence of visual illusions on grasp position" *Experimental Brain Research* **125** 109–114.

Fitts P. M., 1954 "The information capacity of the human motor system in controlling the amplitude of movement" *Journal of Experimental Psychology* **47** 381–391.

Flanagan J. R., Burstedt M. K. O., Johansson R. S., 1999 "Control of fingertip forces in multidigit manipulation" *Journal of Neurophysiology* **81** 1706–1717.

Franklin D. W., Wolpert D. M., 2008 "Specificity of reflex adaptation for task-relevant variability" *Journal of Neuroscience* **28** 14165–14175.

Franz V. H., Gegenfurtner K. R., 2008 "Grasping visual illusions: Consistent data and no dissociation" *Cognitive Neuropsychology* **25** 920–950.

Franz V. H., Gegenfurtner K. R., Bülthoff H. H., Fahle M., 2000 "Grasping visual illusions: No evidence for a dissociation between perception and action" *Psychological Science* **11** 20–25.

Ganel T., Tanzer M., Goodale M. A., 2008 "A double dissociation between action and perception in the context of visual illusions – Opposite effects of real and illusory size" *Psychological Science* **19** 221–225.

Glover S., Dixon P., 2002 "Dynamic effects of the Ebbinghaus illusion in grasping: Support for a planning/control model of action" *Perception & Psychophysics* **64** 266–278.

Goodale M. A., Jakobson L. S., Keillor J. M., 1994a "Differences in the visual control of pantomimed and natural grasping movements" *Neuropsychologia* **10** 1159–1178.

Goodale M. A., Meenan J. P., Bülthoff H. H., Nicolle D. A., Murphy K. J., Raciot C. I., 1994b "Separate neural pathways for the visual analysis of object shape in perception and prehension" *Current Biology* **4** 604–610.

Gordon A. M., Forssberg H., Johansson R. S., Westling G., 1991 "Visual size cues in the programming of manipulative forces during precision grip" *Experimental Brain Research* **83** 477–482.

Grea H., Desmurget M., Prablanc C., 2000 "Postural invariance in three-dimensional reaching and grasping movements" *Experimental Brain Research* **134** 155–162.

Grea H., Pisella L., Rossetti Y., Desmurget M., Tilikete C., Grafton S., Prablanc C., Vighetto A., 2002 "A lesion of the posterior parietal cortex disrupts on-line adjustments during aiming movements" *Neuropsychologia* **40** 2471–2480.

Haffenden A. M., Schiff K. C., Goodale M. A., 2001 "The dissociation between perception and action in the Ebbinghaus illusion: Nonillusory effects of pictorial cues on grasp" *Current Biology* **11** 177–181.

Hesse C., de Grave D. D. J., Franz V. H., Brenner E., Smeets J. B. J., 2008 "Planning movements well in advance" *Cognitive Neuropsychology* **25** 985–995.

Hesse C., Deubel H., 2009 "Changes in grasping kinematics due to different start postures of the hand" *Human Movement Science* **28** 415–436.

Hesse C., Franz V. H., 2009 "Corrective processes in grasping after perturbations of object size" *Journal of Motor Behavior* **41** 253–273.

Hughes C. M. L., Seegelke C., Schack T., 2012 "The influence of initial and final precision on motor planning: individual differences in end-state comfort during unimanual grasping and placing" *Journal of Motor Behavior* **44** 195–201.

Iberall T., Bingham G., Arbib M. A., 1986 "Opposition space as a structuring concept for the analysis of skilled hand movements" in *Generation and Modulation of Action Patterns* Eds H. Heuer and C. Fromm (Berlin: Springer) pp. 158–173.

Jackson S. R., Shaw A., 2000 "The Ponzo illusion affects grip-force but not grip-aperture scaling during prehension movements" *Journal of Experimental Psychology: Human Perception and Performance* **26** 418–423.

Jeannerod M., 1981 "Intersegmental coordination during reaching at natural visual objects" in *Attention and Performance IX* Eds J. Long and A. Baddeley (Hillsdale, NJ: Erlbaum) pp. 153–169.

Jeannerod M., 1984 "The timing of natural prehension movements" *Journal of Motor Behavior* **16** 235–254.

Jeannerod M., 1986 "The formation of finger grip during prehension: A cortically mediated visuomotor pattern" *Behavioural Brain Research* **19** 99–161.

Jeannerod M. 1999 "Visuomotor channels: Their integration in goal-directed prehension" *Human Movement Science* **18** 210–218.

Johansson R. S., Flanagan J. R., 2009 "Coding and use of tactile signals from the fingertips in object manipulation tasks" *Nature Reviews Neuroscience* **10** 345–359.

Johansson R. S., Westling G. R., Backstrom A., Flanagan J. R., 2001 "Eye-hand coordination in object manipulation" *Journal of Neuroscience* **21** 6917–6932.

Kleinholdermann U., Brenner E., Franz V. H., Smeets J. B. J., 2007 "Grasping trapezoidal objects" *Experimental Brain Research* **180** 415–420.

Kopiske K. K., Bruno N., Hesse C., Schenk T., Franz V. H., 2016 "The functional subdivision of the visual brain: Is there a real illusion effect on action? A multi-lab replication study" *Cortex* **79** 130–152.

Lederman S. J., Wing A. M., 2003 "Perceptual judgement, grasp point selection and object symmetry" *Experimental Brain Research* **152** 156–165.

Liu D., Todorov E., 2007 "Evidence for the flexible sensorimotor strategies predicted by optimal feedback control" *Journal of Neuroscience* **27** 9354–9368.

Meulenbroek R. G. J., Rosenbaum D. A., Jansen C., Vaughan J., Vogt S., 2001 "Multijoint grasping movements: Simulated and observed effects of object location, object size, and initial aperture" *Experimental Brain Research* **138** 219–234.

Mon-Williams M., Tresilian J. R., Coppard V. L., Carson R. G., 2001 "The effect of obstacle position on reach-to-grasp movements" *Experimental Brain Research* **137** 497–501.

Oostwoud Wijdenes L., Brenner E., Smeets J. B. J., 2011 "Fast and fine-tuned corrections when the target of a hand movement is displaced" *Experimental Brain Research* **214** 453–462.

Oostwoud Wijdenes L., Brenner E., Smeets J. J., 2014 "Analysis of methods to determine the latency of online movement adjustments" *Behavior Research Methods* **46** 131–139.

Paulignan Y., Jeannerod M., MacKenzie C., Marteniuk R., 1991a "Selective perturbation of visual input during prehension movements. 2. The effects of changing object size" *Experimental Brain Research* **87** 407–420.

Paulignan Y., MacKenzie C., Marteniuk R., Jeannerod M., 1991b "Selective perturbation of visual input during prehension movements. 1. The effects of changing object position" *Experimental Brain Research* **83** 502–512.

Paulun V. C., Kleinholdermann U., Gegenfurtner K. R., Smeets J. B. J., Brenner E., 2014 "Center or side: Biases in selecting grasp points on small bars" *Experimental Brain Research* **232** 2061–2072.

Pavani F., Boscagli I., Benvenuti F., Rabuffetti M., Farnè A., 1999 "Are perception and action affected differently by the Titchener circles illusion?" *Experimental Brain Research* **127** 95–101.

Reilmann R., Gordon A. M., Henningsen H., 2001 "Initiation and development of fingertip forces during whole-hand grasping" *Experimental Brain Research* **140** 443–452.

Rosenbaum D. A., Barnes H. J., Vaughan J., Jorgensen M. J., 1992 "Time course of movement planning – selection of handgrips for object manipulation" *Journal of Experimental Psychology-Learning Memory and Cognition* **18** 1058–1073.

Rosenbaum D. A., Meulenbroek R. J., Vaughan J., Jansen C., 2001 "Posture-based motion planning: Applications to grasping" *Psychological Review* **108** 709–734.

Savelsbergh G. J. P., Whiting H. T. A., Bootsma R. J., 1991 "Grasping tau" *Journal of Experimental Psychology-Human Perception and Performance* **17** 315–322.

Schenk T., 2012 "No dissociation between perception and action in patient DF when haptic feedback is withdrawn" *Journal of Neuroscience* **32** 2013–2017.

Schenk T., Franz V., Bruno N., 2011 "Vision-for-perception and vision-for-action: Which model is compatible with the available psychophysical and neuropsychological data?" *Vision Research* **51** 812–818.

Schot W. D., Brenner E., Smeets J. B. J., 2010 "Posture of the arm when grasping spheres to place them elsewhere" *Experimental Brain Research* **204** 163–171.

Schot W. D., Brenner E., Smeets J. B. J., 2011 "Grasping and hitting moving objects" *Experimental Brain Research* **212** 487–496.

Smeets J. B. J., Brenner E., 1999 "A new view on grasping" *Motor Control* **3** 237–271.

Smeets J. B. J., Brenner E., 2001 "Independent movements of the digits in grasping" *Experimental Brain Research* **139** 92–100.

Smeets J. B. J., Brenner E., 2002 "Does a complex model help to understand grasping?" *Experimental Brain Research* **144** 132–135.

Smeets J. B. J., Brenner E., 2006 "10 years of illusions" *Journal of Experimental Psychology-Human Perception and Performance* **32** 1501–1504.

Smeets J. B. J., Brenner E., 2008 "Grasping Weber's law" *Current Biology* **18** R1089–R1090.

Smeets J. B. J., Brenner E., 2016 "Synergies in grasping" in *Progress in Motor Control – Theories and Translations* Eds M. L. Latash and J. Laczko (Berlin; Heidelberg: Springer) pp. 21–34.

Smeets J. B. J., Brenner E., Biegstraaten M., 2002 "Independent control of the digits predicts an apparent hierarchy of visuomotor channels in grasping" *Behavioural Brain Research* **136** 427–432.

Smeets J. B. J., Glover S., Brenner E., 2003 "Modeling the time-dependent effect of the Ebbinghaus illusion on grasping" *Spatial Vision* **16** 311–324.

Smeets J. B. J., Martin J., Brenner E., 2010 "Similarities between digits' movements in grasping, touching and pushing" *Experimental Brain Research* **203** 339–346.

Smeets J. B. J., Oostwoud Wijdenes L., Brenner E., 2016 "Movement adjustments have short latencies because there is no need to detect anything" *Motor Control* **20** 137–148.

Stelmach G. E., Castiello U., Jeannerod M., 1994 "Orienting the finger opposition space during prehension movements" *Journal of Motor Behavior* **26** 178–186.

Stottinger E., Aigner S., Hanstein K., Perner J., 2009 "Grasping the diagonal: Controlling attention to illusory stimuli for action and perception" *Consciousness and Cognition* **18** 223–228.

Stottinger E., Perner J., 2006 "Dissociating size representation for action and for conscious judgment: Grasping visual illusions without apparent obstacles" *Consciousness and Cognition* **15** 269–284.

Stottinger E., Pfusterschmied J., Wagner H., Danckert J., Anderson B., Perner J., 2012 "Getting a grip on illusions: Replicating Stottinger et al. *Exp Brain Res* (2010) 202:79–88 results with 3-D objects" *Experimental Brain Research* **216** 155–157.

Tresilian J. R., 1998 "Attention in action or obstruction of movement? A kinematic analysis of avoidance behavior in prehension" *Experimental Brain Research* **120** 352–368.

van de Kamp C., Bongers R. M., Zaal F. T. J. M., 2009 "Effects of changing object size during prehension" *Journal of Motor Behavior* **41** 427–435.

van de Kamp C., Bongers R. M., Zaal F. T. J. M., 2010 "Getting hold of approaching objects: In search of a common control of hand-closure initiation in catching and grasping" *Human Movement Science* **29** 518–528.

van de Kamp C., Zaal F. T. J. M., 2007 "Prehension is really reaching and grasping" *Experimental Brain Research* **182** 27–34.

van der Kamp J., Savelsbergh G., Smeets J., 1997 "Multiple information sources guiding the timing of interceptive actions" *Human Movement Science* **16** 787–822.

Vaughan J., Rosenbaum D. A., Meulenbroek R. G., 2001 "Planning reaching and grasping movements: the problem of obstacle avoidance" *Motor Control* **5** 116–35.

Venkadesan M., Valero-Cuevas F. J., 2008 "Neural control of motion-to-force transitions with the fingertip" *Journal of Neuroscience* **28** 1366–1373.

Verheij R., Brenner E., Smeets J. B. J., 2012 "Grasping kinematics from the perspective of the individual digits: A modelling study" *PLoS ONE* **7** e33150.

Verheij R., Brenner E., Smeets J. B. J., 2013a "Gravity affects the vertical curvature in human grasping movements" *Journal of Motor Behavior* **45** 325–332.

Verheij R., Brenner E., Smeets J. B. J., 2013b "Why are the digits' paths curved vertically in human grasping movements?" *Experimental Brain Research* **224** 59–68.

Verheij R., Brenner E., Smeets J. B. J., 2014a "The influence of target object shape on maximum grip aperture in human grasping movements" *Experimental Brain Research* **232** 3569–3578.

Verheij R., Brenner E., Smeets J. B. J., 2014b "Why does an obstacle just below the digits' paths not influence a grasping movement while an obstacle to the side of their paths does?" *Experimental Brain Research* **232** 103–112.

Volcic R., Domini F., 2016 "On-line visual control of grasping movements" *Experimental Brain Research* **234** 2165–2177.

Voudouris D., Brenner E., Schot W. D., Smeets J. B. J., 2010 "Does planning a different trajectory influence the choice of grasping points?" *Experimental Brain Research* **206** 15–24.

Voudouris D., Smeets J. B. J., Brenner E., 2012a "Do humans prefer to see their grasping points?" *Journal of Motor Behavior* **44** 295–304.

Voudouris D., Smeets J. B. J., Brenner E., 2012b "Do obstacles affect the selection of grasping points?" *Human Movement Science* **31** 1090–1102.

Voudouris D., Smeets J. B. J., Brenner E., 2013 "Ultra-fast selection of grasping points" *Journal of Neurophysiology* **110** 1484–1489.

Voudouris D., Smeets J. B. J., Brenner E., 2016 "Fixation biases towards the index finger in almost-natural grasping" *PLoS One* **11** 29.

Westwood D. A., Goodale M. A., 2011 "Converging evidence for diverging pathways: Neuropsychology and psychophysics tell the same story" *Vision Research* **51** 804–811.

Whitwell R. L., Buckingham G., Enns J. T., Chouinard P. A., Goodale M. A., 2016 "Rapid decrement in the effects of the Ponzo display dissociates action and perception" *Psychonomic Bulletin & Review* **23** 1157–1163.

Zaal F. T. J. M., Bongers R. M., 2014 "Movements of individual digits in bimanual prehension are coupled into a grasping component" *PLoS One* **9**.

Zaal F. T. J. M., Bootsma R. J., 1993 "Accuracy demands in natural prehension" *Human Movement Science* **12** 339–345.

9

REACH-TO-GRASP MOVEMENTS

Testing the Cognitive Architecture of Action

Umberto Castiello

The reach-to-grasp movement is performed normally and routinely within the familiar context of everyday living activities. It is the key behavior that allows humans to interact with their environment and it serves as a remarkable experimental test case for probing the cognitive architecture of goal-oriented action. This chapter focuses on experimental evidence that enhances or modifies how we might conceptualize the "cognitive" substrates of prehension. The chapter begins with a consideration on the mechanisms of selection for the control of hand actions. Then, other related aspects of this selection process are pursued. Is action selection a prerequisite of the visual modality? Recent research reviewed here suggests that how reach-to-grasp movements are planned and executed is influenced by information coming from different sensory modalities such as vision, proprioception, audition, taste, and olfaction. How is action selection generated depending on the agent's goal and intention? While there is overwhelming evidence for acting towards objects with an individual intention, the new excitement is in determining how reaching towards and grasping an object is also influenced by the agent's social intention. The chapter closes with some considerations about the predominant role of the cognitive constituents in shaping prehensile behavior and how this might be important for future research developments.

Mechanisms of Selection for the Control of Prehensile Actions

Attention is particularly important in cluttered environments, those with multiple objects that can act as potential obstacles to the intended movement or as separate potential targets. This section examines the attempts to determine the role played by selective mechanisms of attention for the control of action

(Cisek, 2007). It reviews studies that have looked at reach-to-grasp responses to target objects in the presence of other objects (i.e., distractors) within a three-dimensional space (for reviews see Castiello, 1999; Tipper, Howard, & Houghton, 1998). It highlights how motoric aspects of the reach-to-grasp response may be influenced by the distractors, rather than merely addressing the perceptual consequences of the distractors. The studies reviewed here highlight several factors that point out the importance of studying selective processes within three-dimensional environments where attention and action have evolved (Allport, 1987).

If we consider the role that selective attention plays during the control of an action, such as reaching-to-grasp, it may be advanced that attention needs to access object representations that provide information for the coding of volumetric boundaries necessary for positioning the fingers appropriately on the object. Several studies have now shown that selecting a target for grasping in the presence of a distractor object leads to interference effects on movement kinematics (Castiello, 1996; Chieffi, Gentilucci, Allport, Sasso, & Rizzolatti, 1993; Colman, Remington, & Kritikos, 2016; Jackson, Jackson, & Rosicky, 1995; Johansson, Westling, Bäckström, & Flanagan, 2001; Kritikos, Bennett, Dunai, & Castiello, 2000; Mon-Williams, Tresilian, Coppard, & Carson, 2001; Saling, Alberts, Stelmach, & Bloedel, 1998; Tipper, Howard, & Jackson, 1997; Tresilian, 1998; Welsh, Elliott, & Weeks, 1999).

Castiello (1996) developed a double-task procedure (Figure 9.1a,b) capitalizing on the dissociation between overt and covert orienting mechanisms of attention. Overt orienting is accompanied by eye movements, whereas covert orienting occurs while the eye remain still. Participants reached to grasp the target while having to count the number of times that a spotlight illuminated a distractor object. Under these circumstances the characteristics of the distractor did interfere with the kinematics of the reach-to-grasp movement (Figure 9.1a,b). Interference, however, only occurred when covert attention was directed to a lateral non-target, though eye movements were precluded. Apparently, the programming of hand movements can be influenced by attended information in the visual field, and this is independent of eye movements. Distractor interference during prehension movements can even be augmented when the distractor attracts more attention. Support for this last conclusion can be drawn from the study of brain-injured patients. For example, considering that patients with neglect have an attentional bias towards the ipsilesional side, the evidence of interference effects found for distractors positioned on that side suggests that, when distractors receive more attention, interference effects during reaching become more evident (Chieffi et al., 1993).

Using another dual-task paradigm, Bonfiglioli and Castiello (1998) further investigated the role of overt and covert attention in the manifestation of distractor interference effects. In this series of experiments, the target was positioned midsagitally and the distractor was monitored either overtly or covertly.

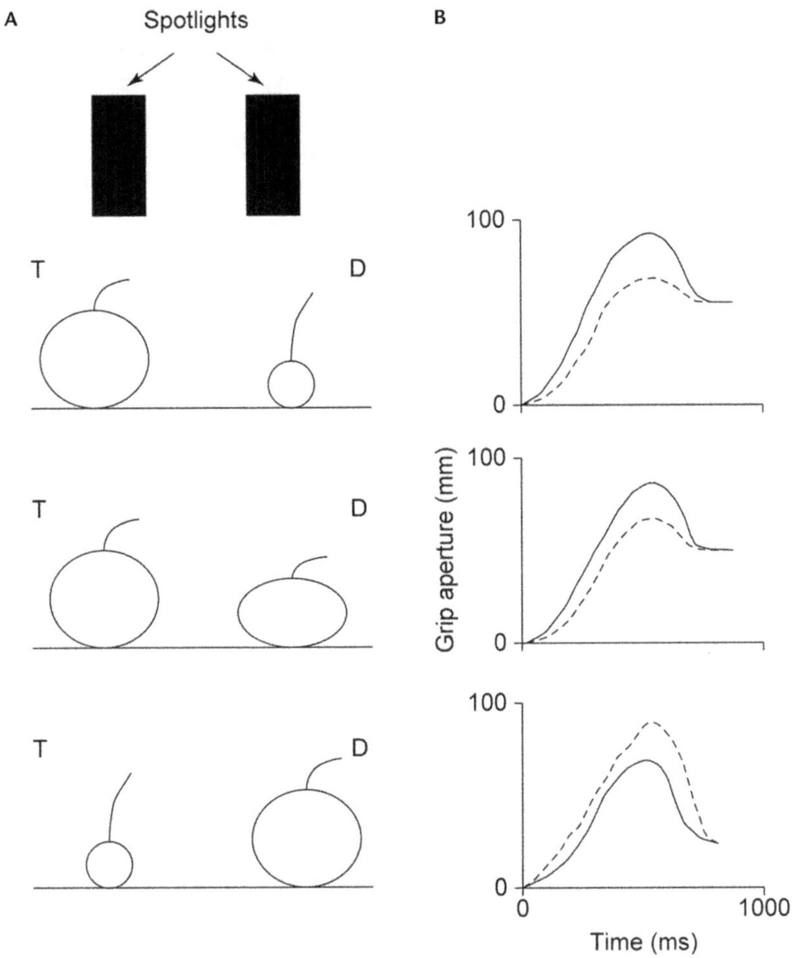

FIGURE 9.1 The dual-task procedure in a grip-aperture task. (A) The dual-task experimental condition, depicted schematically, shows target (T) fruits presented with distractor (D) fruits

Notes: The primary task was to reach and grasp the central target fruit while maintaining gaze fixation of this same fruit. The secondary task was to count the number of times that a laterally placed distractor fruit was highlighted, and then report this count at the end of the trial. A warning tone was given and then the subject was required to start movement when the central target fruit was highlighted. In a single target control condition, which allows baseline assessment, the secondary task did not have to be performed. (B) The amplitude of peak grip aperture to the target was influenced by the type of distractor. If the central target fruit was a cherry (bottom), the amplitude of peak grip aperture was greater when the distractor fruit was an apple (dotted line) than when the cherry was grasped in isolation (solid line). Conversely, the amplitude of peak grip aperture for the grasp of an apple (top) was reduced when the lateral fruit was a mandarin (center) or a cherry (bottom) compared with when the apple was grasped in isolation. Furthermore, when the subject had to perform the secondary "counting" task when the primary and the secondary task were executed upon the same object, interference effects were not revealed (not shown). Modified from Castiello, 1999.

The most consistent finding was that interference effects were evident only when eye movements were fixed on the target and when the distractor was moving. Moreover, interference effects in the grasp, but not the reaching parameters, were observed only when the distractor was smaller than the target and could potentially become the target by a change in illumination. This implies that the intrinsic features of a distractor (such as its size), and not simply the extrinsic feature of location, may elicit competing responses and thus have a selective influence on kinematic parameterization.

A recent study by Similä and McIntosh (2015) was designed to test whether selecting an object for perception prioritizes it for action. In one experiment, participants attended covertly to a flanker on one or other side of a fixated target, prior to reaching for that target, which occasionally jumped, after reach onset, to the attended or non-attended side. Participants corrected their reaches for almost all target jumps. In a second experiment, participants were required to covertly monitor the flanker during reaching. This concurrent perceptual task globally reduced correction behavior, indicating that perception and action share a common attentional resource. Corrections were especially unlikely toward the attended side. The authors assumed that perceptual attention primed an action toward the attended location and that the participant inhibited this primed action. The data thus imply that perceptual selection constrains online action guidance.

Along these lines LeBlanc and Westwood (2014) employed a sequential task paradigm to test the hypotheses that automatic motor plans are created when objects are attended to. They had participants reach to and grasp a target object followed by either a perceptual or motor task associated with a second object. Participants would either have to verbally report the size of the second object or reach out and grasp it. They hypothesized that because sequential tasks are processed as a whole (Henry & Rogers, 1960; Hesse & Deubel, 2010), that peak grip aperture would be scaled towards the size of the second object. Results indicated that the expected interference effects were only produced when participants attended to the size of the second object for a verbal estimation, but not when planning an action to the second object.

Attention, as a mechanism relevant to the representation of peripersonal space, has also been proposed by Tipper and co-workers (Howard & Tipper, 1997; Tipper et al., 1997; Tipper, Lortie, & Baylis, 1992). They advanced the idea that a reaching action is defined by the direction and distance between the origin of the hand and the target. Reaching movements are planned within a hand-centered frame of reference rather than a head or shoulder frame of reference (Soechting & Flanders, 1989). Support for this claim comes from tasks where the location of target and distractor is not known in advance, and hence it is assumed that attention is distributed across the scene. In such situations distractors appear to compete for the control of action, interfering with response to the target. These effects are evident on the spatial path of the hand as it reaches to a target (see Figure 9.2a,b). The most important result of these studies is that the reach path veers towards, or away from the objects or locations which representations should supposedly be inhibited by selective attention. In other words, when attention is anchored on the target object, inhibition acts on the representation of a potential distractor (Tipper et al., 1998): both target and distractor evoke parallel actions, and competition between these simultaneous responses is resolved by inhibition mechanisms (Tipper et al., 1997). In a similar experiment, Welsh et al. (1999) demonstrated that hand movements actually deviated towards and not away from non-target objects. They suggest that this difference may exist because their non-target objects were not physical barriers

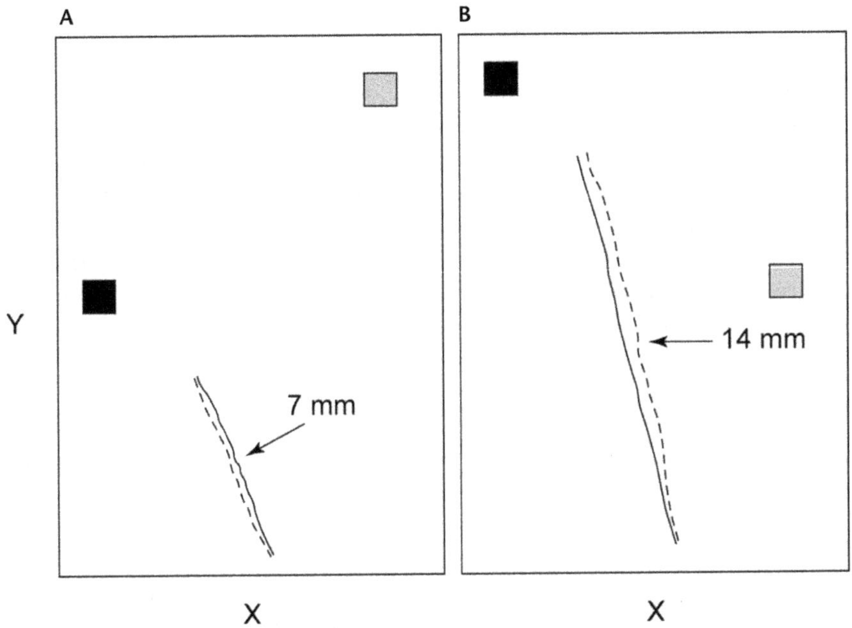

FIGURE 9.2 Evidence for interference effects in a reaching paradigm

Notes: Participants were required to initiate a reach as fast as possible after a cue appeared. Two stimuli, a target, which was either a blue or a green block, and a distractor, which was always a red block, were presented. A color cue for the target was presented on a computer monitor. Participants were instructed to initiate movement towards the target as soon as the cue was presented, but to withhold responses if the cue did not match the target color. The figure represents mean xy coordinates of reach paths to targets (black box) in the presence and absence of a distractor (shaded box). (A) A near-left reach with a far-right distractor is represented. A comparison is shown between the condition where the target is presented alone (dotted line) and the condition where the target is presented with a distractor (solid line). Note that the hand path deviates slightly (up to 7 mm) towards the distractor. (B) A far-left reach with a near-right distractor is represented. Note that in this case, the hand path clearly deviates away from the distractor (up to 14 mm). In a further manipulation, an interval of 300 milliseconds between the color cue and movement initiation was added (Tipper et al., 1997). Under these circumstances, interference effects were not evident. Modified from Castiello, 1999.

to the movement, whereas in many previous studies the non-target objects were intended to be obstacles to movement. Welsh et al. (1999) suggest that a deviation towards the non-target object may occur if actions to both objects are planned in parallel and parts of the non-target location "leak" into the movement plan for the target.

Overall, it is suggested that selective attention is an early mechanism for action selection (Cisek, 2007). Attentional shifts towards the distractor may not only select the relevant object characteristics, but also direct such information to the visuomotor channels for any required modification. Thus, whenever visual attention is allocated to an object, its volumetric and spatial parameters are computed for eventual motor actions such as reach-to-grasping. When visuo-spatial mechanisms are directed towards a distractor object, characteristics of that object may be computed and interfere with those already established for the target object.

A consequence of this argument is that the direction of attention mediates the conversions between sensory and motor systems (Baldauf & Deubel, 2010; Johansson et al., 2001; Stein, 1992). Directing attention towards the signals that are generated in different sensory areas allows binding of different properties for the determination of a specific object (Paillard, 1982). Thus, the direction of attention towards the distractor may select the sensorimotor associations that are required for that object. Interference emerges from competition for computing sensory coordinates for possible conversion to motor output.

Reach-to-Grasp: A Multisensory Experience

One of the classic roles attributed to multisensory integration is that of producing a unitary percept of objects (Newell, 2004). We can recognize a spoon not only by looking at it but also by touching it when it is out of view. We can recognize a species of bird by seeing it, by using visual information about its size, shape, and colors, and also by hearing its song. However, different sensory modalities are used in concert not only to perceive objects but also to represent actions (Fogassi & Gallese, 2004). In this respect, a first fundamental role of multisensory integration is to help us to plan and execute actions. Indeed, in our daily life most of the actions we perform rely on sensory information, and in order to act appropriately, we often have to process in parallel information arriving via more than one sensory modality. The act of kicking a ball, for example, requires the integration of visual, proprioceptive, and tactile modalities. Writing is another example of an action that, to be accomplished accurately, requires the integration of visual, proprioceptive, and tactile information. In addition, recognizing and understanding what other individuals are doing depends on multimodal information (Fogassi & Gallese, 2004). As an example, even without seeing a person pouring water into a glass, we can recognize this action from the sound made by the flowing liquid. Thus, the retrieval in our brain of the representation of a given object, individual, or action greatly benefits from information arriving through different and multiple sensory modalities.

To date, crossmodal links between vision, audition, sense of touch, and proprioception have been extensively documented for grasping an object by using hands (Aziz-Zadeh, Iacoboni, Zaidel, Wilson, & Mazziotta, 2004; Castiello, Bruno, Begliomini, Ansuini, & Grassi, 2010; Etzel, Gazzola, & Keysers, 2008; Gazzola, Aziz-Zadeh, & Keysers, 2006; Goodwin, Jenmalm, & Johansson, 1998; Jenmalm, Dahlstedt, & Johansson, 2000; Johansson & Westling, 1984; Klatzky & Lederman, 1987; Klatzky, Lederman, & Reed, 1987; Klatzky, Pai, & Krotkov, 2000; Patchay, Castiello, & Haggard, 2003; Patchay, Haggard, & Castiello, 2006; Van Beers, Baraduc, & Wolpert, 2002; Zahariev & MacKenzie, 2007).

Here I present a series of kinematic studies based on the interference paradigms mentioned above which demonstrate crossmodal links between vision and other modalities (audition, olfaction, and proprioception) during grasping. The results indicate a strong effect of the information evoked by distractors presented in different modalities on the posture assumed by the hand during a visually guided reach-to-grasp movement. Specifically, effects on hand shaping were found when presenting auditory, olfactory, and proprioceptive information that elicited the representation of objects which differed in size from the to-be-grasped object.

The Role of Tactile Information

The first attempt to investigate the multimodal nature of the visuomotor mechanism underlying grasping has been conducted by Gentilucci, Daprati, and Gangitano (1998). In their experiments, participants reached and grasped a visual target-object presented in different sizes (small, medium, and large target) with one hand, while holding another, an unseen distractor of different sizes (1 cm either smaller or greater than the target) in the other hand. The main finding was that the size of the distractor did affect the kinematics of the grasping executed with the other hand. In particular, maximal hand aperture decreased and increased when the distractor was smaller and larger than the visual target-object, respectively. However, the effect of tactile information was only observed when the visual target-object was small and manipulation was performed using the right hand. This rendered unclear what caused the effect.

In a series of subsequent investigations Patchay et al. (2003, 2006) confirmed and extended these results. By using a similar experimental paradigm, these authors found that maximum hand aperture for the visual target was proportional to the dimension of the distractor, which was manipulated proprioceptively with the other hand (Figure 9.3a,b). Furthermore, analogous patterns were observed when the distractor was manipulated with either the left or the right hand. Noticeably, the effect of tactile information occurred only when the

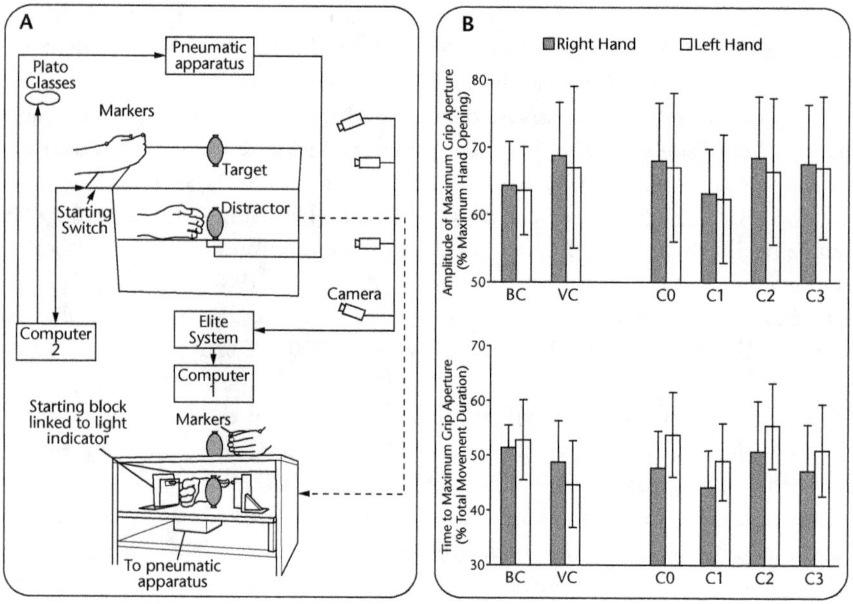

FIGURE 9.3 (A) Schematic representation of the experimental setup. (B) Mean values of amplitude of maximum grip aperture

Notes: (A) Schematic representation of the experimental setup (see the text for a detailed description): transverse view (upper panel) and front view (lower panel) of the work surface, showing the spatially coincident locations of the target and the distractor and the starting positions of both hands. (B) Mean values of amplitude of maximum grip aperture expressed as a percentage of participant's maximum hand opening and time to maximum grip aperture expressed as a percentage of total movement duration for the right and left hands. Error bars are plus and minus one standard deviation. BC, baseline control condition; VC, vibrating control condition; C0, control condition randomized with the different distractor sizes; C1, distractor smaller than target; C2, distractor same size as target; C3, distractor bigger than target. Modified from Patchay et al., 2003.

distractor was actively grasped. The effect was absent when the non-reaching hand received passive tactile and proprioceptive stimulation.

Overall these findings show that exploring a large and a small object activates the movement parameterization which corresponds to the size of that object, i.e., big and small hand aperture, respectively. This signifies that selection of the appropriate "grasp" motor plan for interacting with an object can be based on tactile cues. Therefore, the mechanism underlying the guidance of actions is not only sensitive to the information conveyed via vision but also via the sense of touch.

The Role of Auditory Information

When using hands to manipulate objects and interact with surfaces we generate contact sounds. These auditory events provide important information concerned with the interaction between the moving effectors and the acted-upon object. In particular, the contact sound signals both functional consequences and completion of manipulative actions. For instance, the "crash" sound associated with our hands breaking a walnut can be considered as a contact sound. Upon hearing the "crash," we become aware of having broken the shell, thus, we stop the walnut handling, and we bring the husk to our mouth.

Zahariev and MacKenzie (2007) have focused on the role played by contact sounds for the planning and the execution of hand grasping actions. In their experiments, participants reached towards and grasped a visual target (i.e., a wooden cube) either in the absence or in the presence of a "virtual" contact sound. The contact sound was delivered at the end of the reaching movement, when digits closed around the visual target-object. Their main finding was that reach duration was shorter for the contact sound compared to the no contact sound condition. This result was taken as an evidence of the effect that auditory information might exert on the organization of hand grasping movements. However, the specific reason of why the presence of a contact sound reduced reach duration leaving unchanged hand kinematics was unclear. Furthermore, the delivered contact sound corresponded to the sound of a cork popping out of bottle, a sound which is not normally generated when touching a wooden block. Therefore, the nature of the effect caused by the contact sound remained unexplained.

A subsequent investigation addressed this issue by adopting a similar procedure (Castiello et al., 2010). Here the sound produced by digits, which made contact with objects covered with different materials (i.e., aluminium, paper, string, wool) was recorded. In a first experiment, either one of the recorded contact sounds or a synthetic sound was administered to participants. The task was to reach towards and grasp a visual target-object. The foremost result was that the contact sound did affect movement kinematics. Specifically, both reach duration and the time of hand closure around the visual target decreased when the administered contact sound corresponded to the sound that was actually generated by the forthcoming contact with that visual target. Whereas, when the administered contact sound differed from that associated with the interaction between the hand and the visual target, both reach duration and the time of hand closure around the target increased. Therefore, hearing sounds generated during the end part of action, when the hand touched the visual target-object, had the ability to modulate the "grasp" motor plan selected for that specific target depending on the level of correspondence between the contact sound and the sound produced at touch. Further support to this proposal comes from a second experiment (Castiello et al., 2010). Participants were requested to reach towards and grasp a visual target with the upper and the lower part covered with different materials (e.g., wool and paper, respectively). This task was performed in the presence of a contact sound associated with the material covering one of the two parts of the visual target (e.g., "touching-wool" sound or

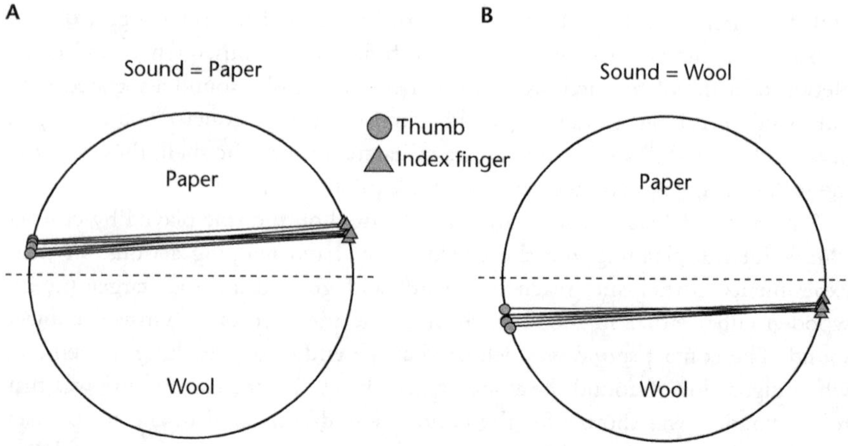

FIGURE 9.4 Graphical representation of contact points for the index finger and thumb in Experiment 2

Notes: The considered measure was in which half of the stimulus object the index finger and thumb contacted the object. This measure was calculated on the basis of the end trajectories of the fingers. A representative example of sound congruent grasp for the wool/paper material combination is presented (participant n. 9). The dashed line indicates the stimulus midline. Modified from Castiello et al., 2010.

"touching-paper" sound). Noticeably, when the presented contact sound was "touching-wool" and "touching-paper," the probability that participants grasped the visual-target object by the wool surface and the paper surface increased above chance, respectively (Figure 9.4a,b).

Altogether, these findings demonstrate that selection of the "grasp" motor plan for interacting with an object can be based on contact sounds. This evidence extends the sensitivity of the mechanism underlying the guidance of actions to the auditory information.

The Role of Chemosensory Information

An aspect which has been largely neglected in terms of the multisensory processes underlying hand grasping movements concerns chemosensory information. To date, only few studies performed in our laboratory considered reach-to-grasp movements performed towards a visual target-object in the presence of an olfactory stimulus (Castiello, Zucco, Parma, Ansuini & Tirindelli, 2006; Tubaldi, Ansuini, Demattè, Tirindelli, & Castiello, 2008; Tubaldi, Ansuini, Tirindelli, & Castiello, 2008). The olfactory stimulus could evoke an object of a smaller or larger dimension than the visual target-object (Figure 9.5a,b). In these circumstances, the maximum distance between the index finger and thumb (i.e., maximum hand aperture) was affected. If the olfactory stimulus evoked an object smaller than the target-object, then maximum hand aperture

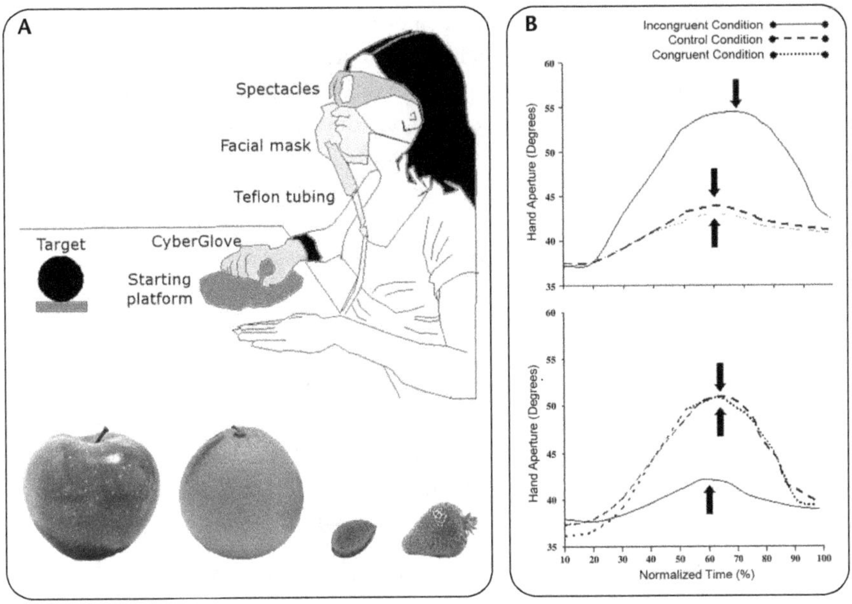

FIGURE 9.5 The experimental setup and the visual targets

Notes: (A) Labels indicate the parts composing the experimental setup. The visual targets defined as "large" were an apple and an orange, whereas those defined as "small" were an almond and a strawberry. (B) Graphical representation for the interaction type of target by condition by time. The panel represents how the time of maximum hand aperture (filled arrow) is delayed for an action toward a small target when olfactory information evokes an object requiring an incongruent large grasp and how it is anticipated for an action toward a large target when olfactory information evokes an object requiring an incongruent small grasp. The time course of maximum hand aperture is expressed in terms relative to the overall reach duration (%).

was smaller than when no odor was delivered. If the olfactory stimulus evoked an object larger than the target-object, then maximum hand aperture was larger than when grasping occurred in the absence of olfactory information (Castiello et al., 2006).

Altogether, the results of these studies indicate that the kinematic patterning of a reach-to-grasp movement was influenced by the "size" of an odor. Crucially, the motor plan evoked by the odor is surprisingly fine grained and when elicited can modulate both the pattern of angular excursion at the level of individual fingers' joints and the degree of synergic movement amongst digits.

More recently, research on the effect of chemosensory information on reach-to-grasp has been extended to flavor (Parma, Ghirardello, Tirindelli, & Castiello, 2011; Parma, Roverato, Ghirardello, Bulgheroni, Tirindelli, & Castiello, 2011). Participants were asked to drink a sip of strawberry juice and then reach and grasp an orange positioned in front of them. Thus, the object elicited by the flavored solution was a strawberry, a fruit that needs a precision grip (i.e., the opposition of the thumb to the index finger) in order to be grasped. Whereas, the visual target was an orange, eliciting a whole-hand grip (i.e., opposition of the thumb to all the other fingers) in order to be grasped. The results indicated either interference or facilitation effects on kinematic parameters such as the maximum hand aperture. In the first instance, maximum hand aperture was significantly smaller when the orange was grasped preceded by a size-incongruent (e.g., strawberry-flavored solution) than when it was grasped preceded by a size-congruent (e.g., apple-flavored solution) stimulation or water (i.e., control condition). In the second instance, maximum hand aperture was more size attuned when the orange was grasped preceded by a size-congruent stimulation (e.g., orange-flavored solution) than when it was grasped preceded by a size-incongruent (e.g., strawberry-flavored solution) or water stimulation. Therefore, the grasp plan evoked by the flavor of a strawberry (i.e., precision grip) influenced the performance of the whole-hand grip necessary to grasp the orange and successfully accomplish the task.

Reach-to-Grasp: An Experimental Window for Understanding Action Goals

Prehension experiments remain a powerful approach for motivating experiments on what constitutes an action goal (Ansuini, Santello, Massaccesi, & Castiello, 2006; Rosenbaum, Meulenbroek, Vaughan, & Jansen, 1999). Research on hand kinematics has proven insightful in revealing how specific kinematic landmarks modulate with respect to object properties, including object size, shape, texture, fragility, and weight. As recently reviewed, all these factors influence the kinematics of grasping (Castiello, 2005). The way an object is grasped, however, does not only depend exclusively on the properties of the object, but it is also influenced by the action's goal. A number of studies on hand kinematics have demonstrated that how the hand approaches an object depends on the manipulative action following object contact and grasping (Ansuini, Giosa, Turella, Altoe', & Castiello, 2008; Ansuini et al. 2006; Armbrüster & Spijkers, 2006; Cohen & Rosenbaum, 2004; Marteniuk, MacKenzie, Jeannerod, Athenes, & Dugas, 1987; Rosenbaum & Jorgensen, 1992; Rosenbaum, Vaughan, Barnes, & Jorgensen, 1992). For instance, Marteniuk et al. (1987) asked subjects to reach for an object and to either fit it into a similarly sized opening or to throw it away. Although the initial task requirements of reaching for the object were identical across the two conditions, kinematic analyses revealed substantial differences. Compared with reaching movements in the "throw condition," reaching movements performed in the "fit condition" revealed lower peak velocities and longer deceleration periods, suggestive of a more accurate approaching phase. More recently, the influence of different consecutive movements on initial reaching and prehension movement was also examined (Armbrüster & Spijkers, 2006). These researchers considered four after-grasp movements differing in direction and accuracy requirements: lifting, raising, throwing, and placing. Their results showed that movement parameters values were affected by the type of consecutive movement. Specifically, peak aperture was larger and peak deceleration was higher when the grasp was followed by either a throwing or a placing rather than a lift movement. Ansuini and colleagues (2006, 2008) added a level of complexity to this analysis considering whether and how each individual fingers' angular excursion varied depending on the accuracy requirements of the action that follows object's grasp. By asking participants to grasp the same object and either lift it and fitting it into a tight or a large niche, they showed that the degree of end-goal accuracy did also affect the shaping of individual digits during the approach phase (Figure 9.6a,b).

Similarly, choice of hand placement on the object has been shown to adapt to the upcoming intention. For example, participants place their thumb and index finger in a higher position when they grasp a bottle when the goal is to pour than when they grasp it with the goal of lifting it (Crajé, Lukos, Ansuini, Gordon, & Santello, 2011; see also Sartori, Straulino, & Castiello, 2011). Continuing on this analysis, it is also hand placement on an object that appears to be

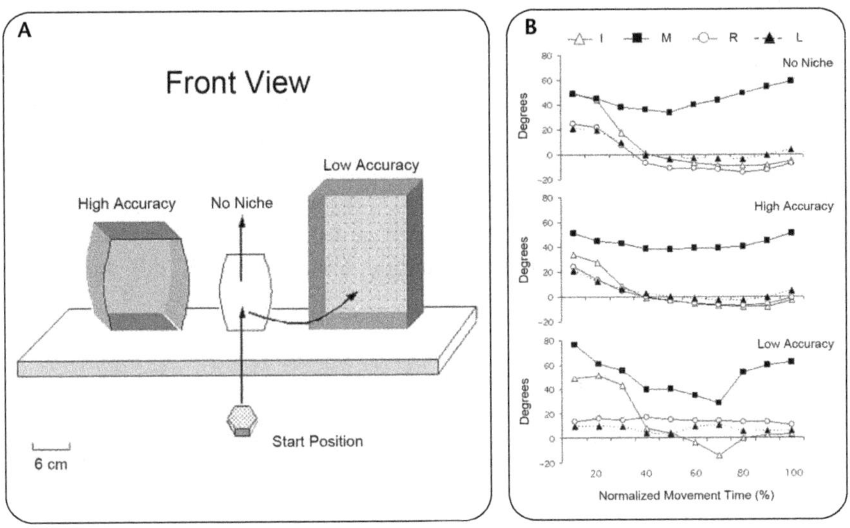

FIGURE 9.6 (A) Workspace and the three experimental conditions. (B) Time course of finger motion during reaching. Index (I), Middle (M), Ring (R) and Little (L) finger

Notes: (A) Experimental setup: workspace (front and top view, respectively) and the three experimental conditions [no-niche is equivalent only to the object lift action (arrow direction)]. (B) Time course of finger motion during reaching. Each trace denotes angular excursion of mcp joints of the index (I), middle (M), ring (R) and little (L) finger (subject 7) during one trial (no. 1) performed in the no-niche, high accuracy, and low accuracy conditions (A, B, and C, respectively). Modified from Ansuini et al., 2006.

sensitive to the planned action following object contact. Cohen and Rosenbaum (2004) asked participants to take hold of a cylindrical object and move it to a new position. They found that the grasp heights on the cylinder were inversely related to the height of this new position. This result was taken as the evidence that actors anticipate the positions they will comfortably adopt upon completing object transport movements. Altogether these findings suggest that the overarching action's goal influence action planning so that, although the to-be-grasped object is the same, different kinematic features are selected.

The "Social" Reach-to-Grasp

For many years cognitive psychology and cognitive neuroscience have focused on individual cognition developing paradigms apt at investigating individual minds in isolation. Applied to social interaction, this "isolation paradigm" approach has led to the paradox of studies in which social interaction is investigated by examining individuals physically isolated in separate compartments that do not allow face-to-face interaction between agents. These "isolation" experiments reflect the underlying assumption that social interaction is ultimately reducible to the understanding of those mental states that people entertain when they interact – or at least, believe to interact, with other agents (e.g., Jacob & Jeannerod, 2005).

Recent findings challenge this perspective by positing that social interaction is deeply rooted in the actions co-acted by agents interacting together (e.g., Becchio, Sartori & Castiello, 2010; Knoblich & Sebanz, 2008). In keeping with this vein, here I review findings from a novel line of research suggesting that social context shapes reach-to-grasp kinematics. I then consider how acting with a partner might influence action planning and control in different social contexts.

Individual Versus Social Intentions

Human beings help one another and request help from one another in many situations. Passing an object represents a basic helping gesture that humans perform in a variety of contexts. In an initial test of the hypothesis that social context influences action planning, Becchio, Sartori, Bulgheroni, and Castiello (2008a) sought to determine whether reach-to-grasp kinematics is sensitive to the social-end goal of passing an object to another person. For the individual condition, participants were requested to reach towards, grasp an object, and move it from one spatial location to another (Figure 9.7a). This condition was contrasted with a social condition in which participants were requested to reach towards and grasp the same object and pass it to a partner (Figure 9.7b). Results revealed that both the "reach-to-grasp" and the "place" phase of the movement were sensitive to the 'social' manipulation (see Figure 9.7c). Critically, no

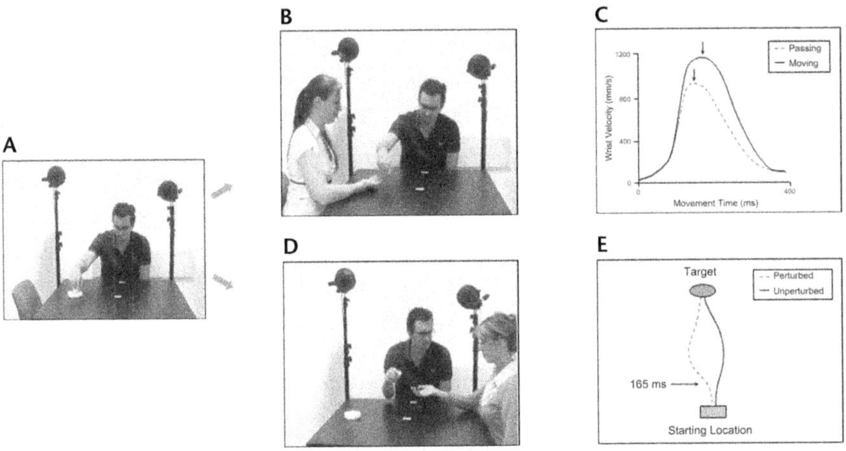

FIGURE 9.7 Reach-to-grasp movement task used to test whether action kinematics is sensitive to the social-end goal of passing an object to another person

Notes: Panel A represents the "individual" condition in which the participants reached towards, grasped the stimulus, and placed it within a concave base. Panel B represents the "social" condition in which following the action of grasping, the stimulus was handed to a partner. Panel C depicts the velocity profiles for the "individual" and the "social" conditions. For the "social" condition the amplitude of peak velocity was lower and it occurred earlier in time than for the "individual" condition. Panel D represents the "social" perturbation paradigm. Although the instruction for the participant was to place the object within the concave base located to her right, the sudden and unexpected "give me the object" request by a partner produced a dramatic deviation of the arm spatial trajectory towards the partner herself (Panel E). Modified from Becchio et al., 2010.

significant differences were found when comparing the individual condition with a condition in which a passive observer witnessed the action.

The same task context – passing an object to a partner – was used by Meulenbroek, Bosga, Hulstijn, and Miedl (2007) to assess the extent to which transfer of kinematic patterns takes place between co-actors involved in transferring objects. First, one of the two actors picked up a cylinder from a nearby location on the table and put it in the middle of the table. Subsequently, the other actor fetched the cylinder and repositioned it in a nearby target area. The authors varied both the size and the weight of the transferred cylinder. Time series analysis of the lifting heights revealed that the actor who fetched the object first showed a systematically larger surprise effect compared with the actor who was asked to transport the object after the first actor had done so. These findings indicate that in the context of a social interaction, kinematic parameters may be picked up during movement observation and integrated into a subsequent motor plan (see also, Ferri, Campione, Dalla Volta, Gianelli, & Gentilucci, 2011; Mason & Mackenzie, 2005).

Surprisingly, the influence of other people's actions is also evident even when it would be more effective for task performance to ignore others' actions (e.g., Sebanz, Knoblich, & Prinz, 2005). Using a perturbation paradigm, Sartori, Becchio, Bulgheroni, and Castiello (2009) investigate to which extent an unexpected social request might influence the kinematics of a preplanned action. Participants were instructed to reach towards, grasp an object, and put it in a container. On 20% of the trials, a perturbation occurred. Specifically, at the moment at which the participant started the action towards the object, unexpectedly a co-experimenter stretched out her right arm and unfolded her hand as to ask for the object (Figure 9.7D). Analysis of spatial trajectories revealed a significant veering in the arm trajectory occurring 165 ms following the perturbation (see Figure 9.7E). Strikingly, in some trials participants totally disregarded the instruction to place the target on the platform, and handed the object towards the experimenter. No perturbation effect was observed when a robotic agent replaced the human co-experimenter or when the perturbation consisted of a human arm movement conveying no communicative intention. These results suggest that, in contrast to a robotic or generic human arm movement, processing a social request had the potency to override the agent's initial motor plan. In line with an ecological perspective to social interaction (Marsh, Richardson, & Schmidt, 2009), it is tempting to interpret these findings as evidence that socially motivated actions (e.g., a request gesture) may act as social affordances that activate an appropriate motor response. Once this request has been processed, the activation of the appropriate response is almost automatic: ignoring the instruction to put the object in a container, participants deviated the trajectory path toward the human agent.

Cooperating and Competing

Cooperation and competition constitute two basic modes of social cognition that involve specific, and often different processes. For example, whereas both cooperation and competition necessitate predicting the action of co-actors, incorporating the action plan of a co-actor with one's own action plan might be beneficial for cooperative, but not for competitive situations. In cooperative contexts this process would allow individuals to adjust the timing of their actions as to achieve a common outcome in real time (Sebanz & Knoblich, 2009). In contrast, in competitive context it might be more beneficial not to incorporate the action plan of an opponent, as this would slow down responses and consequently competitive performance (De Bruijn, Miedl, & Bekkering, 2008).

To test the influence that cooperative and competitive contexts might have on action, Georgiou, Becchio, Glover, and Castiello (2007) analyzed the kinematics of reach-to-grasp movements towards a wooden block in four different conditions: natural individual, fast individual, cooperative, and competitive. For the natural individual condition, participants reached towards, grasped, and moved the stimulus to the center of the working surface at a natural speed comparable to that connoting a cooperative movement. For the individual fast condition, participants performed the same action, but as fast as possible. Therefore at a speed comparable to that characterizing a competitive interaction (Figure 9.8A). For the cooperative and competitive condition participants were assigned in pairs. The cooperation task required them to reach and grasp their respective objects and to cooperate as to form a tower putting one object on top of the other in the middle of the working surface (Figure 9.8B). The competition task was similar except that they had to compete as to place their own object first in the middle of the working surface (Figure 9.8D). Results revealed specific kinematic patterns for cooperation and competition, which were distinct from similar actions performed by each participant in isolation (Figure 9.8C–E). As predicted, for the cooperation, but not for the competition task, a

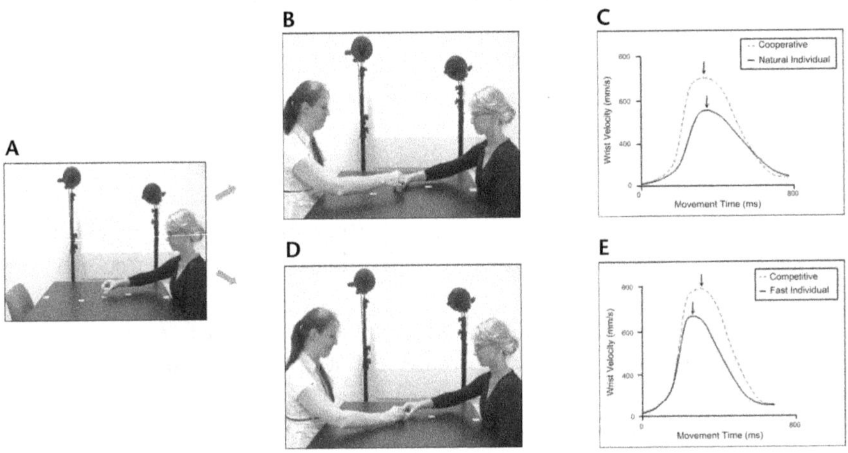

FIGURE 9.8 Experimental setup used for the studies considering cooperative and competitive behaviors

Notes: Panel A represents the individual condition in which participants were requested to reach towards, grasp, and place the stimulus in the middle of the working surface. Panel B represents the "cooperative" task in which the two participants were requested to grasp the object in front of them and form a tower in the middle of the table. The photograph represents the point at which the two participants tend to synchronize the action as to fulfill the task. Panel C represents the velocity profiles obtained for the "cooperative" and the "natural-individual" tasks. Note that deceleration time was longer for the cooperative than for the natural individual condition. The longer time spent in decelerating for the cooperative task suggests that performing an action towards an object with the intention to precisely locate it in a predetermined way for the purpose of cooperation produced a slower feedback guided decelerative phase. Panel D represents the competition task in which participants compete as to put their object in the bottom of the tower first (middle of the table). Panel E represents the velocity profiles obtained for the "competitive" and the "fast-individual" tasks. Note that the amplitude of peak velocity was higher and deceleration time was shorter for the "competitive" than for the "fast-individual" condition. The competition task seemed to elicit a more ballistic type of movement for which no great care at the level of honing phase was required. Modified from Becchio et al., 2010.

high level of correlation between key kinematical parameters of the two participants was found.

In these experiments, the attitude of the actor always matched the attitude of the partner. In real-life social interaction, however, this is not always the case. Suppose to be asked to cooperate with a partner clearly displaying the intention to compete or, vice versa, to have to compete with an opponent displaying the intention to cooperate. Would the incongruent attitude of the partner influence the implementation of your action?

To answer this question, Becchio, Sartori, Bulgheroni, and Castiello (2008b) manipulated the attitude of the partner during the reach-to-grasp phase of a social interaction. Subjects were requested to reach for their respective objects and then either to cooperate or to compete with a partner. The subjects were not aware that the partner was a semi-professional stage actor. For the congruent trials, she assumed an attitude (facial expression and body posture) congruent with the task instructions: cooperative for the cooperative task, competitive for the competitive task. For the incongruent trials, her attitude was manifestly in contrast with the instruction: competitive for the cooperative task, cooperative for the competitive task. Results revealed a significant "mismatch effect" on the agent kinematics: cooperating with a partner displaying the intention to compete rendered the agent's action more competitive. The opposite effect emerged when competing with a partner displaying the intention to cooperate: the kinematic pattern of the agent became similar to a cooperative pattern.

These findings point to the importance of social cues in face-to-face interaction. Facial expression and body posture represent powerful cues for intention which can either support or compromise cooperation. Reversal of kinematic patterning during incongruent trials might indicate that participants used these cues to infer the partner's attitude. This led to a change in their kinematics. In this interpretation, the changes in the agents' kinematics would be mediated by the representation of the partner's attitude. Alternatively, reversal in kinematic patterning might be interpreted as a result of non-intentional interpersonal alignment, by which individuals acting together tend to automatically coordinate their goal-directed actions (Marsh et al., 2009). In this scenario, cooperative and competitive pattern would emerge via a dynamic coupling rather than through cognitive-mediated processes such as mental state attribution.

Communicating with Others

Whereas speech is the most obvious signal that typifies communication in humans, there are many other signals that humans use to communicate. For example, the action of touching one's earlobe – which is per se non-communicative – could become a communicative signal in the context of a poker card game, when two players agree that touching the earlobe means:

"Drop out the current hand." Sartori, Becchio, Bara, and Castiello (2009) employed kinematic methods to investigate whether acting in a communicative context places special demands on action planning. They asked participants to perform the same goal-directed action in two different contexts which were operationalized through an individual and a communicative task. In the individual task, participants were requested to reach towards, grasp, and lift either a blue or a green spherical object according to one of five predetermined sequences. The communicative task was identical to the individual task except that participants executed the sequence with a communicative intent. Each of the sequences of blue and green spheres represented a different meaning in a sort of simplified Morse code. Participants were asked to select a meaning (and thus a sequence) and to communicate it to a partner by lifting the spheres in the predetermined order. Based on a conversion table, the partner had to interpret the meaning of the communicated sequence. Results revealed how the imposition of a communicative meaning was not neutral with respect to action kinematics: although the to-be-grasped object remained the same, movements performed for the "communicative" condition were characterized by a more careful approach to the object compared to the "individual" condition. This finding might be regarded as indicative of an adaptation to a partner: when lifting the object was executed with a purely individual intention, the object could be grasped in whatever orientation without compromising the goal of the action. When the lifting action was performed to show the object to another person, then a more careful determination of contact points might be desirable as to optimize the viewing of the object by the partner. Results obtained for a second experiment, in which a control condition implying the presence of a blindfolded agent was used, confirmed this hypothesis. When the partner was blindfolded and adaptation to the partner was therefore pointless, no communicative effect on movement kinematics was observed.

In this connection, Newman-Norlund, Noordzij, Newman-Norlund, Volman, De Ruiter, Hagoort, and Toni (2009) have demonstrated that in the context of tacit communication game, adaptation of communicative behavior is partner-specific. The mere belief that one is communicating either with a child or an adult influences the way the communicative message is generated. In this experiment participants could not see each other. An interesting question for future research is whether during a face-to-face interaction adaptation of communicative behavior to a specific addressee extends to motor parameterization, i.e., whether in planning a communicative action, participants take into account specific information regarding the partner, such as eye level or head orientation.

Complementary Actions

Suppose that someone hands you a mug by its handle, you will automatically, without giving it a second thought, grab the mug using a whole-hand-grasp (the most appropriate grasping posture in this particular situation). The types of grasps adopted by the two interacting agents are incongruent, but they are nevertheless appropriate and complementary. As a working definition, complementary actions refer here to any form of social interaction wherein two (or more) individuals coordinate and mutually complete their incongruent actions, rather than performing imitative behaviors. In this respect, we can define as complementary affordances all the action possibilities in which suitable motor programs aiming to bring a joint goal to completion are activated (such as grasping and offering a coin when seeing an open hand in sign of request). Depending on its posture and context, therefore, an extended open hand could lead to a donation, to a handshake, or to an infinite number of other actions (Sartori et al., 2009). Activation of a complementary affordance is an important social tool, and it suggests that the automatic, rapid decoding of social cues influences intentional behavior in our everyday interactions, maximizing the efficiency of our responses. These examples illustrate the functional importance of complementary actions in the action–perception domain (Graf, Schütz-Bosbach, & Prinz, 2009), and they prompt us to examine the mechanisms involved in producing those responses.

Since the direct matching between observed and performed actions is thought to occur automatically, when we observe an action, which differs from our intended action we have to inhibit the tendency to imitate (Brass et al., 2005). While the mechanism leading to automatic imitation is relatively well-studied (Heyes, 2011), it is less clear how this automatic tendency is brought under control.

Evidence that task representation plays a pivotal role in shaping our actions has been provided by a series of studies using prehensile actions as a model (Newman-Norlund, Noordzij, Meulenbroek, & Bekkering, 2007; Newman-Norlund, van Schie, van Zuijlen, & Bekkering, 2007; Poljac, van Schie, & Bekkering, 2009; van Schie, van Waterschoot, & Bekkering, 2008) in which participants were explicitly instructed to prepare imitative or complementary actions after viewing a virtual actor grasp a manipulandum using either a precision grip (PG; i.e., opposition between the index finger and thumb) or a whole-hand grasp (WHG; i.e., opposition of the thumb with the other fingers). As expected, participants were faster at preparing their response in imitative contexts if the action to be carried out was congruent with what they had observed. When, instead, they were expected to carry out complementary actions, they responded faster when their action was dissimilar to the one they had just observed. The task representation (imitative vs. complementary) seems then to overrule long-term stimulus-response associations, influencing the way that action–perception coupling takes place. Further evidence concerning this flexible perception–action coupling was produced by a 3D motion capture study (Ocampo & Kritikos, 2010) in which reaching and grasping parameters of congruent responses were found to improve in imitative contexts, and incongruent responses were facilitated in complementary contexts. Consistent with these findings, Longo, Kosobud, and Bertenthal (2008) demonstrated that also the level of action coding can be modified (e.g., toward coding in terms of movements) depending on task requirements. Taken together, these data challenge the idea that action observation automatically leads to imitation in the observer and suggest that, depending on the context, observed actions can prime incongruent responses.

Recently, Sacheli et al. (Sacheli, Candidi, Pavone, Tidoni, & Aglioti, 2012; Sacheli, Tidoni, Pavone, Aglioti, & Candidi, 2013) showed that participants involved in face-to-face interactions can mutually adjust their prehensile actions in time and space even in the absence of instructions to either imitate or perform a complementary response. This demonstrates that priming does not strictly depend on task-constrains, and that humans might indeed be able to actively shift from imitative to complementary actions, thanks to neuro-cognitive processes that still need to be clarified.

Conclusions

As the title suggests, this chapter presents the hand as a theoretical vehicle for understanding the relationality between cognition and the motor system. Not only is the hand a key ingredient in communicative practice in general, but the hand serves as a device for a better understanding of cognitive mechanisms. In other words, the hand, an organ through which we explore our social and objective world, is integral to test the cognitive architecture of action. I believe that unveiling how attentional selection, multisensory integration, individual and social intentionality shape our action not only has important implications for a full understanding of action planning and on-line control, but also for the implementation of "tomorrow's" hands. It might be surprising that attempts to design a robot with functional hands have been met with little success throughout the history of humanoid robot production. And if the hands are the gateway to the world, it is clear that contemporary research is not yet in a position to provide us with any robots with meaningful active relations. Enactive robots do not yet exist. In comparison to animals and humans, the motor skills of today's best robots are limited and inflexible. But, if one were to come about, it would have to behave itself not as a deliberative and precisely calculating machine, but as skillful and dynamic entity in constant "cognitive" adjustment with its environment. A robot of this kind should approximate a reasonable spectrum of different perceptual, cognitive, and motor capabilities. The challenge is to determine how cognitive functions can be integrated into meaningful architectures and to test their functional limits.

References

Allport, A. (1987). Selection-for-action: some behavioural and physiological considerations of attention and action. In H. Heurer & A. F. Sanders (Eds.), *Perspectives on perception and action* (pp. 395–419). Hillsdale, NJ: Lawrence Erlbaum Associates Inc.

Ansuini, C., Giosa, L., Turella, L., Altoe', G., & Castiello, U. (2008). An object for an action, the same object for other actions: effects on hand shaping. *Experimental Brain Research, 185*(1), 111–119. doi:10.1007/s00221-007-1136-4.

Ansuini, C., Santello, M., Massaccesi, S., & Castiello, U. (2006). Effects of end-goal on hand shaping. *Journal of Neurophysiology, 95*, 2456–2465. doi:10.1152/jn.01107.2005.

Armbrüster, C., & Spijkers, W. (2006). Movement planning in prehension: do intended actions influence the initial reach and grasp movement? *Motor Control 10*(4), 311–329. doi:10.1123/mcj.10.4.311.

Aziz-Zadeh, L., Iacoboni, M., Zaidel, E., Wilson, S., & Mazziotta, J. (2004). Left hemisphere motor facilitation in response to manual action sounds. *European Journal of Neuroscience, 19*(9), 2609–2612. doi:10.1111/j.0953-816X.2004.03348.x.

Baldauf, D., & Deubel, H. (2010). Attentional landscapes in reaching and grasping. *Vision Research, 50*(11), 999–1013. doi:10.1016/j.visres.2010.02.008.

Becchio, C., Sartori, L., Bulgheroni, M., & Castiello, U. (2008a). Both your intention and mine are reflected in the kinematics of my reach-to-grasp movement. *Cognition, 106*(2), 894–912. doi:10.1016/j.cognition.2007.05.004.

Becchio, C., Sartori, L., Bulgheroni, M., & Castiello, U. (2008b). The case of Dr. Jekyll and Mr. Hyde: a kinematic study on social intention. *Consciousness and Cognition, 17*(3), 557–564. doi:10.1016/j.concog.2007.03.003.

Becchio, C., Sartori, L., & Castiello, U. (2010). Toward you: the social side of actions. *Current Directions In Psychological Science, 19*(3), 183–188.

Bonfiglioli, C., & Castiello, U. (1998). Dissociation of covert and overt spatial attention during prehension movements: selective interference effects. *Perception & Psychophysics, 60*(8), 1426–1440. doi:10.3758/BF03208003.

Brass, M., Derrfuss, J., & von Cramon, D. Y. (2005). The inhibition of imitative and overlearned responses: a functional double dissociation. *Neuropsychologia, 43*(1), 89–98. doi:10.1016/j.neuropsychologia.2004.06.018.

Castiello, U. (1996). Grasping a fruit: selection for action. *Journal of Experimental Psychology: Human Perception and Performance, 22*(3), 582. doi:10.1037//0096-1523.22.3.582.

Castiello, U. (1999). Mechanisms of selection for the control of hand action. *Trends in Cognitive Sciences, 3*, 264–271. doi:10.1016/S1364-6613(99)01346-7.

Castiello, U. (2005). The neuroscience of grasping. Nature Reviews. *Neuroscience, 6*, 726–736. doi:10.1038/nrn1744.

Castiello, U., Giordano Bruno, L., Begliomini, C., Ansuini, C., & Grassi, M. (2010). When ears drive hands: the influence of contact sound on reaching to grasp. *PLoS ONE, 5*(8), e12240. doi:10.1371/journal.pone.0012240.

Castiello, U., Zucco, A., Parma, V., Ansuini, C., & Tirindelli, R. (2006). Cross-modal interactions between olfaction and vision when grasping. *Chemical Senses, 31*, 665–671. doi:10.1093/chemse/bjl007.

Chieffi, S., Gentilucci, M., Allport, A., Sasso, E., & Rizzolatti, G. (1993). Study of selective reaching and grasping in a patient with unilateral parietal lesion. *Brain, 116*(5), 1119–1137. doi:10.1093/brain/116.5.1119.

Cisek, P. (2007). Cortical mechanisms of action selection: the affordance competition hypothesis. *Philosophical Transactions of the Royal Society of London B: Biological Sciences, 362*(1485), 1585–1599. doi:10.1098/rstb.2007.2054.

Cohen, R. G., & Rosenbaum, D. A. (2004). Where grasps are made reveals how grasps are planned: generation and recall of motor plans. *Experimental Brain Research, 157*(4), 486–495. doi:10.1007/s00221-004-1862-9.

Colman, H., Remington, R., & Kritikos, A. (2016). Grasping remaps the distribution of visuospatial attention and enhances competing action activation. *The Quarterly Journal of Experimental Psychology*, 1–17. doi:10.1080/17470218.2016.1214974.

Crajé, C., Lukos, J. R., Ansuini, C., Gordon, A. M., & Santello, M. (2011). The effects of task and content on digit placement on a bottle. *Experimental Brain Research, 212*(1), 119–124. doi:10.1007/s00221-011-2704-1.

De Bruijn, E. R., Miedl, S. F., & Bekkering, H. (2008). Fast responders have blinders on: ERP correlates of response inhibition in competition. *Cortex, 44*(5), 580–586. doi:10.1016/j.cortex.2007.09.003.

Etzel, J. A., Gazzola, V., & Keysers, C. (2008). Testing simulation theory with cross-modal multivariate classification of fMRI data. *PLoS ONE, 3*(11), e3690. doi:10.1371/journal.pone.0003690.

Ferri, F., Campione, G. C., Dalla Volta, R., Gianelli, C., & Gentilucci, M. (2011). Social requests and social affordances: how they affect the kinematics of motor sequences during interactions between conspecifics. *PLoS ONE, 6(1)*, e15855. doi:10.1371/journal.pone.0015855.

Fogassi, L., & Gallese, V. (2004). Action as binding key to multisensory integration. In G. Calvert, C. Spence, & B. E. Stein (Eds.), *The handbook of multisensory processes* (pp. 425–441). Cambridge, MA: MIT Press.

Gazzola, V., Aziz-Zadeh, L., & Keysers, C. (2006). Empathy and the somatotopic auditory mirror system in humans. *Current Biology, 16*(18), 1824–1829. doi:10.1016/j.cub.2006.07.072.

Gentilucci, M., Daprati, E., & Gangitano, M. (1998). Haptic information differentially interferes with visual analysis in reaching-grasping control and in perceptual processes. *Neuroreport, 9*(5), 887–891.

Georgiou, I., Becchio, C., Glover, S., & Castiello, U. (2007). Different action patterns for cooperative and competitive behaviour. *Cognition, 102*, 415–433. doi:10.1016/j.cognition.2006.01.008.

Goodwin, A. W., Jenmalm, P., & Johansson, R. S. (1998). Control of grip force when tilting objects: effect of curvature of grasped surfaces and applied tangential torque. *The Journal of Neuroscience, 18*(24), 10724–10734.

Graf, M., Schütz-Bosbach, S., & Prinz, W. (2009). *Motor involvement in action and object perception: similarity and complementarity. grounding sociality: neurons, minds, and culture.* New York: Psychology Press.

Henry, F. M., & Rogers, D. E. (1960). Increased response latency for complicated movements and a "memory drum" theory of neuromotor reaction. *Research Quarterly. American Association for Health, Physical Education and Recreation, 31*(3), 448–458. doi:10.1080/10671188.1960.10762052.

Hesse, C., & Deubel, H. (2010). Advance planning in sequential pick-and-place tasks. *Journal of Neurophysiology, 104*(1), 508–516. doi:10.1152/jn.00097.2010.

Heyes, C. (2011). Automatic imitation. *Psychological Bulletin, 137*(3), 463–483. doi:10.1037/a0022288.

Howard, L. A., & Tipper, S. P. (1997). Hand deviations away from visual cues: indirect evidence for inhibition. *Experimental Brain Research, 113*(1), 144–152. doi:10.1007/BF02454150.

Jackson, S. R., Jackson, G. M., & Rosicky, J. (1995). Are non-relevant objects represented in working memory? The effect of non-target objects on reach and grasp kinematics. *Experimental Brain Research, 102*(3), 519–530. doi:10.1007/BF00230656.

Jacob, P., & Jeannerod, M. (2005). The motor theory of social cognition: a critique. *Trends in Cognitive Sciences, 9*(1), 21–25. doi:10.1016/j.tics.2004.11.003.

Jenmalm, P., Dahlstedt, S., & Johansson, R. S. (2000). Visual and tactile information about object-curvature control fingertip forces and grasp kinematics in human dexterous manipulation. *Journal of Neurophysiology, 84*(6), 2984–2997.

Johansson, R. S., & Westling, G. (1984). Roles of glabrous skin receptors and sensorimotor memory in automatic control of precision grip when lifting rougher or more slippery objects. *Experimental Brain Research, 56*(3), 550–564.

Johansson, R. S., Westling, G., Bäckström, A., & Flanagan, J. R. (2001). Eye–hand coordination in object manipulation. *The Journal of Neuroscience, 21*(17), 6917–6932.

Klatzky, R. L., & Lederman, S. J. (1987). The intelligent hand. In: G. H. Bower (Ed.), *The psychology of learning and motivation* (pp. 121–151). San Diego: Academic Press.

Klatzky, R. L., Lederman, S. J., & Reed, C. (1987). There's more to touch than meets the eye: the salience of object attributes for haptics with and without vision. *Journal of Experimental Psychology: General, 116*(4), 356.

Klatzky, R. L., Pai, D. K., & Krotkov, E. P. (2000). Perception of material from contact sounds. *Presence: Teleoperators and Virtual Environments, 9*(4), 399–410. doi:10.1162/105474600566907.

Knoblich, G., & Sebanz, N. (2008). Evolving intentions for social interaction: from entrainment to joint action. *Philosophical Transactions of the Royal Society of London B: Biological Sciences, 363*(1499), 2021–2031. doi:10.1098/rstb.2008.0006.

Kritikos, A., Bennett, K. M. B., Dunai, J., & Castiello, U. (2000). Interference from distractors in reach-to-grasp movements. *The Quarterly Journal of Experimental Psychology: Section A, 53*, 131–151. doi:10.1080/027249800390709.

LeBlanc, K. A., & Westwood, D. A. (2014). Interference in sequential grasping: effects of action and perception. *Journal of Exercise, Movement, and Sport*, 46.

Longo, M. R., Kosobud, A., & Bertenthal, B. I. (2008). Automatic imitation of biomechanically possible and impossible actions: effects of priming movements versus goals. *Journal of Experimental Psychology: Human Perception and Performance, 34*(2), 489–501. doi:10.1037/0096-1523.34.2.489.

Marsh, K. L., Richardson, M. J., & Schmidt, R. C. (2009). Social connection through joint action and interpersonal coordination. *Topics in Cognitive Science, 1*(2), 320–339. doi:10.1111/j.1756-8765.2009.01022.x.

Marteniuk, R. G., MacKenzie, C. L., Jeannerod, M., Athenes, S., & Dugas, C. (1987). Constraints on human arm movement trajectories. *Canadian Journal of Psychology, 41*, 365–378.

Mason, A. H., & MacKenzie, C. L. (2005). Grip forces when passing an object to a partner. *Experimental Brain Research, 163*(2), 173–187. doi:10.1007/s00221-004-2157-x.

Meulenbroek, R. G., Bosga, J., Hulstijn, M., & Miedl, S. (2007). Joint-action coordination in transferring objects. *Experimental Brain Research, 180*(2), 333–343. doi:10.1007/s00221-007-0861-z.

Mon-Williams, M., Tresilian, J. R., Coppard, V. L., & Carson, R. G. (2001). The effect of obstacle position on reach-to-grasp movements. *Experimental Brain Research, 137*(3–4), 497–501. doi:10.1007/s002210100684.

Newell, F. N. (2004). Cross-modal object recognition. In G. Calvert, C. Spence, & B. E. Stein (Eds.), *The handbook of multisensory processes* (pp. 425–441). Cambridge, MA: MIT Press.

Newman-Norlund, R. D., Noordzij, M. L., Meulenbroek, R. G., & Bekkering, H. (2007). Exploring the brain basis of joint action: co-ordination of actions, goals and intentions. *Social Neuroscience, 2*(1), 48–65. doi:10.1080/17470910701224623.

Newman-Norlund, R. D., van Schie, H. T., van Zuijlen, A. M., & Bekkering, H. (2007). The mirror neuron system is more active during complementary compared with imitative action. *Nature Neuroscience, 10*(7). doi:10.1038/nn1911.

Newman-Norlund, S. E., Noordzij, M. L., Newman-Norlund, R. D., Volman, I. A., De Ruiter, J. P., Hagoort, P., & Toni, I. (2009). Recipient design in tacit communication. *Cognition, 111*(1), 46–54. doi:10.1016/j.cognition.2008.12.004.

Ocampo, B., & Kritikos, A. (2010). Placing actions in context: motor facilitation following observation of identical and non-identical manual acts. *Experimental Brain Research, 201*(4), 743–751. doi:10.1007/s00221-009-2089-6.

Paillard, J. (1982). The contribution of peripheral and central vision to visually guided reaching. In D. J. Ingle, M. A. Goodale, & R. J. W. Mansfield (Eds.), *Analysis of visual behavior* (pp. 367–385). Cambridge, MA: MIT Press.

Parma, V., Ghirardello, D., Tirindelli, R., & Castiello, U. (2011). Grasping a fruit: hands do what flavour says. *Appetite, 56*(2), 249–254. doi:10.1016/j.appet.2010.12.013.

Parma, V., Roverato, R., Ghirardello, D., Bulgheroni, M., Tirindelli, R., & Castiello, U. (2011). When flavor guides motor control: an effector independence study. *Experimental Brain Research, 212*, 339–346. doi:10.1007/s00221-011-2733-9.

Patchay, S., Castiello, U., & Haggard, P. (2003). A cross-modal interference effect in grasping objects. *Psychonomic Bulletin & Review, 10*, 924–931. doi:10.3758/BF03196553.

Patchay, S., Haggard, P., & Castiello, U. (2006). An object-centred reference frame for control of grasping: effects of grasping a distractor object on visuomotor control. *Experimental Brain Research, 170*, 532–542. doi:10.1007/s00221-005-0240-6.

Poljac, E., van Schie, H. T., & Bekkering, H. (2009). Understanding the flexibility of action–perception coupling. *Psychological Research PRPF, 73*(4), 578–586. doi:10.1007/s00426-009-0238-y.

Rosenbaum, D. A., & Jorgensen, M. J. (1992). Planning macroscopic aspects of manual control. *Human Movement Science, 11*(1), 61–69. doi:10.1016/0167-9457(92)90050-L.

Rosenbaum, D. A., Meulenbroek, R. G., Vaughan, J., & Jansen, C. (1999). Coordination of reaching and grasping by capitalizing on obstacle avoidance and other constraints. *Experimental Brain Research, 128*(1–2), 92–100. doi:10.1007/s002210050823.

Rosenbaum, D. A., Vaughan, J., Barnes, H. J., & Jorgensen, M. J. (1992). Time course of movement planning: selection of handgrips for object manipulation. *Journal of Experimental Psychology: Learning, Memory, and Cognition, 18*(5), 1058.

Sacheli, L. M., Candidi, M., Pavone, E. F., Tidoni, E., & Aglioti, S. M. (2012). And yet they act together: interpersonal perception modulates visuo-motor interference and mutual adjustments during a joint-grasping task. *PLoS ONE, 7*(11), e50223. doi:10.1371/journal.pone.0050223.

Sacheli, L. M., Tidoni, E., Pavone, E. F., Aglioti, S. M., & Candidi, M. (2013). Kinematics fingerprints of leader and follower role-taking during cooperative joint actions. *Experimental Brain Research, 226*(4), 473–486. doi:10.1007/s00221-013-3459-7.

Saling, M., Alberts, J., Stelmach, G. E., & Bloedel, J. R. (1998). Reach-to-grasp movements during obstacle avoidance. *Experimental Brain Research, 118*(2), 251–258. doi:10.1007/s002210050279.

Sartori, L., Becchio, C., Bara, B., & Castiello, U. (2009). Does the intention to communicate affect action kinematics? *Consciousness and Cognition, 18*(3), 766–772. doi:10.1016/j.concog.2009.06.004.

Sartori, L., Becchio, C., Bulgheroni, M., & Castiello, U. (2009). Modulation of the action control system by social intention: unexpected social requests override preplanned action. *Journal of Experimental Psychology-Human Perception And Performance, 35*(5), 1490–1500. doi:10.1037/a0015777.

Sartori, L., Straulino, E., & Castiello, U. (2011). How objects are grasped: the interplay between affordances and end-goals. *PLoS ONE, 6*(9), e25203. doi:10.1371/journal.pone.0025203.

Sebanz, N., & Knoblich, G. (2009). Prediction in joint action: what, when, and where. *Topics in Cognitive Science, 1*(2), 353–367. doi:10.1111/j.1756-8765.2009.01024.x.

Sebanz, N., Knoblich, G., & Prinz, W. (2005). How two share a task: corepresenting stimulus-response mappings. *Journal of Experimental Psychology: Human Perception and Performance, 31*(6), 1234. doi:10.1037/0096-1523.31.6.1234.

Similä, S. S., & McIntosh, R. D. (2015). Look where you're going! Perceptual attention constrains the online guidance of action. *Vision Research, 110*, 179–189. doi:10.1016/j.visres.2014.06.002.

Soechting, J. F., & Flanders, M. (1989). Sensorimotor representations for pointing to targets in three-dimensional space. *Journal of Neurophysiology, 62*(2), 582–594.

Stein, J. F. (1992). The representation of egocentric space in the posterior parietal cortex. *Behavioral and Brain Sciences, 15*(04), 691–700. doi:10.1017/S0140525X00072605.

Tipper, S. P., Howard, L. A., & Houghton, G. (1998). Action-based mechanisms of attention. *Philosophical Transactions of the Royal Society B: Biological Sciences, 353*(1373), 1385–1393. doi: 10.1098/rstb.1998.0292.

Tipper, S. P., Howard, L. A., & Jackson, S. R. (1997). Selective reaching to grasp: evidence for distractor interference effects. *Visual Cognition, 4*(1), 1–38. doi:10.1080/713756749.

Tipper, S. P., Lortie, C., & Baylis, G. C. (1992). Selective reaching: evidence for action-centered attention. *Journal of Experimental Psychology: Human Perception and Performance, 18*(4), 891.

Tresilian, J. R. (1998). Attention in action or obstruction of movement? A kinematic analysis of avoidance behavior in prehension. *Experimental Brain Research, 120*(3), 352–368. doi:10.1007/s002210050409.

Tubaldi, F. (2010). *Olfaction in action* (Unpublished master's thesis). University of Padova, Italy.

Tubaldi, F., Ansuini, C., Demattè, M. L., Tirindelli, R., & Castiello, U. (2008). Effects of olfactory stimuli on arm-reaching duration. *Chemical Senses, 33*(5), 433–440. doi:10.1093/chemse/bjn010.

Tubaldi, F., Ansuini, C., Tirindelli, R., & Castiello, U. (2008). Odours grab his hand but not hers. *Perception, 37*(12), 1886–1889. doi:10.1068/p6286.

van Beers, R. J., Baraduc, P., & Wolpert, D. M. (2002). Role of uncertainty in sensorimotor control. *Philosophical Transactions of the Royal Society B: Biological Sciences, 357*(1424), 1137–1145. doi:10.1098/rstb.2002.1101.

van Schie, H. T., van Waterschoot, B. M., & Bekkering, H. (2008). Understanding action beyond imitation: reversed compatibility effects of action observation in imitation and joint action. *Journal of Experimental Psychology: Human Perception and Performance, 34*(6), 1493. doi:10.1037/a0011750.

Welsh, T. N., Elliott, D., & Weeks, D. J. (1999). Hand deviations toward distractors evidence for response competition. *Experimental Brain Research, 127*(2), 207–212. doi:10.1007/s002210050790.

Zahariev, M. A., & MacKenzie, C. L. (2007). Grasping at "thin air": multimodal contact cues for reaching and grasping. *Experimental Brain Research, 180*(1), 69–84. doi:10.1007/s00221-006-0845-4.

10

SENSORIMOTOR INTEGRATION ASSOCIATED WITH TRANSPORT-APERTURE COORDINATION AND TOOL-MEDIATED REACHING

Miya K. Rand and Yury P. Shimansky

General Introduction

Reaching movements are one of the most important and frequently performed types of human motor behavior. They have been investigated for more than a century, and the number of related studies is still growing. Due to extensive everyday practice, arm-reaching movements are usually very well coordinated, meaning, according to an optimality approach (Shimansky 2007; Todorov 2004), which is the mainstream theory of movement control, that an accurate internal model (IM) of the arm as well as a controller that generates motor commands have been well developed. To achieve sufficiently high precision of motor task performance, the central nervous system (CNS) needs to acquire and process information about the arm and the target object. Reach to grasp is a typical and usually well-learned motor task that requires high precision of coordination between hand transport and control of grip aperture. This task serves as a convenient experimental model for studying sensorimotor coordination.

While information about the target object (mainly its spatial location and shape) is only visual, information about the arm (its location, velocity, and acceleration) comes from both vision and proprioception. Theoretically, the redundancy of information can help to improve the accuracy of determining the arm's state during movement and thus the precision of task performance. However, to take advantage of that potential redundancy, the CNS needs to realize that both visual and proprioceptive information is related to the same physical object (i.e., the arm). Consequently, a tight correspondence between visual and proprioceptive sensory signals must exist. An internal model of that correspondence is an important part of the entire IM of the controlled object's (i.e., arm's) dynamics. Very often motor tasks are performed using different tools. Tools usually change the dynamics of the controlled object (which involves the arm with the tool) significantly. To perform a motor task using an unfamiliar tool with sufficient precision, both the internal model of the controlled object and the controller need to be adjusted. Studying the characteristics of such adaptation and strategies used by the CNS for it is one of the most important areas of movement control and motor learning. In this chapter we will review control characteristics for both reach-to-grasp movement and reaching movement in tool-use, which are strongly influenced by the CNS's regulation of the precision of sensory information processing.

Visuomotor Coordination: Control of Hand Transport and Grip Aperture

Introduction

Reach-to-grasp movements are highly coordinated both spatially and temporally. Their coordination involves the regulation of the arm transport toward the target object (arm transport component) as well as the organization of the finger movements required to grasp and subsequently manipulate the object (grasp component, Jeannerod 1981, 1984). Traditionally, arm transport is thought to be planned based on analyzing visually the spatial properties of the target object such as its distance and direction, whereas the grasp is thought to be planned based on the visual analysis of the object's intrinsic properties such as its size and shape (Jeannerod et al. 1995). As a reach toward the target object progresses, the two components become more interdependent (Saling et al. 1996; Timmann et al. 1996), while the spatial organization of the fingers required to fit the object's shape becomes more stereotypic (Santello and Soechting 1998). Thus, processing visual information of both the object and the hand during reaching plays an important role in the successful control of reach-to-grasp movement.

Eye-Hand Coordination during Reach-to-Grasp Movement

Eye movements during reach-to-grasp arm movements are controlled to provide effective flow of visual information for the control of this manual action. Thus, eye movements during reach-to-grasp movements are well coordinated spatially and temporally with arm transport and grasping. Johansson and colleagues (2001) studied gaze locations during a sequential motor behavior that consists of reaching for a rectangular bar, grasping, lifting, and bringing it to another location. Prior to reach initiation and during reaching toward the bar, gazes were generally fixated on the would-be-grasp point on the bar. In other words, gazes were directed not to the hand but to the target object. During the final phase of reaching while the fingertips were still approaching the bar, gazes were either kept at the would-be-grasp point or started shifting to another location where the bar would be transported to. By the time of the onset of bar lift after completing grasping, gazes had left the grasp point of the bar in most trials. Thus, the gazes were generally fixated on the grasp object prior to and during arm reaching. Such gaze behavior is useful for gathering visual information about the object's properties for planning arm control before the movement onset. Fixating the gaze on the object during arm transport is similar to the case of reaching to a target (a phenomenon known as gaze anchoring, Neggers and Bekkering 2000; Rand and Stelmach 2010), which is useful to control reaching actions because it places the object in the fovea as the hand approaches it, and allows for effective use of visual feedback about both the object and the approaching hand to guide the hand to the object (Elliott 1988; Paillard 1982; Sanders and Knill 2004).

At the same time, eye movements are not well coordinated with reach-to-grasp movements when the same task is performed in the dark (i.e., without vision) in a memory-guided manner (Flanagan et al. 2008). Participants were able to complete the required task in the dark with a slower speed than in the case of full vision. However, gaze fixations were scattered over a large area and not directed to the would-be-grasp point, the bar, or the location where the bar would be transported to. Furthermore, the number of fixations was halved when performing in the dark compared to performance in the light. Thus, the tight spatial and temporal eye-hand coordination observed under full vision was lost during the memory-guided reach-to-grasp movements.

Dependence of Basic Kinematic Parameters on Vision Availability

While only a few studies examined the effects of presence or absence of haptic, auditory, and graphic cues relaying information about a finger contract with the target object on transport and grasp components of reach-to-grasp movements (e.g., Zahariev and MacKenzie, 2007), the role of vision in controlling reach-to-grasp movements has been studied extensively in the past by comparing performance under various degrees of availability of visual feedback (Jeannerod 1999). Generally, duration of the arm transport increases with full or partial removal of visual feedback during reach-to-grasp movement. The increase often occurs in the later part of arm transport where the arm is decelerating and/or grip aperture is closing toward the object (e.g., Berthier et al. 1996; Churchill et al. 2000; Gentilucci et al. 1994; Jakobson and Goodale 1991; Rand et al. 2007). Regarding the grasp component, maximum grip aperture during arm transport also increases in response to full or partial removal of visual feedback (e.g., see same references as above).

When full vision is not available, reach-to-grasp movements are controlled by using visual information that is available prior to movement initiation. Thus, extensive neurocomputational utilization of the internal representations of target object location and hand motion as well as proprioceptive information about the hand is required to compensate for the lack of visual information during coordination of arm transport with finger aperture regulation. The lengthening of reach duration without visual feedback is partly due to the increase in the neurocomputational load and partly due to the CNS applying a strategy of increasing the accuracy of reaching by reducing the speed. The increased maximum grip aperture without visual feedback reflects a strategy of increasing safety margin for grasping. Thus, visual feedback during movement execution plays an important role in the reduction of uncertainty about the properties of the target object and the arm/finger movements being executed, thereby allowing effective control of reach-to-grasp movements.

Effects of Mixing Vision and No-Vision Trials

With full vision, the size of maximum grip aperture during arm transport is carefully scaled relative to the target size (Smeets and Brenner 1999) as well as the speed of arm transport (Rand et al. 2006b; Wallace et al. 1990; Wing et al. 1986). Even though the maximum aperture size increases when vision is removed as described above, sequencing of two types of trials (with vs. without vision) in a set of successive trials further influences this regulation. The first factor is the homogeneity of successive trials. When trials of the same type are repeated in succession (i.e., a blocked condition), maximum grip aperture increases in without-vision trials compared to trials with visual feedback (Jakobson and Goodale 1991; Rand et al. 2007). Conversely, when the two types of vision-related trials are mixed in a randomized order, the size of maximum grip aperture is always increased as if all trials were performed without vision (Jakobson and Goodale 1991), apparently revealing the CNS's strategy to prepare the control of grasp formation for the worst-case scenario (i.e., without-vision trial) and use it during both types of trials in order to simplify the overall control of reach-to-grasp movements. This phenomenon, however, seems to happen only when transport duration is relatively short, so that the CNS cannot take advantage of the available visual feedback. Conversely, when transport duration is long enough, the aperture size is regulated depending on the availability of vision (Fukui and Inui 2006). Indeed, the critical nature of the availability of ample time for controlling grip aperture in heterogeneous successive trials is not limited to vision manipulation (Rand et al. 2004).

Other factors related to the effects of trial sequencing are repetition and predictability of the two vision-related types of trials (with vs. without vision). Goodale and colleagues compared three different arrangements of trials of different vision-related types: blocked, alternated, and randomized orders (Whitwell et al. 2008; Whitwell and Goodale 2009). With blocked trials, the trial series was homogeneous with respect to the trial type (i.e., all trials within the same block were either with vision or without vision) and also predictable regarding the availability of vision. With alternating trials, the trial series was heterogeneous with respect to the trial type, but predictable. With random trials, not only the trial series was heterogeneous, but the next trial type was unpredictable. The results showed that the difference of the maximum grip aperture between the with- and without-vision trials was larger under the homogeneous condition (i.e., blocked condition) than under the heterogeneous conditions (i.e., alternated and randomized conditions), whereas that difference was similar between the two heterogeneous conditions. This means that repetition of the same vision-related trial type more critically influences the aperture size than the predictability of vision-related type of upcoming trials. Therefore, to regulate the size of grip aperture, the CNS utilizes sensorimotor memory accumulated based on past trials rather than predictive knowledge about the availability of visual feedback in an upcoming trial. This feature appears stronger for the more optimized reach-to-grasp movements (i.e., performed using the dominant hand, Tang et al. 2014).

Transport and Aperture Coordination (TAC) and its Model

It has been suggested that the grasp formation (the grasp component) and kinematics of the arm transport (the transport component) during reach-to-grasp movements are coordinated by establishing a temporal linkage between them (Hoff and Arbib 1993; Jeannerod 1999; Jeannerod et al. 1995), so that the peak of aperture (and hence the initiation of aperture closure) occurs at a certain fixed timing in relation to the overall movement duration. In contrast, there has been another hypothesis that those two components are linked spatially; in particular, that grip aperture is modulated based on the distance between the hand and the target object (Haggard and Wing 1991, 1995). In more recent studies, the hand-target distance where aperture closure began during reaching was shown to be much more invariant than the duration of aperture closure during reaching (e.g., Alberts et al. 2002; Rand et al. 2004; Rand and Stelmach 2005; Wang and Stelmach 2001). These empirical observations have led to the view that the onset of aperture closure is triggered by spatial information based on the hand-target distance. Nevertheless, when we examined reach-to-grasp movements in a large range of reaching speed (slow, normal, fast, and maximal speed conditions), the hand-target distance at aperture closure initiation did vary across the conditions, although that distance remained relatively invariant within each speed condition (Rand et al. 2006b). Thus, hand-target distance is a spatial parameter that plays a dominant, although not exclusive role in the initiation of aperture closure during reach-to-grasp movements.

In search for an answer to the question of how the initiation of aperture closure during arm transport is triggered by the CNS, we used an approach based on the theoretical concept of control law (e.g., Naslin 1969) to elucidate the relationship between arm movement parameters and a decision to initiate aperture closure. A control law describes the dependence of control action (in this case the initiation of aperture closure during reaching) on the parameters of the motor plant, which in this case includes the dynamics of the arm and its relationship with the reach target. To this end, we have formulated a hypothesis that the initiation of aperture closure is governed by a certain control law, i.e., a function defined on certain state parameters of arm-target dynamics. Specifically, it is presumed that finger closure is initiated during the reach when the distance to target (D) crosses a threshold (D_{thr}) that is a function of grip aperture (G), hand velocity (V_{w}), and hand acceleration (A_{w}) measured at the time of finger closure initiation:

$$D = D_{\text{thr}} (G, V_{\text{w}}, A_{\text{w}}), \tag{1}$$

where D_{thr} is the distance-to-target threshold. It is assumed that aperture closure is not initiated while $D > D_{\text{thr}}$. This we called a transport-aperture coordination (TAC) model for grasp initiation.

For the aforementioned experimental data with a large range of speed conditions, the hand-target distance where aperture closure began indeed depended on hand velocity, peak aperture and hand acceleration (Figure 10.1, Rand et al. 2006b).

Next, we fitted the experimental data to the above model by a linear approximation. Specifically, the function D_{thr} $(G, V_{\text{w}}, A_{\text{w}})$ was presented as

$$D_{\text{thr}} (G, V_{\text{w}}, A_{\text{w}}) = D_0 + k_G G + k_{Vw} V_w + k_{Aw} A_w, \tag{2}$$

where D_0, k_G, k_{Vw}, and k_{Aw} are constant coefficients that are unknown parameters requiring identification. The unknown coefficients of the model of the control law were identified based on the standard method, namely by minimizing the least square deviation of the model's prediction from the actual, experimentally measured hand-target distance at which aperture closure began (i.e., aperture closure distance). The results confirmed the existence of a relationship among the four movement parameters (Rand et al. 2006b) described by the above model, meaning that a control law that triggers the aperture closure during arm transport is determined based on grip aperture, wrist velocity, and acceleration initiation.

This way of describing transport-aperture coordination is useful for quantifying effects of different experimental conditions, which often can be presented as a general shift of the hand–target distance threshold and interpreted as an increase or decrease in safety margin. An important advantage of this approach is that such shifts of safety margin cannot be expressed based on a difference in the value of just one kinematic parameter, because an individual parameter usually varies within a wide range between different trials and participants.

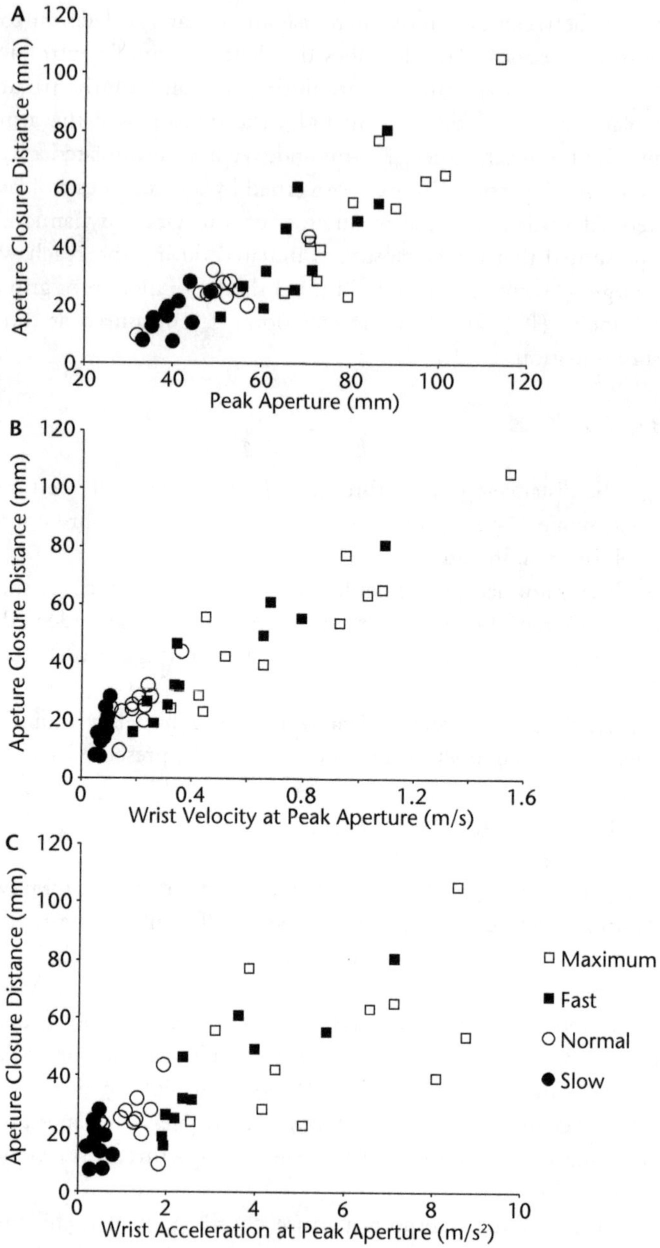

FIGURE 10.1 Scatter plots for hand-target distance for grasp initiation (i.e., aperture closure distance) as a function of the amplitude of aperture (A), the wrist velocity (B), and wrist acceleration (C) at the time of maximum aperture

Notes: Mean values for each participant are plotted for four speed conditions of reach-to-grasp movement (slow: filled circles; normal: open circles; fast: filled squares; and maximum: open squares). From Rand et al. 2006b.

Next, we will review the experiments that applied this model for examining the effects of vision on transport-aperture coordination (TAC) for aperture closure initiation.

Dependence of TAC for Aperture Closure Initiation on Vision

In the first experiment (Rand et al. 2007), participants reached for, grasped, and lifted a cylindrical target, which was placed 30 cm from the starting position along the midline of the trunk (front target condition) or at a 45° angle to the left of the midline of the trunk (left target condition). There were four visibility-related conditions where the target or the hand or both were either visible or not visible ("target visible, hand visible," "target not visible, hand visible," "target visible, hand not visible," and "target not visible, hand not visible").

To examine whether the relationship between D, G, V_w, and A_w (Eq. 2) estimated in trials from the "target visible, hand visible" condition was significantly different from the relationships in the other conditions, a residual error analysis was performed. Residual errors were calculated for each participant for each condition using the TAC model for grasp initiation as follows:

1 based on values from all participants and all trials, a multiple linear regression analysis was applied for the aperture closure distance (D) as a function of the three parameters (G, V_w, and A_w) for the "target visible, hand visible" condition combined with the front-target condition;
2 the intercept constant (k_0) and slope coefficients (k_G, k_{Vw}, and k_{Aw}) were determined from the regression analysis involving the three parameters (G, V_w, and A_w);
3 using the identified values of k_0, k_G, k_{Vw}, and k_{Aw}, the residual error (E) was calculated as $E = k_0 + k_G G + k_{Vw} V_w + k_{Aw} A_w - D$ for all trials for each condition separately. Next, these residual errors were compared across the three conditions.

With this calculation, the mean residual error across trials for the "target visible, hand visible" condition for the front target is zero because the residual errors for this condition are based on the coefficients obtained from the multiple regression analysis (Figure 10.2a). When either the hand or target was not visible, the residual errors shifted in the negative direction compared to that under full visibility. This means that when either the hand or target was not visible, the aperture closure distance was systematically increased compared to its value for the same amplitude of peak aperture, hand velocity, and acceleration under full visibility. This result implies that the hand-target distance threshold was increased, indicating an increase in the distance-related safety margin for aperture closure initiation when the hand or target is not visible. Under "target not visible, hand not visible" condition, the reduction of residual errors was maximal, indicating

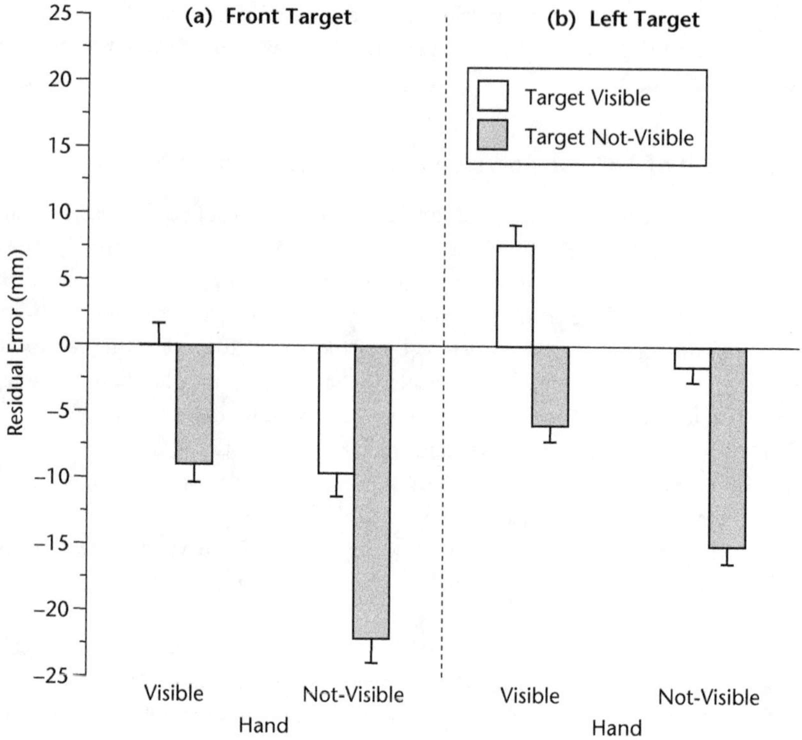

FIGURE 10.2 The effect of target object- and hand-visibility manipulation for the front (a) and left (b) target conditions

Notes: Mean residual errors for different visibility conditions are plotted. Mean residual error values for the front and left target conditions are calculated based on the "target visible, hand visible, front-target" condition (a). Error bars represent standard error. Modified from Rand et al. 2007.

that the distance-related safety margin for aperture closure initiation was further increased. Thus, transport-aperture coordination reflected in the distance-related safety margin is highly sensitive to the extent of visibility of the hand and the target object.

Additionally, the distance-related safety margin is also sensitive to subtle changes in task demands. A small extension of the wrist joint was needed in the last phase of arm transport to the front target in order to grasp it, whereas there was no need for such an extension when reach-to-grasp with the right hand was made to the left target. As a result, residual errors were generally shifted in the positive direction in reaches made to the left target (Figure 10.2b) compared to reaches made to the front target (Figure 10.2a), indicating that the distance-related safety margin was decreased for slightly simpler reach-to-grasp movements to the left target.

Dependence of Aperture Closure Initiation on Vision in Movement Disorders

We have applied a similar analysis to examine whether distance-related safety margin for aperture closure initiation is shifted in Parkinson's disease (PD) patients (Rand et al. 2006a, 2010a). PD is a progressive neurological disorder caused by the degeneration of dopamine-producing neurons in the basal ganglia of the brain, which leads to various difficulties in producing movements (such as slowness, tremor, rigidity, and postural instability). Since PD patients' reaching movements are usually slower than those of healthy individuals, it is difficult to assess the effect of PD on transport-aperture coordination based on only one kinematic parameter. However, a data analysis using the TAC model that describes the relationship among four movement parameters (hand-target distance (D), grip aperture (G), hand velocity (V_w), hand acceleration (A_w)) measured at the time of finger closure initiation can precisely reveal the effect of PD on transport-aperture coordination for aperture closure initiation.

In one experiment, reach-to-grasp movements were made to cylindrical target objects with full availability of vision in both PD patients and age-matched healthy controls (Rand et al. 2006a). For analysis using the TAC model (Eq. 1), we first applied a linear regression analysis to the associated four parameters measured from the controls and obtained the intercept and slope coefficients. Subsequently, we used these coefficients to calculate the residual errors based on the four parameters measured in PD patients. Next, we compared the mean residual errors of control participants (which was zero) with those of the patients. The results showed that, even though the patients showed slower transport speed and shorter hand-target distance at the time of aperture closure than the controls, residual errors did not significantly differ from those of the controls. That means that the hand-target distance threshold did not significantly differ between the patients and controls, implying that the distance-related safety margin for aperture closure initiation is unaffected by PD when vision is fully available.

In the next study, we examined the effects of PD on distance-related safety margin for aperture closure initiation during more complex trunk-assisted reach-to-grasp movements (Rand et al. 2010a). During the experiment, the target was placed beyond the reach of the participants' extended arms, and thus, a forward motion of the trunk had to be included to extend the reach to the target (with-trunk condition). Due to the involvement of the trunk's forward motion during reaching, coordination among different body segments (finger, arm, and trunk) becomes more complex than that without trunk involvement (Marteniuk and Bertram 2001; Wang and Stelmach 2001). In another condition, the target was placed within the reach of the extended arm, requiring no trunk motion (without-trunk condition). Furthermore, the availability of vision was also manipulated in four different ways ("target visible, hand visible," "target not visible, hand visible," "target visible, hand not visible," "target not visible, hand not visible") to examine the effects of visibility on the distance-related safety margin.

We applied a similar TAC model to the experimental data. Due to the increased complexity of the motor plant because of trunk involvement, two more parameters (trunk velocity (V_t) and trunk acceleration (A_t)) were added to the four original parameters as the TAC model of aperture closure initiation:

$$D = D_{thr}\,(G,\ V_w,\ A_w,\ V_t,\ A_t). \tag{3}$$

Based on this model, residual errors were calculated using the method described in the previous section. Namely, for each group and each trunk condition, intercept and slope coefficients resulting from multiple linear regression analysis applied to data of the "target visible, hand visible" condition were used to calculate residual errors of each of four vision-related conditions. The results showed that in both groups, the residual errors shifted further in the negative direction under with-trunk condition (Figure 10.3b) compared to without-trunk condition (Figure 10.3a), and thus the distance-related safety margin increased for reach-to-grasp that included trunk motion. When the hand or the target object was not visible, both groups increased the distance-related safety margin for grasp initiation compared to the case of full visibility.

The distance-related safety margin was the greatest under "target not visible, hand not visible" condition. The increase in the distance-related safety margin due to the absence of target object visibility or hand visibility was accentuated in the PD patients compared to that in the controls. The above findings imply that, when full vision is not available, PD patients have significant limitations regarding neural computations required for efficient utilization of the internal representations of target object location and hand motion as well as proprioceptive information about the hand to compensate for the lack of visual information during performance of complex, multi-segment goal-directed movements.

Expansion of the TAC Model to Include the Entire Movement

Since reach-to-grasp movements are performed very frequently, the skill of reaching to grasp must be very well learned, meaning that the movement is highly optimized. If the control system is optimal, there has to be tight coordination between independently controlled processes that are required to finish simultaneously with each other and have a definite final state. In the case of reach-to-grasp movements, there are two such processes: hand transport and grasping, the final state of which is determined by the target's location and size, respectively. Based on these considerations, we extended the transport-aperture coordination model for aperture-closure initiation to a similar, more

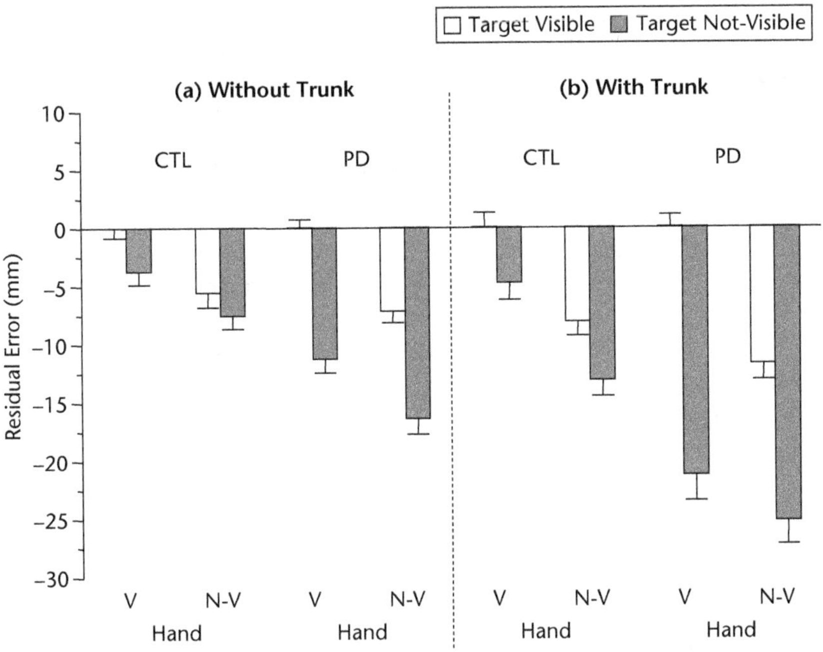

FIGURE 10.3 The effect of target object- and hand-visibility manipulation for the without-trunk (a) and with-trunk (b) conditions

Notes: Mean residual errors across all trials and all participants are plotted for all conditions. The manipulations of the target object and hand visibility include the "visible" conditions (V) and "not visible" conditions (N-V). Trunk involvement manipulations include the "without-trunk" condition (a) and "with-trunk" condition (b). Mean residual error values have been calculated for each trunk-related condition separately based on the "target visible, hand visible" condition. Error bars represent standard error. CTL and PD refer to healthy controls and Parkinson's disease patients, respectively. Modified from Rand et al. 2010a.

general equation (model) that fits entire reach-to-grasp movements for both aperture-opening and aperture-closure phases. On the basis of an optimal control theory (Naslin 1969), transport-aperture coordination were described (Rand et al. 2008, 2010b) by the following equation (Model 1).

$$D = D(A_w, V_w, G, A_g, V_g), \tag{4}$$

where V_g and A_g are grip aperture velocity and acceleration, respectively, and the other variables are the same as above (i.e., hand-target distance (D), grip aperture (G), hand velocity (V_w), hand acceleration (A_w), see Eq. 1). These six parameters completely describe the (one-dimensional) dynamics of hand transport and that of finger aperture. This equation can be viewed as a generalization of Eq. 1 from the point of aperture-closure initiation (where $V_g = 0$) to the entire movement. If the control of transport-aperture coordination is optimal (or almost optimal), the equation system representing the quantitative model of motor coordination is supposed to be correct for every point in time throughout the movement towards the target object.

Difference in TAC Precision Between Aperture-Opening and Closure Phases

We fitted the above six experimental parameters measured in an experiment where reach-to-grasp movements were made to a cylindrical target with full vision under four transport speed conditions (slow, normal, fast, maximal) combined with two target distance conditions (near vs. far target) to Model 1 (Rand et al. 2010b). Similar to the above case (Eq. 1), a linear approximation of Model 1 was used to fit the data during the aperture-opening phase and aperture-closure phase, separately, described as follows.

1. based on values from all data points of all trials in each condition and each participant, a multiple linear regression analysis was applied to the hand-target distance (D) as a function of the five parameters (A_w, V_w, G, A_g, V_g);
2. the intercept (k_0) and slope coefficients (k_{Aw}, k_{Vw}, k_G, k_{Ag}, k_{Vg}), were calculated;
3. by using the above coefficients, the residual error (E) was calculated by using the equation $E = k_0 + k_{Aw}\,Aw + k_{Vw}\,Vw + k_G\,G + k_{Ag}\,A_g + k_{Vg}\,V_g - D$ for all data points of all trials of each condition and in each participant.
4. the magnitude (root mean square [RMS]) of the residual errors across all trials was calculated for each condition and participant, and the error magnitudes were compared across conditions.

The results showed that the magnitude of residual errors during the aperture-closure phase (Figure 10.4b) was small across all experimental conditions and substantially smaller than that during the aperture-opening phase (Figure 10.4a), indicating that TAC was highly precise during the aperture-closure phase and consistent across different trials. Even when inter-individual variability was taken into account, the residual-error magnitude for the aperture-closure phase remained small (Figure 10.4b, black columns), indicating that TAC during aperture-closure phase was consistent across different participants. These results suggest that movement control is highly optimized during that phase. This is understandable because significant inaccuracy in transport-aperture coordination during it is likely to result in a costly error of target acquisition (e.g., upsetting and perhaps even damaging the target object).

For the aperture-opening phase (Figure 10.4a), in contrast, the magnitude of residual errors was generally large across all conditions. The magnitude was substantially larger for the far-target condition compared to the near-target condition, and also when inter-individual variability was taken into account (black

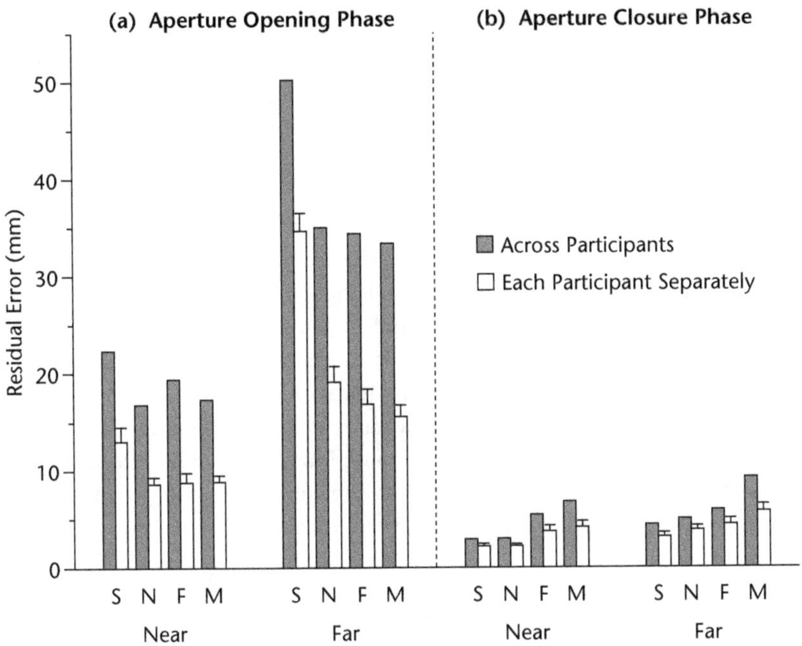

FIGURE 10.4 Residual error magnitude of TAC approximation with Model 1 across trials for the aperture-opening phase (a) and the aperture closure phase (b)

Notes: The magnitude (root mean square [RMS]) of the residual errors is plotted for each condition. S, N, F and M refer to low, normal, fast, and maximal condition, respectively. "Near" and "far" refer to the near- and far-transport distance conditions, respectively. The model's coefficients were calculated based on a dataset comprising all participants, all trials, and all data points within a movement phase (black columns) and based on a dataset comprising all trials and all data points for each participant separately (white columns). Error bars represent standard error. Modified from Rand et al. 2010b.

columns) compared the case in which it was not (white columns). These results indicate imprecise TAC during that phase, suggesting that movement control is less optimized.

Furthermore, the effects of transport speed manipulation were opposite between the aperture-opening and closure phases. The magnitude of residual errors during the aperture-closure phase was greater under faster-speed conditions (Figure 10.4b) which was consistent with the principle of speed-accuracy tradeoff (Fitts 1954; Schmidt et al. 1979). In contrast, the magnitude of residual errors during the aperture-opening phase was smaller for the faster-speed conditions (Figure 10.4a). This indicates that the precision of TAC was relaxed for slower movements during the initial part of reach-to-grasp movements, which was inconsistent with the principle of speed-accuracy tradeoff. These opposing movement-speed effects reveal a two-phase control strategy employed by the CNS: the precision of TAC was relaxed during the early movement phase and increased during the final phase (Shimansky and Rand 2013). We also observed a similar two-phase strategy in planar reaching movements (Rand and Shimansky 2013).

Dependence of the Characteristics of Each Phase on Vision

The theoretical insights regarding the two-phase strategy of reaching movement control can be used for predicting how the availability of visual information during the movement affects parameters of those phases. For example, without vision, the rate of sensory information inflow is considerably smaller compared to that under full vision. Consequently, it can be expected that without vision the final phase (during which relatively high precision of state estimation is required) will be prolonged and become more distinctive, because the time for sensory information integration must become longer for having the same precision of state estimation. The safety margin for successful grasping is also expected to be increased because of greater uncertainty with respect to hand-target relationship, so grasp initiation should occur at a longer distance to the target.

The above predictions can be verified based on the fact that the final phase of reach-to-grasp movements coincides with the aperture-closure phase. It turns out that the above predictions are readily supported by experimental data (Rand et al. 2007, 2010a). First, aperture closure time increases when the target is not visible compared to the condition of target visibility. It is also longer when the hand is not visible compared to that when the hand is visible (Rand et al. 2007). Second, aperture-closure distance becomes longer (i.e., grasp initiation occurs at a longer distance from the target) when the target is not visible. However, the aperture-closure distance does not significantly differ between the conditions of hand visibility and invisibility. Third, hand-target distance-related safety margin increases when either the target or the hand is not visible (Figure 10.2, Figure 10.3).

Conclusions

Reach-to-grasp coordination is one of the best examples of two-phase strategy utilization, where the aperture-closure phase is actually the final phase of movement control. The initiation of that phase is determined by the amount of sensory information available during the movement and requirements for the precision of target acquisition: the greater the required precision, the more information has to be accumulated and processed to meet that requirement.

Vision plays an important role in the control of reach-to-grasp movements. Namely, visual information helps to decrease uncertainty about the properties of the target object as well as the arm and finger configuration and motion. Generally, any reduction of the quality of vision results in slowness of reach-to-grasp movement and wider grip aperture during reach. However, the regulation of grip aperture size depending on the availability of vision is subject to other factors, such as availability of time and repetition of "vision" or "no-vision" condition in successive trials. Availability of vision also influences transport-aperture coordination. When vision is removed, the hand-target distance-related safety margin for aperture closure initiation during arm transport, which is determined by a function of grasp magnitude, arm velocity, and acceleration, is increased.

Tool Use: Implicit and Explicit Hand Representations

Introduction: Reaching with a Tool

Physical interactions with the environment often involve utilization of various tools for performing a wide variety of tasks. To accomplish a motor task that involves reach-type movements with a tool, one has to control the movement of the effective part of the tool (e.g., the head of a hammer, the cursor on a computer monitor, etc.). Thus, the CNS has to take into account the kinematic and dynamic relationship between the body part (e.g., hand) holding the tool and the effective part of the tool. Tool-use actions related to reaching can be categorized into at least two different types. One type corresponds to using mechanical tools, such as a hammer, a tennis racket, etc., which directly extend the reach of the hand(s). The tool's motion is governed by the laws of mechanics and can be sensed both through proprioception (e.g., one can feel the inertial, gravitational, and friction forces of the hammer) and vision (cf. Heuer et al. 2013). The other type of tool-use actions related to reaching is characterized by using electronic tools, which includes special means or channels for delivering feedback information about hand and tool motion to accomplish the reaching task, such as controlling a cursor on a computer monitor by moving a mouse with the hand. In this case, the kinematic and dynamic relationship between the hand and the effective part of the tool (i.e., cursor) is flexibly determined by the related computational program, and thus lacks mechanical transparency. If a cursor is displayed at a distance or even on a different plane, an additional coordinate transformation to link the hand and cursor positions is required. Consequently, in order to control reaching movements with a tool (both mechanical and electronic ones), one needs to learn the tool-determined visuomotor transformation through practice. Since electronic tools are frequently used in our daily life and occupational behavioral activities, it is important to understand how tool-use actions are learned and performed skillfully. The following sections describe visuomotor processes involved in learning of using electronic tools where the cursor on a monitor is controlled by moving the hand.

Implicit and Explicit Components of Learning to Reach with Electronic Tools

To excel in reaching with an electronic tool, one needs to acquire knowledge of tool-determined visuomotor transformation, the relation between the input (i.e., hand movement) and the output (i.e., cursor movement) of the tool, through learning. Once it is acquired, one can use that knowledge to adjust hand movements to control cursor movements as intended. Visuomotor transformation paradigms have been often used to study this type of learning. One of the most typical paradigms involves visuomotor rotation (Figure 10.5) as a tool-determined visuomotor transformation (e.g., Cunningham 1989; Heuer et al. 2013; Krakauer et al. 2000). Usually the task goal is to bring the cursor displayed on the monitor's screen (which is the feedback of hand movements) from the starting position to the target location by making a reaching movement on a horizontal plane. The direction of cursor motion is rotated by a certain angle (e.g., 45°) relative to the direction of hand motion. Through practice, participants need to learn the relationship between the cursor and hand motion in order to bring the cursor to the target. This type of learning can be mediated by implicit adaptation and/or explicit learning (e.g., Heuer et al. 2013; Mazzoni and Krakauer 2006; Morehead et al. 2015; Rand and Rentsch 2015; Taylor et al. 2014). Explicit learning (characterized by conscious awareness) is accomplished by making preplanned adjustments of reaching direction based on explicit knowledge of the transformation acquired through practice. For example, if one becomes aware of a 45° visuomotor rotation, one can

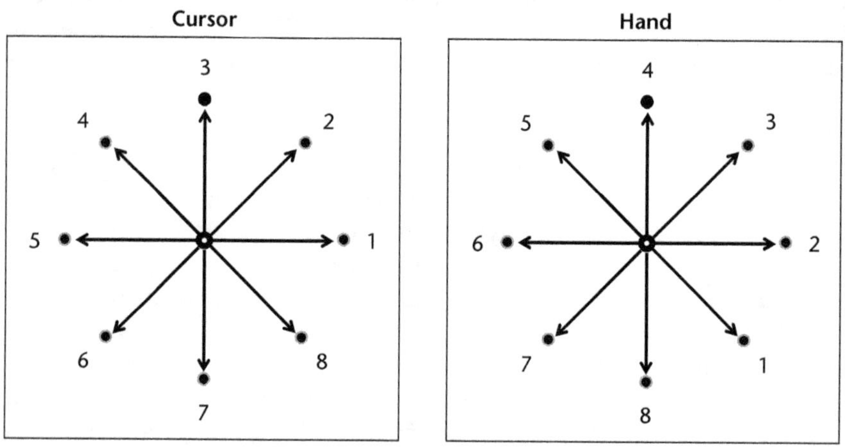

FIGURE 10.5 Rotational transformation

Notes: The left plot shows the typical target arrangement where a feedback cursor moves from the central starting point to the targets. The right plot shows the associated movements of the hand. Numbers refer to the different target positions and the associated correct end positions of the hand under a 45° visuomotor rotation.

intentionally adjust the reaching direction by –45° to make the cursor move towards the target. Conversely, implicit adaptation (without conscious awareness) is thought to be accomplished by (subconscious) development of an internal model of the visuomotor transformation through practice, during which the CNS gradually recalibrates the cursor-hand relationship based on that internal model to adjust the direction of reaching. Either one or both of these processes can compensate for the applied visuomotor rotation to achieve the goal. Both processes can run concurrently. Different neural substrates are thought to be involved in implicit and explicit adaptive processes (Granek et al. 2013; Pisella et al. 2009; Rand and Rentsch 2015).

Factors that Influence Implicit and Explicit Processes of Adaptation to Visuomotor Rotation

The extent of the involvement of implicit and explicit processes in learning a visuomotor rotation are largely altered by different factors related to experimental tasks, such as the amount of rotation (Granek et al. 2013; Heuer and Hegele 2008; Morehead et al. 2015; Rentsch and Rand 2014; Werner and Bock 2007), type of visual feedback (Hinder et al. 2010; Rand and Rentsch 2016; Rentsch and Rand 2014; Shabbott and Sainburg 2010; Taylor et al. 2014) and target visibility during reaching (Taylor and Ivry 2011). Other factors are individual specificities and aging. There are large inter-individual differences in the amount of explicit knowledge acquired through practice (Heuer and Hegele 2008), but participants who acquire greater explicit knowledge of the visuomotor rotation generally learn better the rotation with implicit and explicit processes combined (Heuer and Hegele 2008; Werner and Bock 2007). In terms of the aging factor, explicit learning is reduced in older adults compared to young adults, but implicit adaptation is not influenced by aging (Heuer et al. 2013).

These observations prompted a question: why do some individuals acquire through practice one type (e.g., explicit) of knowledge of tool-determined visuomotor transformation more than other individuals? One possible factor that alters the level of explicit knowledge acquired during practice is the difference in the magnitude of biases of visual and proprioceptive percepts of hand position during reaching movements under a visuomotor rotation. For example, if there is a stronger bias of the proprioceptively perceived hand position toward the visually perceived position of the feedback cursor, these perceived hand and cursor positions are less separated under the visuomotor rotation, thereby making it difficult to discriminate between them. As a result, acquisition of explicit knowledge of the visuomotor rotation would be impeded. A similar logic may be also applied to the level of implicit knowledge of the rotation acquired through practice.

Currently, however, there is no firm evidence supporting existence of separate implicit and explicit percepts of hand position. Based on the above-described ideas, Rand and Heuer have attempted examining implicit and explicit measures of perceived hand position under visuomotor rotation. Before reviewing their studies, we briefly summarize current understanding of multisensory integration in the next section.

Integration of Visually and Proprioceptively Sensed Hand Position Estimates

Without a tool, both visual and proprioceptive information serve to monitor hand position based on a prior assumption that the hand is just one object perceived through the two different sensory modalities. Signals coming from the two sensory sources are integrated to obtain a single estimate of hand position (van Beers et al. 1999). The mainstream current view on how two sources are integrated follows from the principle of an optimal integration, where the CNS weights each source of information to minimize the uncertainty in perceived position by minimizing the statistical variance of the combined perceptual estimate (e.g., Cheng et al. 2007; Ernst and Banks 2002; Ernst and Bülthoff 2004). According to that principle, the CNS assesses the precision of hand position estimates made based on visual and proprioceptive information, and the combination weights for the different sources of information are determined as inverse variances of the corresponding estimate error. For example, when the visual estimate of hand position is more (or less) precise, the variance of that estimate is low (or high). In the integration of these two sources, the visual estimate has a higher (or lower) weight and the proprioceptive estimate has a lower (or higher) weight, so that the variance of the combined estimate is lower than the variance of each single-sensory domain estimate. The benefit of such fusion of sensory estimates through an optimal integration is the maximal precision of the combined hand position estimate.

When a tool is used, visual information specifying the position of the effective part of the tool (e.g., cursor) and proprioceptive information specifying the position of the hand are spatially separated. They are related to each other only through the tool-specific hand-tool visuomotor transformation. However, even though the percept of the hand position and that of the cursor position refer to different objects, they are strongly biased toward each other due to their high inter-correlation across movements. This kind of inter-relation between visual and proprioceptive sensory signals, which does not result in a fused percept, but in distinct, although highly inter-correlated perceptual estimates, has been referred to as coupling (Ernst 2006; Parise et al. 2012). The idea of coupling can be also conceived in terms of weighted combinations of different sensory signals similar to sensory fusion of percepts related to the same object (Debats et al. 2017).

Experimental Paradigm to Study Implicit and Explicit Judgments of Hand Positions

Vision-proprioceptive integration of perceived hand position for reaching movements has been explored usually through explicit measurement of hand position percept (e.g., Izawa et al. 2012; Ladwig et al. 2013; Synofzik et al. 2008; van Beers et al. 1999; Zbib et al. 2016). However, Rand and Heuer have recently developed both explicit and implicit measures of judged hand positions to study the nature of coupling between vision and proprioception related to hand positions during tool use (Rand and Heuer 2013, 2016). Next, we briefly describe their methods.

During the experimental studies, seated participants held a stylus with their right hand and made reaching movements on a digitizer, while looking at the feedback cursor presented on the screen of a monitor (Figure 10.6a). The essence of the task was that in each trial, the target was briefly displayed at a randomly selected location in an area spanned from 60° left to 60° right of the 12 o'clock position (Figure 10.6b, 1st panel). Next, the participants made sequentially a forward-reaching movement from the central position to the remembered target location and a return movement back to the central position. Only during the forward movement (Figure 10.6b, 2nd panel), visual feedback was displayed; it was rotated by a certain angle (randomly selected in each trial from the following set: {−30°, −18°, −6°, 6°, 18°, 30°} or other similar sets depending on the particular experimental setup) relative to the direction of hand motion. Thus, the participants had to modify their movements online to bring the cursor to the remembered target location. Consequently, there was a difference between the cursor and hand positions (i.e., directions from the center) at the end of the forward movement (Figure 10.6b, 2nd panel). After the forward movement, the subsequent return movement was made to the central position without visual feedback (Figure 10.6b, 3rd panel).

After the return movement, the participants made an explicit judgment of the direction of either the remembered final cursor or the hand position at the end of the forward movement. For the judgment of cursor direction (Figure 10.6b, 4th panel, top), a short straight-line segment (served as a spatial direction marker) was moved on the monitor's screen at a constant speed clockwise or counter-clockwise (like a clock arm). The participant instructed the examiner to stop the line at the direction that matched the cursor direction at the end of the forward movement. For the judgment of hand direction (Figure 10.6b, 4th panel, bottom), the participant moved the stylus counter-clockwise or clockwise and stopped the stylus at the direction that matched the hand direction at the end of the forward movement. One half of the set of trials included the cursor-direction judgment and the other half included the hand-direction judgment.

For the analysis of *explicit judgments of sensed hand (or cursor) direction*, first the angular deviation of the explicitly judged hand (or cursor) direction from the

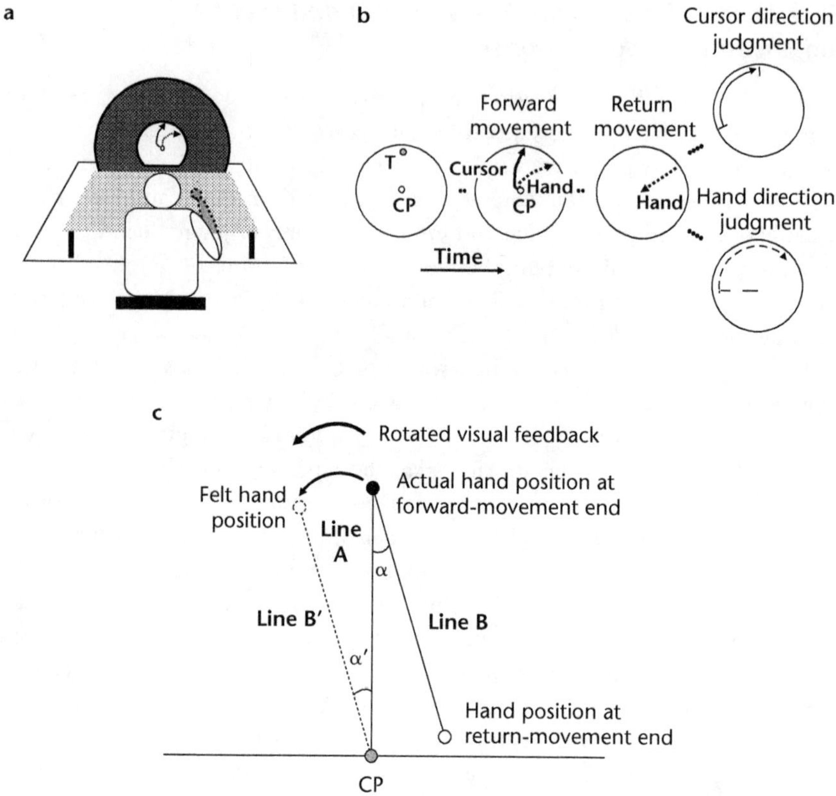

FIGURE 10.6 Behavioral task of a forward and return movement and analysis. (a) Experimental setting. (b) A schematic task procedure. (c) Implicit measure of hand direction

Notes: (b) CP and T refer to the central position and a target, respectively. Dashed arrow and solid arrow refer to the hand and cursor movements, respectively. (c) An illustration of relationship between the actual hand position (black circle) at the end of the forward movement and its felt hand position (dotted white circle), which is estimated based on the shift of the hand position at the end of the return movement (solid white circle) from the central position (CP, grey circle). Modified from Rand and Heuer 2013.

actual hand (or cursor) direction at the end of the forward movement was measured in each trial. Next, based on all trials of hand-direction (or cursor-direction) judgment made by each participant, the angular deviations were subjected to a linear regression in relation to visual feedback rotations. The slope parameter specified the strength of sensory coupling in terms of the biases of the judgments of hand (or cursor) directions relative to the visual-feedback rotation. The slope parameter was used as an explicit measure of the bias of sensed hand (or cursor) direction toward the direction of the cursor (or hand).

For the analysis of *implicit judgments of sensed hand direction*, first the angular deviation of the direction of the return movement from the direction of the forward movement was measured in each trial, that is, $\alpha' = \alpha$ (Figure 10.6c). This was used as an estimate of the rotation of the felt hand position relative to the actual hand position, namely, an implicit measure. This measuring technique exploits the existence of error propagation in successive aiming movements (Bock and Eckmiller 1986; Heuer and Sangals 1998; Heuer and Sülzenbrück 2012), in particular the propagation of errors that originate from visually induced deviations between the physical and the sensed positions of the hand (Holmes et al. 2004; Holmes and Spence 2005; Rossetti et al. 1995). In this case, the visually induced angular deviations of the sensed position of the hand from the actual position (angle α' in Figure 10.6c) occur in the forward movement due to the rotated visual feedback, and they are estimated from the angular error of the return movement (angle α in Figure 10.6c). Next, the angular deviations (α') from all trials of each participant were subjected to the same linear regression analysis in relation to the visual feedback rotations. The slope parameter was used as an implicit measure of the bias of the sensed hand direction toward the direction of the cursor.

Explicit Judgments of Hand and Cursor Directions

Explicit measures of the judged hand and cursor directions were analyzed to obtain features of coupling between visual and proprioceptive percepts in tool-use actions. The average angular deviation between the actual cursor direction and the explicitly judged (seen) cursor direction was characterized by a slightly negative slope in its linear relation to the feedback rotation angle (Figure 10.7a, cursor-explicit), indicating that the judged cursor direction was slightly biased toward the hand direction. In contrast, the average angular deviation between the actual hand direction and the explicitly judged (felt) hand direction had a steep positive slope in relation to the rotation angle (Figure 10.7a, hand-explicit), indicating that the judged hand direction was heavily biased toward the cursor direction. Thus, there are asymmetric mutual biases between visual and proprioceptive percepts with a strong visual dominance (Figure 10.7a, Rand and Heuer 2013, 2016). The feature of a strong visual dominance over proprioception is often found in tool-use actions as attention tends to be directed to the visual information (Collins et al. 2008; Reed et al. 2010), whereas conscious awareness of hand position becomes limited (Müsseler and Sutter 2009). Moreover, proprioceptive input tends to be functionally neglected during tool use (Heuer and Rapp 2012).

Another difference between explicit hand- and cursor-direction judgments is that the inter-trial variability of the former usually was about twice as large as that of the latter (Figure 10.7b, hand-explicit vs. cursor-explicit, Rand and

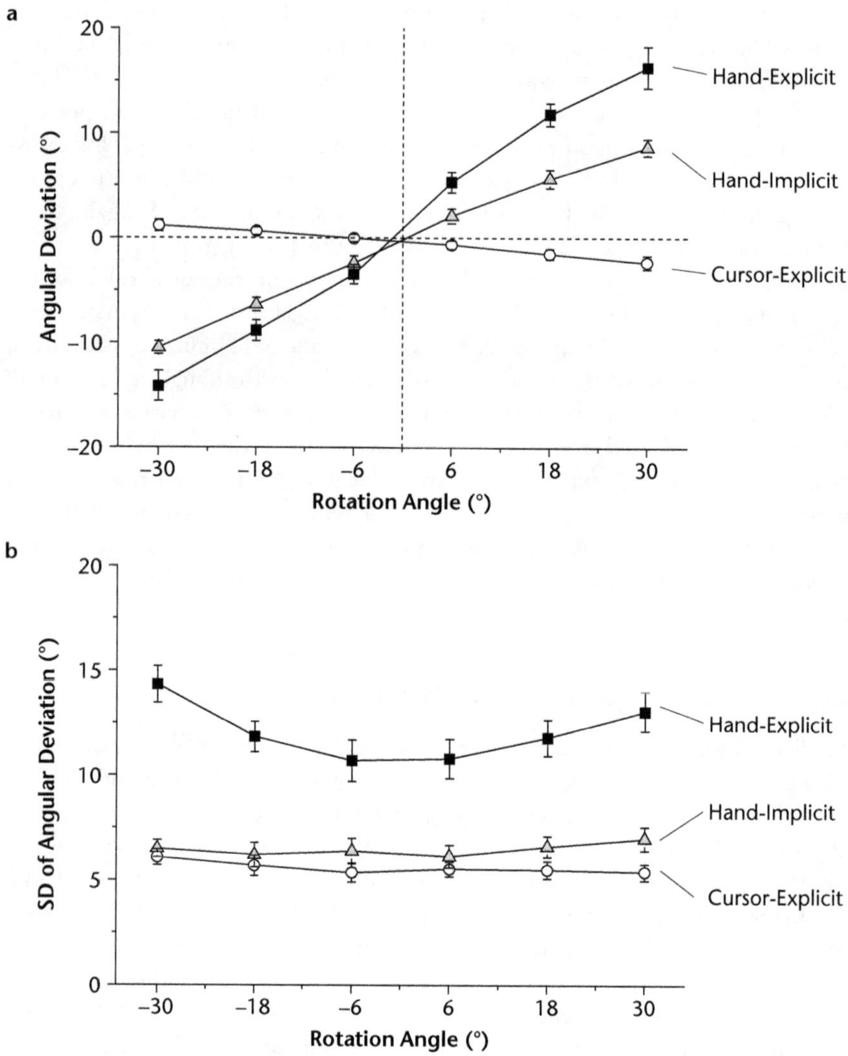

FIGURE 10.7 Implicit and explicit judgments. (a) Angular deviations between the judged direction and the actual direction. (b) Inter-trial variability (measured as SD) of judgments

Notes: Average values (± SE) across participants are plotted against visual feedback rotation angles. Modified from Rand and Heuer 2016.

Heuer 2013, 2016), indicating that the reliability of hand-direction judgments is substantially lower compared to that of cursor-direction judgments. From the perspective of the optimal integration principle, this reliability difference also contributed to the strong visual dominance in coupling between visual and proprioceptive information.

Characteristics of Implicit and Explicit Judgments of Hand Directions

Different characteristics between implicit and explicit judgments of hand directions have been observed (Rand and Heuer 2013, 2016, 2017, 2018). The main differences are (1) the bias of judged hand direction toward the cursor direction was about twice as strong for the explicit judgment compared with the implicit judgment (Figure 10.7a, hand-explicit vs. hand-implicit) and (2) that inter-trial variability of explicit judgments were substantially larger than that of implicit judgments (Figure 10.7b, hand-explicit vs. hand-implicit). Furthermore, the two types of judgments were uncorrelated.

Further differences between the implicit and explicit judgments of hand directions were found in the biases of judged hand direction toward the cursor direction by examining various factors, namely, effects of aging, effects of adaptation, effects of increasing reliability of proprioceptive signals at the end of the forward movement, effects of trial type repetition (whether a trial with explicit hand-direction judgment was immediately preceded by a trial of the same type or the alternative type, i.e., with explicit cursor-direction judgment), and effects of increasing the frequency of trials with explicit hand-direction judgment from low to even and from even to high with respect to the frequency of trials with explicit cursor-direction judgment (Table 10.1a). On the other hand, similarities between the implicit and explicit judgments of hand direction were also found in the biases of judged hand direction toward the cursor direction when the following two sets of comparisons were performed between trial subseries arranged differently with respect to the frequency and order of two types of trials (including a test of explicit hand-direction judgment or explicit cursor-direction judgment). In one set of comparisons, trials of the above two types were dispersed randomly within a subseries and the frequency of trials with explicit hand-direction judgment varied between the subseries from low to high with respect to the frequency of trials with cursor-direction judgment (Table 10.1b). In the other set of comparisons, the manner of arrangement (blocked, alternated, or randomized) of the two trial types within a subseries varied between different subseries (Table 10.1b).

Taken together, despite some commonalities found between the implicit and explicit judgments of hand directions, a number of discrepant features were revealed between them. These discrepancies support the notion of two different neural representations of hand direction (e.g., Dijkerman and de Haan, 2007; de Vignemont, 2010; Head and Holmes 1911; Paillard 1991), similar to cognitive (perceptual, explicit) and motor action-oriented (implicit) representations of visual stimuli (e.g., Bridgeman et al. 1979; Milner and Goodale 1995, 2008). Different representations of hand direction can also be based on different combinations of the available sources of information. For example, the explicit

TABLE 10.1 Differences and commonalities between implicit and explicit judgments of hand direction: Characteristics of bias of judged hand direction toward the cursor direction

Characteristics and factors examined	*Bias of explicit judgment*	*Bias of implicit judgment*
a Differences		
Strength of bias	Large	Small
Inter-trial variability of judged direction	Large	Small
Aging effect: older adults compared with young adults	Increase	No change
Altered reliability effect: increasing reliability of proprioceptive information	Decrease	No change
Effect of adaptation to a visuomotor rotation on inter-individual variability of bias strength compared with no adaptation	Decrease	No change
Effect of trial type repetition: bias strength in trials for which the preceding trial was of the same type, i.e., included explicit *hand*-direction judgment, compared to the bias in trials for which the preceding trial was of the alternative type, i.e., with explicit *cursor*-direction judgment	Increase	Decrease
Trial frequency effect: even frequency of trials with explicit hand-direction judgments (hand-50%:cursor-50%) compared with low frequency (hand-20%:cursor-80%).	No change	Decrease
Trial frequency effect: high frequency of trials with explicit hand-direction judgment (hand-80%:cursor-20%) compared with even frequency (hand-50%:cursor-50%).	Decrease	No change
b Commonalities		
Trial frequency effect: high frequency of trials with explicit hand-direction judgement (hand-80%:cursor-20%) compared with low frequency (hand-20%:cursor-80%).	Decrease	Decrease
Sequence effect regarding trial homogeneity: blocked (homogeneous) trial series compared with alternated (inhomogeneous) trial series with respect to two types of trials (explicit hand- and cursor-direction judgments).	No change	No change
Sequence effect regarding trial type predictability: randomized (unpredictable) trial-type order compared with alternated (predictable) trial-type order with respect to two types of trials (explicit hand- and cursor-direction judgments).	Decrease	Decrease

Note
A bias decrease (or increase) suggests a strengthened (or weakened) internal representation of hand movement direction.

representation could rely more on signals from various muscle and joint receptors, whereas the implicit representation could rely more on corollary discharge (Sperry 1950).

Contribution of Implicit and Explicit Percepts of Hand Positions to Learning of the Tool-Determined Visuomotor Transformation

What role do the implicit and explicit percepts of hand direction play in learning of the tool-determined visuomotor transformation? It seems reasonable to assume that the explicit measure taps a representation being used for explicit learning of the visuomotor transformation and that the implicit measure taps a representation being used for implicit learning. Then, the learning process would likely benefit when perceptual biases become weaker, so that the direction of hand (tool control input) and the direction of cursor (tool control output) become easier to discriminate. Conversely, the learning would likely suffer when perceptual biases become stronger, so that it is harder to discriminate between the directions of hand and cursor motion.

In support for the above hypothesis, the age-related strengthening of bias of the explicit measure of the sensed hand direction toward the cursor direction (Table 10.1a) was found matching well with the observed age-related deficits in both explicit discrimination between the directions of hand and cursor (Rand et al. 2013) and explicit learning of visuomotor rotation (Heuer and Hegele 2008; Heuer et al. 2013). Furthermore, the absence of age-related changes in the bias of the implicit measure (Table 10.1a) again matches well with the observed absences of age-related deficits both in implicit discrimination between the directions of hand and cursor (Rand et al. 2013) and implicit learning of visuomotor rotation (Heuer and Hegele 2008; Heuer et al. 2013). Despite these empirical observations, however, it still remains unclear how the implicit and explicit perceptual biases of hand direction toward the cursor direction are modulated during learning of a visuomotor transformation.

Recent studies showed that explicitly judged hand directions are changed relative to their actual hand directions after learning to reach under a visuomotor rotation (Izawa et al., 2012; Ostry and Gribble 2016; Synofzik et al. 2008; Zbib et al. 2016). Namely, as a result of learning, an adaptive change in the internal representation of the cursor-hand relationship is made, and thereby the judged direction of the hand is shifted toward the direction of the cursor (proprioceptive recalibration, Zbib et al. 2016), thus increasing the explicit perceptual bias. Furthermore, our recent study of sensory coupling (Rand and Heuer 2017) indicates that the proprioceptive recalibration resulting from the learning of a visuomotor rotation affects the explicit perceptual bias of hand direction toward the cursor direction, but not the implicit perceptual bias.

Summary

For both reach-to-grasp and tool-use experimental paradigms, movement control characteristics are strongly influenced by the CNS's regulation of the precision of sensory information processing. That precision is determined by the statistical reliability of sensory signals of different modalities and internal representations as information sources. Indeed, the precision of estimation of hand or tool spatial location and motion velocity as well as the location and shape of the target object can be crucial for successful performance of the motor task. That precision is maximized by optimal integration of visual and proprioceptive information as well as information derived from the internal models of the hand, the tool, and the target object. At the same time, during the initial phase of reaching movement, the precision of sensory information processing is not yet so critical and, therefore, it can be lowered to decrease the cost of information processing. Neurophysiological studies are required for understanding the neural mechanisms employed by the CNS for the regulation of the precision of sensory information processing and cross-modal sensory integration.

Acknowledgment

A part of the research reported in this chapter has been supported by the German Research Foundation (DFG), grant Ra 2183/1.

References

Alberts J.L., Saling M., Stelmach G.E. (2002) Alterations in transport path differentially affect temporal and spatial movement parameters. *Exp Brain Res* 143:417–425.

Berthier N.E., Clifton R.K., Gullapalli V., McCall D.D., Robin D. (1996) Visual information and object size in the control of reaching. *J Mot Behav* 28:187–197.

Bock O., Eckmiller R. (1986) Goal-directed arm movements in absence of visual guidance: evidence for amplitude rather than position control. *Exp Brain Res* 62:451–458.

Bridgeman B., Lewis S., Heit G., Nagle M. (1979) Relation between cognitive and motor-oriented systems of visual position perception. *J Exp Psychol Hum Percept Perform* 5:692–700.

Cheng K., Shettleworth S.J., Huttenlocher J., Rieser J.J. (2007) Bayesian integration of spatial information. *Psychol Bull* 133:625–637.

Churchill A., Hopkins B., Rönnqvist L., Vogt S. (2000) Vision of the hand and environmental context in human prehension. *Exp Brain Res* 134:81–89.

Collins T., Schicke T., Röder B. (2008) Action goal selection and motor planning can be dissociated by tool use. *Cognition* 109:363–371.

Cunningham H.A. (1989) Aiming error under transformed spatial mappings suggests a structure for visual-motor maps. *J Exp Psychol Hum Percept Perform* 15:493–506.

Debats N.B., Ernst M.O., Heuer H. (2017) Perceptual attraction in tool-use: evidence for a reliability-based weighting mechanism. *J Neurophysiol* 117:1569–1580.

de Vignemont, F. (2010) Body schema and body image – pros and cons. *Neuropsychologia* 48:669–680.

Dijkerman, H.C., De Haan, E.H.F. (2007) Somatosensory processes subserving perception and action. *Behav Brain Sci* 30:189–239.

Elliott D. (1988) The influence of visual target and limb information on manual aiming. *Can J Psychol* 42:57–68.

Ernst M.O. (2006) A Bayesian view on multimodal cue integration. In: Knoblich G., Thornton I.M., Grosjean M., Shiffrar M. (eds) *Human body perception from the inside out.* Oxford University Press, Oxford, pp. 105–131.

Ernst M.O., Banks M.S. (2002) Humans integrate visual and haptic information in a statistically optimal fashion. *Nature* 415:429–433.

Ernst M.O., Bülthoff H.H. (2004) Merging the senses into a robust percept. *Trends Cogn Sci* 8:162–169.

Fitts, P.M. (1954) The information capacity of the human motor system in controlling the amplitude of movement. *J Exp Psychol* 47:381–391.

Flanagan J.R., Terao Y., Johansson R.S. (2008) Gaze behavior when reaching to remembered targets. *J Neurophysiol* 100:1533–1543.

Fukui T., Inui, T. (2006) The effect of viewing the moving limb and target object during the early phase of movement on the online control of grasping. *Hum Mov Sci* 25:349–371.

Gentilucci M., Toni I., Chieffi S., Pavesi G. (1994) The role of proprioception in the control of prehension movements: a kinematic study in a peripherally deafferented patients and in normal subjects. *Exp Brain Res* 99:483–500.

Granek J.A., Pisella L., Stemberger J., Vighetto A., Rossetti Y., Sergio L.E. (2013) Decoupled visually-guided reaching in optic ataxia: differences in motor control between canonical and non-canonical orientations in space. *PLoS One* 8:e86138.

Haggard P., Wing A.M. (1991) Remote responses to perturbation in human prehension. *Neurosci Lett* 122:103–108.

Haggard P., Wing A.M. (1995) Coordinated responses following mechanical perturbation of the arm during prehension. *Exp Brain Res* 102:483–494.

Head H., Holmes H.G. (1911) Sensory disturbances from cerebral lesions. *Brain* 34:102–254.

Heuer H., Hegele M. (2008) Adaptation to visuo-motor rotations in younger and older adults. *Psychol Aging* 23:190–202.

Heuer H., Rapp K. (2012) Adaptation to novel visuo-motor transformations: further evidence of functional haptic neglect. *Exp Brain Res* 218:129–140.

Heuer H., Sangals J. (1998) Task-dependent mixtures of coordinate systems in visuomotor transformations. *Exp Brain Res* 119:224–236.

Heuer H., Sülzenbrück S. (2012) The influence of the dynamic transformation of a sliding lever on aiming errors. *Neurosci* 207:137–147.

Heuer H., Hegele M., Rand M.K. (2013) Age-related variations in the control of electronic tools. In: Schlick C., Frieling E., Wegge J. (eds) *Age-differentiated work systems.* Springer, Heidelberg, pp. 369–390.

Hinder M.R., Riek S., Tresilian J.R., de Rugy A., Carson R.G. (2010) Real-time error detection but not error correction drives automatic visuomotor adaptation. *Exp Brain Res* 201:191–207.

Hoff B., Arbib M.A. (1993) Models of trajectory formation and temporal interaction of reach and grasp. *J Mot Behav* 25:175–192.

Holmes N.P., Crozier G., Spence C. (2004) When mirrors lie: "visual capture" of arm position impairs reaching performance. *Cogn Affect Behav Neurosci* 4:193–200.

Holmes N.P., Spence C. (2005) Visual bias of unseen hand position with a mirror: spatial and temporal factors. *Exp Brain Res* 166:489–497.

Izawa J., Criscimagna-Hemminger S.E., Shadmehr R. (2012) Cerebellar contributions to reach adaptation and learning sensory consequences of action. *J Neurosci* 32:4230–4239.

Jakobson L.S., Goodale M.A. (1991) Factors affecting higher-order movement planning: a kinematic analysis of human prehension. *Exp Brain Res* 86:199–208.

Jeannerod M. (1981) Intersegmental coordination during reaching at natural visual objects. In: Long J., Baddeley A. (eds.) *Attention and performance IX*. Erlbaum, Hillsdale, pp. 153–168.

Jeannerod M. (1984) The timing of natural prehension movements. *J Mot Behav* 16:235–254.

Jeannerod M. (1999) Visuomotor channels: their integration in goal directed prehension. *Hum Mov Sci* 18:201–218.

Jeannerod M., Arbib M.A., Rizzolatti G., Sakata H. (1995) Grasping objects: the cortical mechanisms of visuomotor transformation. *Trends Neurosci* 18:314–320.

Johansson R.S., Westling G., Bäckström A., Flanagan J.R. (2001) Eye-hand coordination in object manipulation. *J Neurosci* 21:6917–6932.

Krakauer J.W., Pine Z.M., Ghilardi M.F., Ghez C. (2000) Learning of visuomotor transformations for vectorial planning of reaching trajectories. *J Neurosci* 20: 8916–8924.

Ladwig S., Sutter C., Müsseler J. (2013) Intra- and intermodal integration of discrepant visual and proprioceptive action effects. *Exp Brain Res* 231:457–468.

Marteniuk R.G., Bertram C.P. (2001) Contributions of gait and trunk movements to prehension: perspectives from world- and body-centered coordination. *Motor Control* 2:151–165.

Mazzoni P., Krakauer J.W. (2006) An implicit plan overrides an explicit strategy during visuomotor adaptation. *J Neurosci* 26:3642–3645.

Milner A.D., Goodale M.A. (1995) *The visual brain in action*. Oxford University Press, Oxford.

Milner A.D., Goodale M.A. (2008) Two visual systems re-viewed. *Neuropsychologia* 46:774–785.

Morehead J.R., Qasim S.E., Crossley M.J., Ivry R. (2015) Savings upon re-aiming in visuomotor adaptation. *J Neurosci* 35:14386–14396.

Müsseler J., Sutter C. (2009) Perceiving one's own movements when using a tool. *Conscious Cogn* 18:359–365.

Naslin P. (1969) *Essentials of optimal control*. Boston Technical Publishers, Cambridge.

Neggers S.F.W., Bekkering H. (2000) Ocular gaze is anchored to the target of an ongoing pointing movement. *J Neurophysiol* 83:639–651.

Ostry D.J., Gribble P.L. (2016) Sensory plasticity in human motor learning. *Trends Neurosci* 39:114–123.

Paillard J. (1982) The contribution of peripheral and central vision to visually guided reaching. In: Ingle D., Goodale M., Mansfield R. (eds.) *Analysis of visual behaviour*. MIT Press, Cambridge, pp. 367–385.

Paillard J. (1991) Motor and representational framing of space. In: Paillard J. (ed.) *Brain and space*. Oxford University Press, Oxford, pp. 163–182.

Parise C.V., Spence C., Ernst M.O. (2012) When correlation implies causation in multisensory integration. *Curr Biol* 22:46–49.

Pisella L., Sergio L., Blangero A., Torchin H., Vighetto A., Rossetti Y. (2009) Optic ataxia and the function of the dorsal stream: contributions to perception and action. *Neuropsychologia* 47:3033–3044.

Rand M.K., Heuer H. (2013) Implicit and explicit representations of hand position in tool use. *PLoS One* 8:e68471.

Rand M.K., Heuer H. (2016) Effects of reliability and global context on explicit and implicit measures of sensed hand position in cursor-control tasks. *Front Psychol* 6:2056.

Rand M.K., Heuer H. (2017) Contrasting effects of adaptation to a visuomotor rotation on explicit and implicit measures. *Psychol Res* DOI: 10.1007/s00426-017-0931-1.

Rand M.K., Heuer H. (2018) Dissociating explicit and implicit measures of sensed hand position in tool use: Effect of relative frequency of judging different objects. *Atten Percept Psychophys* 80:211–221. DOI: 10.3758/s13414-017-1438-y.

Rand M.K., Rentsch S. (2015) Gaze locations affect explicit process but not implicit process during visuomotor adaptation. *J Neurophysiol* 113:88–99.

Rand M.K., Rentsch S. (2016) Eye-hand coordination during visuomotor adaptation with different rotation angles: effects of terminal visual feedback. *PloS One* 11: e0164602.

Rand M.K., Shimansky Y.P. (2013) Two-phase strategy of neural control for planar reaching movements. I. XY coordination variability and its relation to end-point variability. *Exp Brain Res* 225:55–73.

Rand M.K., Stelmach G.E. (2005) Effect of orienting the finger opposition space on the control of reach-to-grasp movements. *J Mot Behav* 37:65–78.

Rand M.K., Stelmach G.E. (2010) Effects of hand termination and accuracy constraint on eye-hand coordination during sequential two-segment movements. *Exp Brain Res* 207:197–211.

Rand M.K., Lemay M., Squire L.M., Shimansky Y.P., Stelmach G.E. (2007) Role of vision in aperture closures control during reach-to-grasp movements. *Exp Brain Res* 181:447–460.

Rand M.K., Lemay M., Squire L.M., Shimansky Y.P., Stelmach G.E. (2010a) Control of aperture closure initiation during reach-to-grasp movements under manipulations of visual feedback and trunk involvement in Parkinson's disease. *Exp Brain Res* 201:509–525.

Rand M.K., Shimansky Y., Hossain A.B.M., Stelmach G.E. (2008) Quantitative model of transport-aperture coordination during reach-to-grasp movements. *Exp Brain Res* 188:263–274.

Rand M.K., Shimansky Y.P., Hossain A.B.M.I., Stelmach G.E. (2010b) Phase dependence of transport-aperture coordination variability reveals control strategy of reach-to-grasp movements. *Exp Brain Res* 207:49–63.

Rand M.K., Shimansky Y., Stelmach G.E., Bloedel J.R. (2004) Adaptation of reach-to-grasp movement in response to force perturbations. *Exp Brain Res* 154:50–65.

Rand M.K., Smiley-Oyen A.L., Shimansky Y.P., Bloedel J.R., Stelmach G.E. (2006a) Control of aperture closure during reach-to-grasp movements in Parkinson's disease. *Exp Brain Res* 168:131–142.

Rand M.K., Squire L.M., Stelmach G.E. (2006b) Effect of speed manipulation on the control of aperture closure during reach-to-grasp movements. *Exp Brain Res* 174:74–85.

Rand M.K., Wang L., Müsseler J., Heuer H. (2013) Vision and proprioception in action monitoring by young and older adults. *Neurobiol Aging* 34:1864–1872.

Reed C.L., Betz R., Garza J.P., Roberts R.J. Jr. (2010) Grab it! Biased attention in functional hand and tool space. *Atten Percept Psychophys* 72:236–245.

Rentsch S., Rand M.K. (2014) Eye-hand coordination during visuomotor adaptation with different rotation angles. *PLoS One* 9:e109819.

Rossetti Y., Desmurget M., Prablanc C. (1995) Vector coding of movement: vision, proprioception, or both? *J Neurophysiol* 74:457–463.

Saling M., Mescheriakov S., Molokanova E., Stelmach G.E., Berger M. (1996) Grip reorganization during wrist transport: the influence of an altered aperture. *Exp Brain Res* 108:493–500.

Sanders J.A., Knill D.C. (2004) Visual feedback control of hand movements. *J Neurosci* 24:3223–3234.

Santello M., Soechting J.F. (1998) Gradual molding of the hand to object contours. *J Neurophysiol* 79:1307–1320.

Schmidt R.A., Zelaznik H.N., Hawkins B., Frank J.S., Quinn J.T. (1979) Motor output variability: a theory for the accuracy of rapid motor acts. *Psychol Rev* 86:415–451.

Shabbott B.A., Sainburg R.L. (2010) Learning a visuomotor rotation: simultaneous visual and proprioceptive information is crucial for visuomotor remapping. *Exp Brain Res* 203:75–87.

Shimansky Y.P. (2007) Role of optimization in simple and learning based adaptation and its biologically plausible mechanisms. In: Williams T.O. (ed.) *Biological cybernetics research trends*. Nova Science Publishers, Hauppauge, pp. 95–164.

Shimansky Y.S., Rand M.K. (2013) Two-phase strategy of controlling motor coordination determined by task performance optimality. *Biol Cybern* 107:107–129.

Smeets J.B.J., Brenner E. (1999) A new view on grasping. *Motor Control* 3:237–271.

Sperry R.W. (1950) Neural basis of the spontaneous optokinetic response produced by visual inversion. *J Comp Physiol Psychol* 43:482–489.

Synofzik M., Lindner A., Thier P. (2008) The cerebellum updates predictions about the visual consequences of one's behavior. *Curr. Biol* 18:814–818.

Tang R., Whitwell R.L., Goodale M.A. (2014) Explicit knowledge about the availability of visual feedback affects grasping with the left but not the right hand. *Exp Brain Res* 232:293–302.

Taylor J.A., Ivry R.B. (2011) Flexible cognitive strategies during motor learning. *PLoS Comput Biol* 7:e1001096.

Taylor J.A., Krakauer J.W., Ivry R.B. (2014) Explicit and implicit contributions to learning in a sensorimotor adaptation task. *J Neurosci* 34:3023–3032.

Timmann D., Stelmach G.E., Bloedel J.R. (1996) Grasping component alterations and limb transport. *Exp Brain Res* 108:486–492.

Todorov E. (2004) Optimality principles in sensorimotor control. *Nat Neurosci* 7:907–915.

Van Beers R.J., Sittig A.C., Denier van der Gon J.J. (1999) Integration of proprioceptive and visual position-information: an experimentally supported model. *J Neurophysiol* 81:1355–1364.

Wallace S.A., Weeks D.L., Kelso J.A.S. (1990) Temporal constraints in reaching and grasping behavior. *Hum Mov Sci* 9:69–93.

Wang J., Stelmach G.E. (2001) Spatial and temporal control of trunk-assisted prehensile actions. *Exp Brain Res* 136:231–240.

Werner S., Bock O. (2007) Effects of variable practice and declarative knowledge on sensorimotor adaptation to rotated visual feedback. *Exp Brain Res* 178:554–559.

Whitwell R.L., Goodale M.A. (2009) Updating the programming of a precision grip is a function of recent history of available feedback. *Exp Brain Res* 194:619–629.

Whitwell R.L., Lambert L.M., Goodale M.A. (2008) Grasping future events: explicit knowledge of the availability of visual feedback fails to reliably influence prehension. *Exp Brain Res* 188:603–611.

Wing A.M., Turton A., Fraser C. (1986) Grasp size and accuracy of approach in reaching. *J Mot Behav* 18:245–260.
Zahariev M.A. and MacKenzie C.L. (2007) Grasping at "thin air": multimodal contact cues for reaching and grasping. *Exp Brain Res* 180:69–84.
Zbib B., Henriques D.Y.P., Cressman E.K. (2016) Proprioceptive recalibration arises slowly compared to reach adaptation. *Exp Brain Res* 234:2201–2213.

11

DEXTEROUS MANIPULATION

Bridging the Gap between Hand Kinematics and Kinetics

Marco Santello

List of Abbreviations

aIPS	anterior part of intraparietal sulcus
BOLD	blood oxygenation level dependent
CNS	central nervous system
CoP	center of pressure
cTBS	continuous theta burst
EEG	electroencephalography
GFR	grip force rate
M1	primary motor cortex
PMd	pre-motor dorsal area
PMv	premotor ventral areas
S1	primary somatosensory cortex
T_{com}	compensatory torque
TMS	transcranial magnetic stimulation

11.1 Introduction

Skilled manipulation has been extensively studied over the past four decades. To further our understanding of the neural mechanisms underlying this sophisticated motor behavior, scientists have used a wide range of experimental approaches, including motion capture, electromyography, imaging, neuromodulation, and computational models. Although each of these approaches has provided significant insights, often the focus has been on specific components of manipulation control. In this chapter, I provide a brief review of research on grasp kinematics and kinetics, to then discuss the functional continuum between these two domains. I will use this premise as the rationale for studying dexterous manipulation by integrating experimental approaches and behavioral models that can capture such continuum. The goal of this integration is to enable the characterization of interrelated behavioral phenomena – including sensorimotor memory, online feedback, planning and execution, learning and generalization –, and inference of the underlying neural mechanisms. Examples of clinical and robotics applications of insights provided by research on neural control of the hand are also described. The chapter concludes with open questions and directions for future research. When appropriate and due to space limitations, the reader is referred to review article and chapters in this book.

11.2 Why Study Grasping and Manipulation?

Object manipulation, although not unique to humans (see review by Schieber & Santello, 2004; see Part II), is nevertheless uniquely 'human' from the perspectives of breadth and sophistication of manipulative behaviors – buttoning a shirt, drawing, changing the orientation of a smart phone in the hand without dropping it, and playing a musical instrument, are but a few examples of tasks that require accurate sensorimotor transformations *and* a biomechanical structure that can support a wide range of coordinated movements within and across digits. Interestingly, what makes the hand an incredibly versatile sensorimotor system is also what makes understanding the underlying control mechanisms particularly challenging. Such challenges are exemplified by the wide array of experimental approaches and theoretical frameworks – some of which are described in this book – that have been used and developed by researchers over the years.

Why has the hand attracted such a great interest in the neuroscience community? One reason is that our hands play a key role in activities of daily living, and that injury – ranging from peripheral neuropathies to neurodegenerative disorders and loss of the hand – can have a significant impact on quality of life. However, besides the clinical significance of understanding neural control of the hand, another reason is that the hand is an ideal model system for addressing key questions in neuroscience. For example, the classical problem of how the central nervous system coordinates motion across a large number of degrees of freedom has been studied by applying the concept of kinematic hand synergies and identifying commonalities among grasp patterns – defined by multivariate analysis of joint angles – used to grasp imagined and real objects (Santello et al., 1998; Santello et al. 2002; for review see Santello et al., 2013). A similar framework has been used to quantify the extent to which motor unit pairs or populations of intrinsic and extrinsic hand muscles may receive common input during grasping (e.g., Johnston et al., 2009; Winges & Santello, 2004; Winges et al., 2008). More recently, studies of sensorimotor learning have started to examine manipulation tasks to investigate adaptation and related phenomena, i.e., generalization and retention (Fu et al., 2011; Fu et al., 2014; Fu & Santello, 2015).

This is an interesting extension of behavioral paradigms that, until recently, have focused on arm movements and reaching tasks using perturbations based on force fields or visuomotor rotations (for review see Wolpert et al., 2011). In these studies, what makes the hand unique relative to reaching tasks is that – besides differences between the hand and arm in mechanical and neuromuscular features – neural principles of sensorimotor learning of hand-object interactions may differ, to some extent, from those described for arm movements, the former being characterized by visual contextual cues, i.e., object geometry (Fu & Santello, 2012; Fu & Santello, 2015; see section 11.7). Another example is the problem of multimodal sensory integration, whereby the question of how multiple streams of sensory information from different modalities must be effectively integrated to successfully manipulate an object (e.g., Sarlegna et al., 2010) – that is, while the object is being manipulated, it might be necessary to integrate visual feedback of the object orientation with tactile and proprioceptive inputs relaying information about what digits are touching the object and how much force is being exerted.

11.3 Grasp Kinematics

The act of reaching towards a tool or an object is often motivated, and followed by a manipulation, through which the object position or orientation is changed. Below I briefly review studies addressing neural mechanisms and principles underlying the coordination of reach-to-grasp movements.

11.3.1 Neural Mechanisms

As described in Parts II and III of this book, during object grasping and manipulation the central nervous system (CNS) relies on visual object attributes such as size, shape, and orientation to shape the hand in preparation for a grasp. Electrophysiological and imaging studies have demonstrated the role of anterior part of intraparietal sulcus (aIPS) and premotor ventral (PMv) areas for accurate hand shaping (Davare et al., 2011; Gallivan et al., 2011; Schettino et al., 2015). Non-human primate studies have further identified visuomotor neurons within aIPS and 'canonical' neurons within PMv whose activation depends on grasp type. Different PMv cell populations are thought to encode the motor repertoire required for different grasp actions (Castiello, 2005; Murata et al., 1997; Umilta et al., 2007).

With regard to experimental evidence obtained from human subjects, patients with posterior parietal cortex lesions experience difficulties in shaping the fingers according to intrinsic object features (Rizzolatti, Fogassi, & Gallese, 2002). In healthy subjects, transient reversible disruption of PMv and bilateral aIPS areas through transcranial magnetic stimulation (TMS) prior to object lift-off leads to increased variability in hand shaping (Binkofski et al., 1998;

Jeannerod et al., 1994). Furthermore, virtual lesion studies using TMS have implicated aIPS in the online control of grasping action (Davare et al., 2006; (Davare et al., 2007). Specifically, when the object orientation was changed following movement onset, disruption of contralateral aIPS impaired adjustments in grip aperture (Hamilton & Grafton, 2006; Rice et al., 2006; Tunik et al., 2007; Tunik et al., 2005). Moreover, deficits in grasp kinematics were observed when a TMS pulse was delivered within 65 ms of object perturbation, suggesting the role of aIPS in visually based detection of error in grip aperture scaling (Tunik et al., 2005). This finding suggests that aIPS plays an important role for integrating visual feedback about object size with the motor command for online adjustment of grasp kinematics during movement execution. However, it is not known whether the online correction of grasp aperture influenced subsequent application of digit forces on the object for grasp and manipulation. This would be crucial when performing manipulation tasks requiring fine modulation of digit forces to position for successful task performance, such as it would be necessary in the task described below (section 11.6).

11.3.2 Coordination of Reach-to-Grasp Movements

Early work on reach-to-grasp movements focused on identifying proximal (shoulder, elbow, and wrist joints) and distal (finger joints, fingertip positions) movement parameters that might be sensitive to object features, as these would ultimately impact how the object is manipulated, e.g., size and shape, or features of the manipulation task itself. The origins of research on grasp kinematics can be traced to early work by Jeannerod (Jeannerod, 2009) examining the spatial and temporal characteristics of finger span modulation to object size. This work on precision grip inspired many more grasp kinematics studies (reviewed in Parts II and III of this book), and led to investigations of how whole-hand shape changes as a function of the geometry of imagined or real objects (Santello & Soechting, 1998; Santello et al., 1998; Santello et al., 2002). More recent investigations have re-examined how the control of reach and grasp kinematics might be controlled as a unit, both from a behavioral and neural standpoint (see Chapter 6 by Rouse and colleagues). A novel view on the functional role of hand pre-shaping, which was first proposed by Smeets and colleagues, posits that the modulation of finger span and fingertip trajectories during the reach is functionally linked to the control of contact points (reviewed in Chapter 8 by Smeets and Brenner).

11.3.3 Choice of Digit Placement

I would argue that contact points are indeed an important component of manipulation, even though it should be considered as an 'intermediate' control variable. Specifically, the trajectories of fingertips during the reach determine where the object is grasped. So, from the perspective of studies of reach kinematics, contact point distribution could be considered as the variable the system is trying to control. While this view is likely to be correct, it should be emphasized that ensuring that the object is grasped at desired points does not, by itself, guarantee that the subsequent phases of manipulation are attained as planned. That is, the ultimate goal of reach-to-grasp goes beyond the act of grasping per se: the ultimate goal of transporting the hand to an object is manipulation, which in turn requires accurate spatial and temporal distribution of forces among the digits involved in the grasp. However, the relation between contact point and force distributions has only recently started to be investigated. Below I briefly review literature on grasp kinetics to introduce the phenomenon of digit force-to-position modulation.

11.4 Grasp Kinetics

11.4.1 Neural Mechanisms

Electrophysiological studies in monkeys demonstrated that activity of pre-motor dorsal area (PMd) and primary motor cortex (M1) neurons is modulated during exertion of grasp forces (Rice et al., 2006; Tunik et al., 2005). Patients with an acute injury of parietal region of the brain (Hendrix et al., 2009; Hepp-Reymond et al., 1999) are unable to predictively scale their grip forces during self-induced modulation of loads on a hand-held object during arm movements. In neurologically intact individuals, BOLD-related cortical activity was significantly larger in right intraparietal cortex when subjects applied a small versus larger force using a precision grip (Nowak et al., 2003). For object manipulation tasks, learning to scale digit forces based on object properties (i.e., weight) also seems to engage a fronto-parietal network (Ehrsson et al., 2001). Specifically, when object weight was kept constant across successive lifts, virtual lesion of PMd resulted in disruption of predictive scaling of grip forces based on arbitrary visuomotor associations or delayed the onset of load force scaling (Dafotakis et al., 2008; Nowak et al., 2005). Virtual lesions of M1 (Chouinard et al., 2005; Davare et al., 2006; Nowak et al., 2009) resulted in disruption of grip force scaling based on memory of prior lifts, and virtual lesion of contralateral aIPS (Chouinard et al., 2005; Nowak et al., 2009) led to an overshooting of peak rates of grip and load forces, suggesting a possible interference with the internal representation of object weight. Thus, aIPS, PMd, and M1 play an important role in the storage and/or retrieval of finger force scaling appropriate for constrained grasping and manipulation.

When object property is changed unexpectedly, several nodes within the fronto-parietal network have been shown to be involved in modulating digit forces for grasping and manipulation. Neurophysiological and imaging studies have identified the involvement of premotor areas, sensorimotor cortices, and posterior parietal cortex (Davare et al., 2007, 2006). Specifically, BOLD-related activity in anterior intraparietal region within one (Chouinard et al., 2005; Ehrsson et al., 2003; Jenmalm et al., 2006; Schmitz et al., 2005; Shadmehr & Holcomb, 1997; van Nuenen et al., 2012) or both (Jenmalm et al., 2006) hemispheres increased during unpredictable changes in object weight suggesting that these areas are involved in monitoring the difference between predicted and actual object weight, inducing corrections in digit forces, and may help in updating the internal representation of object properties for digit force scaling during successive lifts (Schmitz et al., 2005). 'Virtual' lesion of aIPS *prior to object contact* using single-pulse TMS impaired scaling of grip forces following lift onset to stabilize the object whose weight was changed unexpectedly between trials (Ehrsson et al., 2001; Jenmalm et al., 2006; Schmitz et al., 2005). However, and as noted above in relation of grasp kinematics studies, the contribution of aIPS in online control of digit forces in relation to digit position remains to be addressed.

11.4.2 Behavioral Studies

Johansson and his colleagues pioneered the study of grasp kinetics in the mid-80s by combining behavioral and electrophysiological approaches. This pioneering work – which has had a significant impact in the field of motor neuroscience – revealed, for the first time, fundamental sensorimotor control processes responsible for grasp force modulation to object properties, as well as how tactile mechanoreceptors influence force control (for review see Johansson & Flanagan, 2009). For example, sensorimotor control of finger forces is now known to involve predictive mechanisms that depend on visual cues about object properties and/or stored internal representations, i.e., *sensorimotor memory*, of previous manipulations with the same or similar objects. Incorrect predictions in the form of object slips when the surface is more slippery than expected, or inability to lift the object at the expected time due to an underestimation of object weight, are updated within a few lifts (Dafotakis et al., 2008). The brain is believed to use feedback information from tactile skin afferents that encode information about texture, curvature, and weight of the object, direction of the applied fingertip force to sense, correct, and update motor predictions for forthcoming lifts, if necessary (Gordon et al., 1993; Westling & Johansson, 1984). Specifically, tactile feedback signals are used to monitor task progression and compared with signals expected at distinct *sensorimotor control points*, i.e., initial object contact or onset of object lift (Birznieks et al., 2001; Johansson & Birznieks, 2004). A mismatch triggers corrective responses to update the motor plan, i.e., upscaling of finger forces to enable lift-off of a heavier than expected object. This work on precision grip also launched many other studies on multi-finger grasping addressing a number of questions, including the sensorimotor mechanisms underlying the distribution of finger forces as a function of object properties, i.e., mass, mass distribution, and grip configuration (Li et al., 1998; Santello & Soechting, 2000; for review, see Zatsiorsky & Latash, 2004, 2008).

It should be noted that, until recently, virtually all of these studies have focused on an experimental approach that constrains where subject grasp objects (we will refer to this type of grasp as 'constrained grasping'). This constraint has been imposed mostly by technical limitations associated with the need of mounting sensors to objects at fixed locations. For example, precision grip studies have used objects with thumb and index finger sensors facing each other (e.g., Gordon et al., 1993; Westling & Johansson, 1984; for simplicity, we refer to this configuration as 'collinear' locations). Similarly, multi-finger studies have used manipulanda with multiple force/torque sensors at fixed thumb-finger configurations (Santello & Soechting, 2000).

11.5 Bridging the Gap between Grasp Kinematics and Kinetics

11.5.1 Choice of Contact Points

As pointed out above, studies of grasp kinematics and kinetics have provided significant insights into neural control of grasping and manipulation. However, virtually all of these studies have focused on *either* grasp kinematics *or* kinetics. Specifically, the above-reviewed studies of grasp kinematics did not quantify the extent to which hand pre-shaping or contact point distributions may affect coordination of digit forces from contact to onset of object manipulation. Conversely, studies of grasp kinetics have quantified in great detail between-digit force coordination, but have done so without taking into account how hand pre-shaping during the reach may impact contact point distributions, and therefore digit force control. Thus, these studies used objects with sensors that have to be grasped at pre-defined locations on the object. As discussed in greater detail below, constraining where to grasp an object captures only a subset of hand-object interactions we perform in everyday activities, as we normally choose contact points depending on object geometry, task demands, and our knowledge of object dynamics.

Clearly, the focus of many studies on either grasp kinematics or kinetics had left a critical gap, given that these two dimensions should really be conceived as belonging to a functional continuum. One of the first studies pointing out the functional link between where the object is grasped and execution of manipulation was performed by Rosenbaum and his colleagues (Cohen & Rosenbaum, 2004; Rosenbaum et al., 1996; Rosenbaum et al., 1992). This work highlighted the fact that choice of contacts on an object is not only driven by its geometry or physical properties, but also by the upcoming task. Thus, by virtue of what those authors defined "end-state comfort effect," the height at which an object is grasped reflects the avoidance of an uncomfortable posture at the end of the manipulation – so, one would grasp an object at the bottom if this were to be placed on a higher shelf, and so forth. Similarly, it has been shown that the way we grasp a bottle is sensitive to the planned manipulation, e.g., on whether the bottle is grasped to be lifted or to pour liquid in a glass (Crajé et al., 2011).

11.5.2 Grasp Kinematics-Kinetics Continuum

An extension of this framework would predict that choice of contact points should also be sensitive not only to the features of the upcoming task, but also to the extent to which object dynamic properties are known. In a study of whole-hand grasping, subjects were asked to grasp a cylindrical object anywhere along its graspable surface. The cylindrical object was mounted on a base to which the experimenter added a mass in a blocked or random fashion across trials. When the mass was added to the left or right compartment, it created a counterclockwise or clockwise torque. Subjects were instructed to lift the object as straight as possible, which required them to learn to anticipate the external torque at the time of lift onset. Not surprisingly, subjects learned to minimize object roll within the first three trials but only when the direction of the external torque was predictable on a trial-to-trial basis (Lukos et al., 2007). Importantly, however, subjects spontaneously changed contact point distribution as a function of torque direction, such that the thumb was positioned higher and lower than the index finger when the added mass was on the thumb or finger side, respectively. Furthermore, a follow-up study found that grasp kinematics and forces (inferred by the amount of object roll) were differentially sensitive to implicit learning of object dynamics and explicit visual or verbal cues about object's mass distribution (Lukos et al., 2008). These findings extended the above-described observations by Rosenbaum and colleagues on the task-dependent sensitivity of where adults choose to grasp objects. At the same time, a novel conclusion of these studies was that choice of contact points should be integrated with anticipatory force control mechanisms to enable successful object manipulation. This conclusion, however, was based on biomechanical and theoretical considerations, that is: subjects changed digit placement *and* successfully learned to minimize object roll when the direction of the external torque was predictable. Therefore, they must have changed *both* digit position *and* forces. Nevertheless, the object used by Lukos et al. (2007, 2008) could not be sensorized to measure forces from all digits, and therefore the strategies used to modulate forces to position could not be addressed.

11.6 Digit Force-to-Position Modulation as a Fundamental Mechanism for Dexterous Manipulation

11.6.1 Choice of Digit Placement: Behavioral Consequences

To understand how digit forces and position are coordinated in an anticipatory fashion, Fu et al. (2010) designed a sensorized object to be grasped with the thumb and index finger. The design of the object and the use of torque/force sensors allowed to measure digit forces and accurately reconstruct the center of pressure (CoP) of each digit, and therefore address the limitations of previous work. Subjects were asked to perform a similar task as the one used by Lukos et al. (2007), but introduced a new comparison between two grasping conditions, *constrained* and *unconstrained* grasping. For the unconstrained grasping condition, subjects could choose to grasp the object anywhere along the vertical grasp surfaces, whereas for the constrained grasping condition subjects had to grasp the object at predetermined contact points on the object that were placed at the same height relative to the object's base (Figure 11.1). Both grasping conditions required subjects to learn to generate a compensatory torque, T_{com} (same magnitude but opposite direction to the external torque) in an anticipatory fashion, i.e., before object lift onset. However, we should note that generating a torque by grasping the object at collinear points (constrained condition) can only be accomplished by generating asymmetrical tangential (vertical) forces with thumb and index finger (Salimi et al., 2000). In contrast, we expected that – based on findings by Lukos et al. (2007, 2008) – in the unconstrained grasping condition subjects would spontaneously adopt different contact point distributions as a function of object center of mass. In turn, these contact point distributions would have to be accompanied by different digit force distributions. This design allowed us for the first time to address how humans learn to coordinate digit positions and forces for dexterous manipulation on a trial-to-trial basis.

It was found that subjects could learn anticipatory control of the compensatory torque within the first three trials in both grasp conditions. However, the unconstrained condition was characterized by trial-to-trial variability in digit center of pressure. Interestingly, this digit position variability persisted even after T_{com} had been learned (trials 4–10). The key observation of this study was that digit position variability was compensated for by modulating the distribution of digit load (vertical) forces *on a trial-by-trial basis* (Figure 11.2), this phenomenon being prevalent for unconstrained grasping where digit position variability was largest.

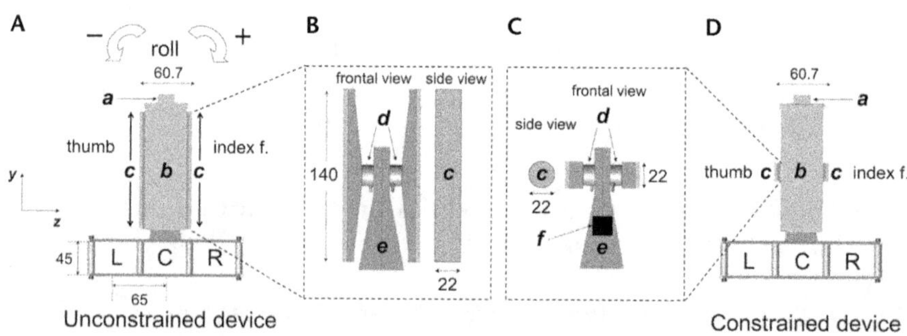

FIGURE 11.1 Grasp devices for studying constrained and unconstrained grasping

Notes: Sensorized devices used to study constrained (A,B) and unconstrained (C,D) grasping (adapted from Fu et al., 2010).

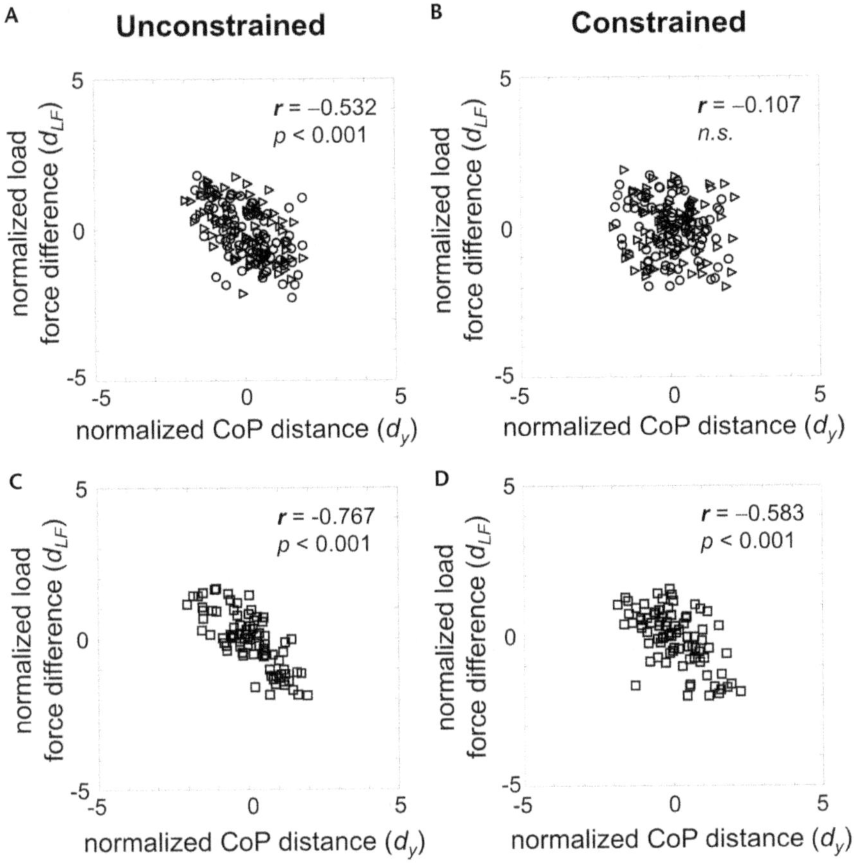

FIGURE 11.2 Digit load force distribution and digit position covariation

Notes: Digit load force distribution (difference between thumb and index finger load force) is modulated as a function of digit position (vertical distance between thumb and index finger center of pressure) in unconstrained grasping (A: circle and triangle symbols denote left and right center of mass, respectively; C: centered center of mass). For constrained grasping, digit load force to position covariation is found only for the centered center of mass, but not for left and right center of mass (D and B, respectively). Data are from all subjects (trials 4–10) and are expressed in normalized form. Adapted from Fu et al. (2010).

11.6.2 Theoretical Considerations

Digit force-to-position modulation is a critically important phenomenon for manipulation and highlights several theoretical features of the underlying control mechanisms. First, force-to-position modulation is necessary for ensuring that manipulation can be successfully performed. Specifically, generating the same forces learned through previous manipulations (i.e., using sensorimotor memory as demonstrated for constrained grasping) (Johansson & Westling, 1984; Westling

& Johansson, 1984) – but applied at different contact points – would prevent generation of the torque required to lift the object straight: this problem is equivalent to spilling water from a glass every time you grasp it if you ignored the fact that the glass is not grasped exactly at the same points.

Second, the ability to modulate digit forces to position on each trial implies that the CNS is factoring in where the digits are placed relative to each other *and* can determine, in a relatively short time (from contact to object lift onset, ~400–500 ms after contact), the digit force distribution that is appropriate to digit position. Feedback of digit placement could be mediated by vision of fingertip trajectories and proprioception of hand opening-closing throughout the reach and contact, as well as tactile sensing which might contribute to provide information about relative digit position. This feedback-based mechanism would not replace the memory-driven mechanism that allows to retrieve and use manipulative forces that have been used in the past, but rather integrate it and triggered by sensing a mismatch between expected and actual digit position. Interestingly, haptic feedback of the object size alone is sufficient to enable digit force-to-position modulation, even though this modulation is slower than when visual and haptic feedback can be used (Fu & Santello, 2014).

Third, the CNS's ability to choose the digit force distribution that is appropriate for a given digit contact distribution implies that learning a dexterous manipulation task like the one used by Fu and colleagues (2010) builds a 'high-level' task representation (in this example, T_{com}) which drives the selection of digit forces until the learned torque is attained. At this time point, the signals responsible for lifting the object can be released. It should be noted that such high-level task representation is necessary when the low-level effector representations (i.e., digit forces and/or position) can vary across trials, as it happens with unconstrained grasping. In this scenario, a high-level representation would allow flexible recombination of low-level variables (this is further discussed in section 11.7).

11.6.3 Differential Contributions of Feedforward and Feedback Control Mechanisms to Constrained and Unconstrained Grasping

To further explore the extent to which digit force control might depend on sensory feedback of digit placement, a follow-up study compared grip force rates (GFR) during the loading phase, i.e., from contact to object lift onset, in unconstrained versus constrained grasping (Mojtahedi et al., 2015). To tease out a potential contribution of feedback to force control, the shape of GFR was analyzed using two complementary techniques. The first approach consisted of fitting a Gaussian function to GFR. The goodness of the fit was used as a metric to quantify the extent to which GFR would approximate a bell-shaped profile, a feature that is typically interpreted as indicative of feedforward control in arm movements (Ghez & Gordon, 1987; Ghez et al., 1995; Gordon et al., 1995; Jeannerod, 1984; (Johansson & Westling, 1988) and force control in grasping (Gordon et al., 1993; Jenmalm & Johansson, 1997; Johansson & Westling, 1988). A weak strength in the Gaussian function fit would denote a GFR profile that departs from a bell-shaped time course. The second approach consisted of applying a time-frequency domain analysis (continuous wavelet transformation) to quantify the correlation of GFR with a Mexican Hat function. A strong correlation at a particular time point and frequency would denote a greater similarity between a bell-shaped profile and GRF, hence a greater contribution of feedforward control, and vice versa for weaker correlations. The main findings of this study were that both Gaussian function fitting and continuous wavelet transform analysis revealed a greater contribution of feedback mechanisms for digit force control in unconstrained than constrained grasping (Figure 11.3A and B, respectively). These findings are consistent with the proposition that these two grasp contexts differ in terms of force control mechanisms despite aiming at the same task goal.

11.7 Dexterous Manipulation: High- vs. Low-Level Representations of Control Variables

11.7.1 What Is Being Learned as We Learn to Manipulate an Object?

Support for the notion of 'high-level' representation of dexterous manipulation being built despite variability of digit placement was provided by a follow-up study (Fu et al., 2011). This work addressed the extent to which significantly greater changes in digit placement than those found during trial-to-trial reach-to-grasp movements would still be accompanied by digit force-to-position modulation to ensure successful performance. If so, this phenomenon could be interpreted as further evidence for subjects building a central, high-level task representation that is robust to large changes in the organization of peripheral effectors, i.e., digit and force distributions. To test this hypothesis, the authors asked subjects to learn the same task described above consisting of minimizing object roll using unconstrained grasping (Fu et al., 2010), but using either two digits (thumb and index finger; 2*d* grip), or three digits (thumb, index, and middle fingers; 3*d* grip). After subjects learned the task using either grip type, subjects who learned the task using a 2*d* grip were asked to now use a 3*d* grip, whereas those who had learned the task using a 3*d* grip were asked to use a 2*d* grip. Both groups were able to successfully transfer the learned torque to the (unpracticed) new grip type, despite significant biomechanical differences between the two grip configurations, e.g., shift in finger center of pressure and re-distribution of finger forces when going from 2*d* to 3*d* (Fu et al., 2011). These results confirmed the notion of digit force-to-position

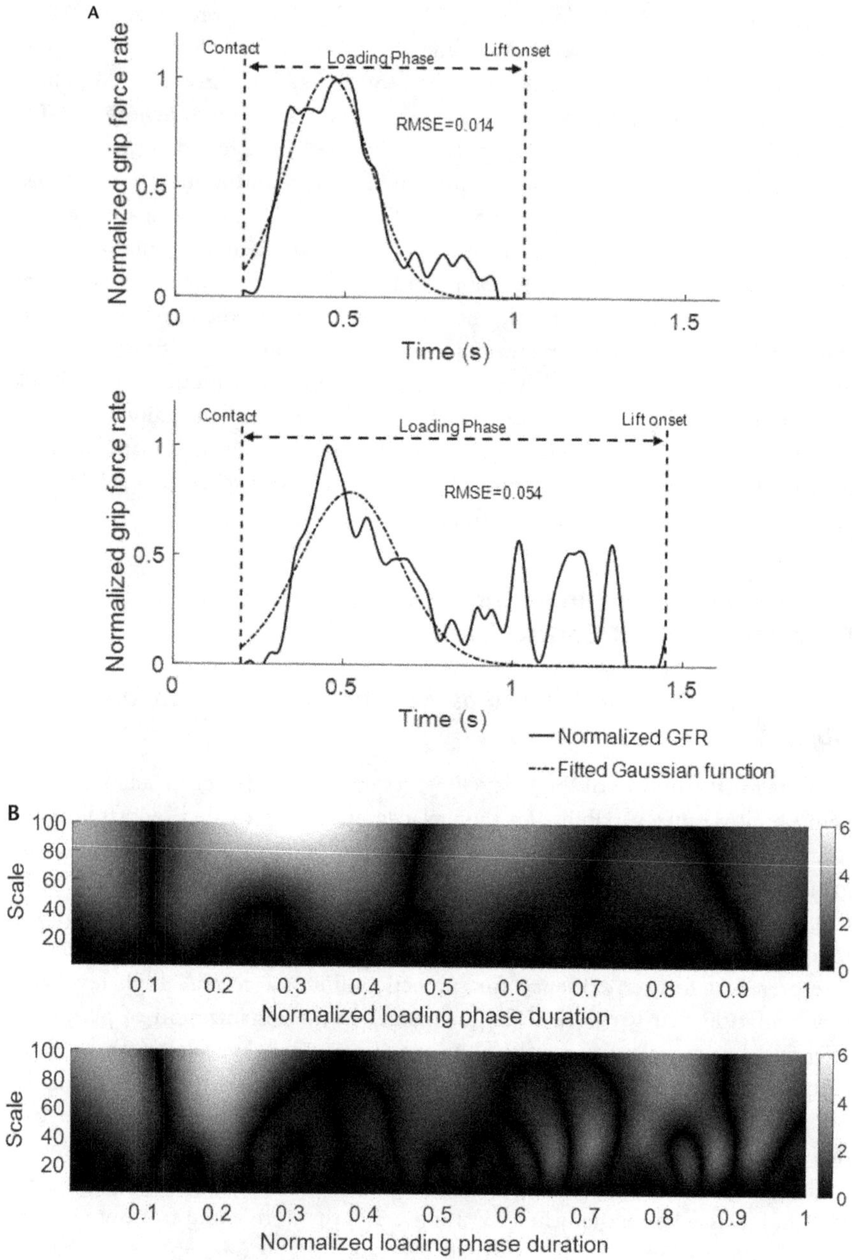

FIGURE 11.3 Assessing feedforward and feedback control mechanisms using feature extraction methods on grip force rate

Notes: A: Normalized grip force rate (GFR) measured during loading phase (contact to object lift onset) of constrained and unconstrained grasping (top and bottom plots, respectively). The dashed trace is the Gaussian function that generated the best fit to GFR and is shown together with

modulation as a fundamental and robust mechanism for ensuring that a given manipulation task can be executed despite small or large trial-to-trial changes in digit position.

11.7.2 Generalizing Learned Manipulation to Different Grasp Contexts

The above example of full transfer of learned manipulation to a different context (grip configuration) raises the question of whether manipulation can always be transferred to different contexts, or conversely whether there might be factors that may prevent such transfer. To address this question, we performed a series of studies manipulating the contexts subjects were required to transfer learned manipulation to. The first study examined whether subjects could perform the learned manipulation for a given torque direction when faced with performing the manipulation for an opposite torque direction (Zhang et al., 2010). Here, subjects were asked to rotate the object (without lifting it) – hence changing the direction of the external torque on the T-shaped object from left to right, or vice versa – after having learned the compensatory torque in the previous block of trials. We found that subjects failed to transfer the learned torque to the new object orientation on the very first lift. However, after multiple exposure to object rotations, subjects gradually improved in their ability to generate the torque in the appropriate direction. Interestingly, this improvement was mostly driven by learning how to correctly position the digits on the rotated object, whereas subjects could not attain digit force distributions appropriate for the rotated object (Zhang et al., 2010).

To gain further insight into the mechanisms underlying this dual adaptation to opposite grasp contexts (external torque direction), we recently re-examined this task while recording electroencephalography (EEG) (Fine et al., 2017). Our analysis focused on the sensitivity of sensorimotor α and β, and medial frontal θ frequency bands to performance error, i.e., object tilt occurring either in the initial learning of manipulation through consecutive lifts or every time that the object was lifted in the opposite context. Specifically, we predicted single-trial EEG data from a computational learning model that can adapt to multiple contexts simultaneously based on a switching mechanism (Lee & Schweighofer,

---

corresponding root-mean-squared error (RMSE) of the fit. Note the approximately six-fold higher RMSE for unconstrained than constrained grasping, denoting a less bell-shaped GFR profile for the former condition. B: Continuous wavelet transformation (CWT) applied to grip force rate measured during the time-normalized loading phase of constrained and unconstrained grasping (top and bottom, respectively). The CWT coefficients computed on the constrained task are larger at the higher scale (lower pseudo-frequencies) than the unconstrained task, but CWT coefficients from the unconstrained task are larger at lower scales (higher pseudo-frequencies) than the constrained task. Data in A and B are from the same subject manipulating the inverted-T object shape with a right center of mass. Adapted from Mojtahedi et al. (2015).

2009). Interestingly, we found that neural oscillations could discriminate the source of behavioral error. Specifically, only medial frontal θ was predicted by a component of the model state that adapts to performance errors caused by a context switch. In contrast, α and β were predicted by a model state that was updated from context-independent performance errors. These findings point to neural mechanisms accounting for grasp context switch and supporting dual adaptation to different grasp contexts (Fine et al., 2017).

Other examples of failure to adapt to different grasp contexts have been provided by several studies (e.g., Ingram et al., 2010). For example, subjects are able to use visual cues about object properties (e.g., mass distribution of a U- or L-shaped object) to predict the appropriate force distribution for manipulation. However, following repeated hand-object interactions, they are unable to use the same cues when asked to grasp the U-shaped object at a different handle or the L-shaped object at a different orientation, despite strong salient shape cues (Fu & Santello, 2012, 2015). These results point to a conflict between vision-based motor planning and sensorimotor memory, which would then limit the extent to which a learned manipulation can be transferred to a different context.

In sum, the findings discussed in this section suggest that context-specific learning of motor actions limits generalization of learned dexterous manipulation. Furthermore, when grasping is performed at unconstrained contacts, digit position and forces appear to be differentially sensitive to these learning generalization constraints.

11.8 Choice of Digit Placement in Dexterous Manipulation Engages Different Neural Mechanisms

The above-reviewed behavioral evidence suggest that the phenomenon of digit force-to-position modulation is important for dexterous manipulation and is mediated by different control mechanisms. Nevertheless, behavioral evidence alone is insufficient to demonstrate or reveal the underlying neural mechanisms. Below I review ongoing work aiming at understanding information processing among cortical areas and its sensitivity to grasp context.

11.8.1 Digit Force and Position Planning

We have recently used EEG to determine the extent to which neural oscillations may capture grasp-context dependent and independent cortical activation (McGurrin et al., 2017). An important advantage of using EEG for comparing constrained and unconstrained grasping is that it allowed us to examine not only cortical activation during grasp execution, but also during planning. Recall that participants can be certain of digit position and concurrent force distributions in constrained but not unconstrained grasping, whereby control of the former and the latter is dominated by memory and online feedback, respectively. Based on the functional role assigned to medial frontal θ activity (for review see Cavanagh et al., 2010; Cavanagh & Shackman, 2015), we hypothesized that this neural oscillation frequency would be greater during planning unconstrained than constrained grasping. This hypothesis was confirmed by larger planning medial frontal θ activity in unconstrained grasping, and also by the ability of this neural oscillation frequency to predict longer reaction times. We also found greater post-movement β synchronization following reach onset and during grasp execution in constrained than unconstrained grasping, suggesting more certainty in feedforward estimations of digit forces in this grasp context (Tan et al., 2016).

11.8.2 Cortical Information Processing Is Sensitive to Grasp Context

Our theoretical framework predicts that M1 would have a dual role according to whether digit force control can be primarily driven by memory versus online feedback of digit placement. In the former case, M1 has been shown to be involved in storing and retrieval of sensorimotor memory of digit forces (Chouinard et al., 2005; Jenmalm et al., 2006; Nowak et al., 2005). In the latter scenario, M1 would be involved in modulating digit force to position based on inputs from primary somatosensory cortex (S1). We have tested this framework by applying continuous theta burst (cTBS) to M1 after subjects learned to manipulate the inverted T-object shape in an unconstrained or constrained grasp context (McGurrin et al., 2015). We also applied cTBS to S1 for the unconstrained grasp context. As predicted, we found grasp-context and cortical-area specific effects of virtual lesions to M1 and S1 on digit position or forces. Specifically, in unconstrained grasping, M1 integrity is critical not only for storing and retrieving memory of learned digit forces and position, but also for coordinating control of digit forces based on feedback of digit placement. Furthermore, S1 provides M1 with online feedback of digit position necessary for trial-to-trial digit force modulation.

11.9 Conclusions

This chapter focused on the need to integrate – conceptually and experimentally – investigations of grasp kinematics and kinetics. An example of such integration has been provided by the experimental model of unconstrained grasping. This model has proved to be very useful in gaining insights into how choice of digit placement and more 'natural' grasp contexts affect not only grasp behavior and sensorimotor learning but also the underlying neural mechanisms. Further work is needed to improve our understanding of how high- and low-level neural representations of grasp variables are built and the factors that facilitate or interfere their generalization to novel contexts.

Acknowledgment

This work was partially supported by National Science Foundation Grant BCS-1455866.

References

Binkofski, F., Dohle, C., Posse, S., Stephan, K. M., Hefter, H., Seitz, R. J., & Freund, H. J. (1998). Human anterior intraparietal area subserves prehension: a combined lesion and functional MRI activation study. *Neurology, 50*(5), 1253–1259.

Birznieks, I., Jenmalm, P., Goodwin, A. W., & Johansson, R. S. (2001). Encoding of direction of fingertip forces by human tactile afferents. *The Journal of Neuroscience: The Official Journal of the Society for Neuroscience, 21*(20), 8222–8237.

Castiello, U. (2005). The neuroscience of grasping. *Nature Reviews. Neuroscience, 6*(9), 726–736. https://doi.org/10.1038/nrn1744.

Cavanagh, J. F., Frank, M. J., Klein, T. J., & Allen, J. J. B. (2010). Frontal theta links prediction errors to behavioral adaptation in reinforcement learning. *NeuroImage, 49*(4), 3198–3209. https://doi.org/10.1016/j.neuroimage.2009.11.080.

Cavanagh, J. F., & Shackman, A. J. (2015). Frontal midline theta reflects anxiety and cognitive control: meta-analytic evidence. *Journal of Physiology Paris, 109*(1–3), 3–15. https://doi.org/10.1016/j.jphysparis.2014.04.003.

Chouinard, P. A., Leonard, G., & Paus, T. (2005). Role of the primary motor and dorsal premotor cortices in the anticipation of forces during object lifting. *The Journal of Neuroscience: The Official Journal of the Society for Neuroscience, 25*(9), 2277–2284. Retrieved from www.ncbi.nlm.nih.gov/pubmed/15745953.

Cohen, R., & Rosenbaum, D. (2004). Where grasps are made reveals how grasps are planned: generation and recall of motor plans. *Experimental Brain Research, 157*(4). https://doi.org/10.1007/s00221-004-1862-9.

Crajé, C., Lukos, J. R., Ansuini, C., Gordon, A. M., & Santello, M. (2011). The effects of task and content on digit placement on a bottle. *Experimental Brain Research, 212*(1), 119–124. https://doi.org/10.1007/s00221-011-2704-1.

Dafotakis, M., Sparing, R., Eickhoff, S. B., Fink, G. R., & Nowak, D. A. (2008). On the role of the ventral premotor cortex and anterior intraparietal area for predictive and reactive scaling of grip force. *Brain Research, 1228*, 73–80. https://doi.org/10.1016/j.brainres.2008.06.027.

Davare, M., Andres, M., Clerget, E., Thonnard, J.-L., & Olivier, E. (2007). Temporal dissociation between hand shaping and grip force scaling in the anterior intraparietal area. *Journal of Neuroscience, 27*(15), 3974–3980.

Davare, M., Andres, M., Cosnard, G., Thonnard, J.-L., & Olivier, E. (2006). Dissociating the role of ventral and dorsal premotor cortex in precision grasping. *Journal of Neuroscience, 26*(8), 2260–2268. https://doi.org/10.1523/JNEUROSCI.3386-05.2006.

Davare, M., Kraskov, A., Rothwell, J. C., & Lemon, R. N. (2011). Interactions between areas of the cortical grasping network. *Current Opinion in Neurobiology, 21*(4), 565–570. https://doi.org/10.1016/j.conb.2011.05.021.

Ehrsson, H. H., Fagergren, E., & Forssberg, H. (2001). Differential fronto-parietal activation depending on force used in a precision grip task: an fMRI study. *Journal of Neurophysiology, 85*(6), 2613–2623.

Ehrsson, H. H., Fagergren, A., Johansson, R. S., & Forssberg, H. (2003). Evidence for the involvement of the posterior parietal cortex in coordination of fingertip forces for grasp stability in manipulation. *Journal of Neurophysiology, 90*(5), 2978–2986.

Fine, J. M., Moore, D., & Santello, M. (2017). Neural oscillations reflect latent learning states underlying dual-context sensorimotor adaptation. *NeuroImage, 163*, 93–105. https://doi.org/10.1016/j.neuroimage.2017.09.026.

Fu, Q., Choi, J. Y., Gordon, A. M., Jesunathadas, M., & Santello, M. (2014). Learned manipulation at unconstrained contacts does not transfer across hands. *PLoS ONE, 9*(9). https://doi.org/10.1371/journal.pone.0108222.

Fu, Q., Hasan, Z., & Santello, M. (2011). Transfer of learned manipulation following changes in degrees of freedom. *The Journal of Neuroscience: The Official Journal of the Society for Neuroscience, 31*(38), 13576–13584. https://doi.org/10.1523/JNEUROSCI.1143-11.2011.

Fu, Q., & Santello, M. (2012). Context-dependent learning interferes with visuomotor transformations for manipulation planning. *Journal of Neuroscience, 32*(43), 15086–15092. https://doi.org/10.1523/JNEUROSCI.2468-12.2012.

Fu, Q., & Santello, M. (2014). Coordination between digit forces and positions: interactions between anticipatory and feedback control. *Journal of Neurophysiology, 111*(7), 1519–1528. https://doi.org/10.1152/jn.00754.2013.

Fu, Q., & Santello, M. (2015). Retention and interference of learned dexterous manipulation: interaction between multiple sensorimotor processes. *Journal of Neurophysiology, 113*(1), 144–155. https://doi.org/10.1152/jn.00348.2014.

Fu, Q., Zhang, W., & Santello, M. (2010). Anticipatory planning and control of grasp positions and forces for dexterous two-digit manipulation. *Journal of Neuroscience, 30*(27), 9117–9126. https://doi.org/10.1523/JNEUROSCI.4159-09.2010.

Gallivan, J. P., McLean, D. A., Valyear, K. F., Pettypiece, C. E., & Culham, J. C. (2011). Decoding action intentions from preparatory brain activity in human parieto-frontal networks. *The Journal of Neuroscience: The Official Journal of the Society for Neuroscience, 31*(26), 9599–9610. https://doi.org/10.1523/JNEUROSCI.0080-11.2011.

Ghez, C., & Gordon, J. (1987). Trajectory control in targeted force impulses – I. Role of opposing muscles. *Experimental Brain Research, 67*(2), 225–240. https://doi.org/10.1007/BF00248545.

Ghez, C., Gordon, J., & Ghilardi, M. F. (1995). Impairments of reaching movements in patients without proprioception. II. Effects of visual information on accuracy. *Journal of Neurophysiology, 73*(1), 361–372.

Gordon, A. M., Westling, G., Cole, K. J., & Johansson, R. S. (1993). Memory representations underlying motor commands used during manipulation of common and novel objects. *Journal of Neurophysiology, 69*(6), 1789–1796.

Gordon, J., Ghilardi, M. F., & Ghez, C. (1995). Impairments of reaching movements in patients without proprioception. I. Spatial errors. *Journal of Neurophysiology, 73*(1), 347–360. Retrieved from www.ncbi.nlm.nih.gov/pubmed/7714577.

Hamilton, A., & Grafton, S. T. (2006). Goal representation in human anterior intraparietal sulcus. *The Journal of Neuroscience: The Official Journal of the Society for Neuroscience, 26*(4), 1133–1137. https://doi.org/10.1523/JNEUROSCI.4551-05.2006.

Hendrix, C. M., Mason, C. R., & Ebner, T. J. (2009). Signaling of grasp dimension and grasp force in dorsal premotor cortex and primary motor cortex neurons during reach to grasp in the monkey. *Journal of Neurophysiology, 102*(1), 132–145. https://doi.org/10.1152/jn.00016.2009.

Hepp-Reymond, M., Kirkpatrick-Tanner, M., Gabernet, L., Qi, H. X., & Weber, B. (1999). Context-dependent force coding in motor and premotor cortical areas. *Experimental Brain Research. Experimentelle Hirnforschung. Expérimentation Cérébrale, 128*(1–2), 123–133.

Ingram, J. N., Howard, I. S., Flanagan, J. R., & Wolpert, D. M. (2010). Multiple grasp-specific representations of tool dynamics mediate skillful manipulation. *Current Biology, 20*(7), 618–623. https://doi.org/10.1016/j.cub.2010.01.054.

Jeannerod, M. (1984). The timing of natural prehensile movements. *Journal of Motor Behaviour, 16*(3), 235–254. https://doi.org/10.1080/00222895.1984.10735319.

Jeannerod, M. (2009). The study of hand movements during grasping: a historical perspective. In D. A. Nowak & J. Hermsdörfer (Eds.), *Sensorimotor control of grasping: physiology and pathophysiology* (pp. 127–140). Cambridge: Cambridge University Press. https://doi.org/10.1017/CBO9780511581267.011.

Jeannerod, M., Decety, J., & Michel, F. (1994). Impairment of grasping movements following a bilateral posterior parietal lesion. *Neuropsychologia, 32*(4), 369–380.

Jenmalm, P., & Johansson, R. S. (1997). Visual and somatosensory information about object shape control manipulative fingertip forces. *The Journal of Neuroscience: The Official Journal of the Society for Neuroscience, 17*(11), 4486–4499. Retrieved from www.ncbi.nlm.nih.gov/pubmed/9151765.

Jenmalm, P., Schmitz, C., Forssberg, H., & Ehrsson, H. H. (2006). Lighter or heavier than predicted: neural correlates of corrective mechanisms during erroneously programmed lifts. *The Journal of Neuroscience: The Official Journal of the Society for Neuroscience, 26*(35), 9015–9021. https://doi.org/10.1523/JNEUROSCI.5045-05.2006.

Johansson, R. S., & Birznieks, I. (2004). First spikes in ensembles of human tactile afferents code complex spatial fingertip events. *Nature Neuroscience,* 7(2), 170–177. https://doi.org/10.1038/nn1177.

Johansson, R. S., & Flanagan, J. R. (2009). Coding and use of tactile signals from the fingertips in object manipulation tasks. *Nature Reviews. Neuroscience, 10*(5), 345–359. https://doi.org/10.1038/nrn2621.

Johansson, R. S., & Westling, G. (1984). Roles of glabrous skin receptors and sensorimotor memory in automatic control of precision grip when lifting rougher or more slippery objects. *Experimental Brain Research, 56*(3), 550–564. https://doi.org/10.1007/BF00237997.

Johansson, R. S., & Westling, G. (1988). Coordinated isometric muscle commands adequately and erroneously programmed for the weight during lifting task with precision grip. *Experimental Brain Research, 71*(1), 59–71. https://doi.org/10.1007/BF00247522.

Johnston, J. A., Winges, S. A., & Santello, M. (2009). Neural control of hand muscles during prehension. *Advances in Experimental Medicine and Biology, 629*, 577–596. https://doi.org/10.1007/978-0-387-77064-2_31.

Lee, J.-Y., & Schweighofer, N. (2009). Dual adaptation supports a parallel architecture of motor memory. *Journal of Neuroscience, 29*(33), 10396–10404. https://doi.org/10.1523/JNEUROSCI.1294-09.2009.

Li, Z. M., Latash, M. L., & Zatsiorsky, V. M. (1998). Force sharing among fingers as a model of the redundancy problem. *Experimental Brain Research, 119*(3), 276–286. https://doi.org/10.1007/s002210050343.

Lukos, J., Ansuini, C., & Santello, M. (2007). Choice of contact points during multidigit grasping: effect of predictability of object center of mass location. *The Journal of Neuroscience: The Official Journal of the Society for Neuroscience, 27*(14), 3894–3903. https://doi.org/10.1523/JNEUROSCI.4693-06.2007.

Lukos, J. R., Ansuini, C., & Santello, M. (2008). Anticipatory control of grasping: independence of sensorimotor memories for kinematics and kinetics. *The Journal of Neuroscience: The Official Journal of the Society for Neuroscience, 28*(48), 12765–12774. https://doi.org/10.1523/JNEUROSCI.4335-08.2008.

McGurrin, P., Fine, J., & Santello, M. (2017). Context-dependent brain dynamics during grasping and dexterous manipulation. *Soc. Neurosci. Abst.* 406.02, *47th Meeting of the Society for Neuroscience*, Washington DC, USA: November 2017.

McGurrin, P., Parikh, P., & Santello, M. (2015). The role of primary motor cortex in the coordination of digit force and position for dexterous object manipulation. *Soc. Neurosci. Abst.* 244.22, *45th Meeting of the Society for Neuroscience*, Chicago IL, USA: October 2015.

Mojtahedi, K., Fu, Q., & Santello, M. (2015). Extraction of time and frequency features from grip force rates during dexterous manipulation. *IEEE Transactions on Biomedical Engineering, 62*(5), 1363–1375.

Murata, A., Fadiga, L., Fogassi, L., Gallese, V., Raos, V., Rizzolatti, G., & Object, G. R. (1997). Object representation in the ventral premotor cortex (area f5) of the monkey. *Journal of Neurophysiology, 78*(4), 2226–2230.

Nowak, D. A., Berner, J., Herrnberger, B., Kammer, T., Grön, G., & Schönfeldt-Lecuona, C. (2009). Continuous theta-burst stimulation over the dorsal premotor cortex interferes with associative learning during object lifting. *Cortex; a Journal Devoted to the Study of the Nervous System and Behavior, 45*(4), 473–482. https://doi.org/10.1016/j.cortex.2007.11.010.

Nowak, D. A., Hermsdörfer, J., & Topka, H. (2003). Deficits of predictive grip force control during object manipulation in acute stroke. *Journal of Neurology, 250*(7), 850–860. https://doi.org/10.1007/s00415-003-1095-z.

Nowak, D. A., Voss, M., Huang, Y.-Z., Wolpert, D. M., & Rothwell, J. C. (2005). High-frequency repetitive transcranial magnetic stimulation over the hand area of the primary motor cortex disturbs predictive grip force scaling. *The European Journal of Neuroscience, 22*(9), 2392–2396. https://doi.org/10.1111/j.1460-9568.2005.04425.x.

Rice, N. J., Tunik, E., & Grafton, S. T. (2006). The anterior intraparietal sulcus mediates grasp execution, independent of requirement to update: new insights from transcranial magnetic stimulation. *The Journal of Neuroscience: The Official Journal of the Society for Neuroscience, 26*(31), 8176–8182. https://doi.org/10.1523/JNEUROSCI.1641-06.2006.

Rizzolatti, G., Fogassi, L., & Gallese, V. (2002). Motor and cognitive functions of the ventral premotor cortex. *Current Opinion in Neurobiology, 12*(2), 149–154.

Rosenbaum, D. A., Van Heugten, C. M., & Caldwell, G. E. (1996). From cognition to biomechanics and back: the end-state comfort effect and the middle-is-faster effect. *Acta Psychologica, 94*(1), 59–85. https://doi.org/10.1016/0001-6918(95)00062-3.

Rosenbaum, D. A., Vaughan, J., Barnes, H. J., & Jorgensen, M. J. (1992). Time course of movement planning: selection of handgrips for object manipulation. *Journal of Experimental Psychology: Learning, Memory, and Cognition, 18*(5), 1058–1073. https://doi.org/10.1037/0278-7393.18.5.1058.

Salimi, I., Hollender, I., Frazier, W., & Gordon, A. M. (2000). Specificity of internal representations underlying grasping. *Journal of Neurophysiology, 84*(5), 2390–2397.

Santello, M., Baud-Bovy, G., & Jörntell, H. (2013). Neural bases of hand synergies. *Frontiers in Computational Neuroscience,* 7(April), 23. https://doi.org/10.3389/fncom.2013.00023.

Santello, M., Flanders, M., & Soechting, J. F. (1998). Postural hand synergies for tool use. *The Journal of Neuroscience, 18*(23), 10105–10115. https://doi.org/citeulike-article-id:423192.

Santello, M., Flanders, M., & Soechting, J. F. (2002). Patterns of hand motion during grasping and the influence of sensory guidance. *The Journal of Neuroscience, 22*(4), 1426–1435. https://doi.org/22/4/1426 [pii].

Santello, M., & Soechting, J. F. (1998). Gradual molding of the hand to object contours. *Journal of Neurophysiology, 79*(3), 1307–1320. https://doi.org/papers2://publication/uuid/A2B015E6-8AE1-4A49-ACE2-9042E4BE8C17.

Santello, M., & Soechting, J. F. (2000). Force synergies for multifingered grasping. *Experimental Brain Research, 133*(4), 457–467. https://doi.org/10.1007/s002210000420.

Sarlegna, F. R., Baud-Bovy, G., & Danion, F. (2010). Delayed visual feedback affects both manual tracking and grip force control when transporting a handheld object. *Journal of Neurophysiology, 104*(2), 641–653. https://doi.org/10.1152/jn.00174.2010.

Schettino, L. F., Adamovich, S. V., Bagce, H., Yarossi, M., & Tunik, E. (2015). Disruption of activity in the ventral premotor but not the anterior intraparietal area interferes with on-line correction to a haptic perturbation during grasping. *Journal of Neuroscience, 35*(5), 2112–2117. https://doi.org/10.1523/JNEUROSCI.3000-14.2015.

Schieber, M. H., & Santello, M. (2004). Hand function: peripheral and central constraints on performance. *Journal of Applied Physiology, 96*(6), 2293–2300. https://doi.org/10.1152/japplphysiol.01063.2003.

Schmitz, C., Jenmalm, P., Ehrsson, H. H., & Forssberg, H. (2005). Brain activity during predictable and unpredictable weight changes when lifting objects. *Journal of Neurophysiology, 93*(3), 1498–1509. https://doi.org/10.1152/jn.00230.2004.

Shadmehr, R., & Holcomb, H. H. (1997). Neural correlates of motor memory consolidation. *Science (New York, N.Y.), 277*(5327), 821–825.

Tan, H., Wade, C., & Brown, P. (2016). Post-movement beta activity in sensorimotor cortex indexes confidence in the estimations from internal models. *Journal of Neuroscience, 36*(5), 1516–1528. https://doi.org/10.1523/JNEUROSCI.3204-15.2016.

Tunik, E., Frey, S. H., & Grafton, S. T. (2005). Virtual lesions of the anterior intraparietal area disrupt goal-dependent on-line adjustments of grasp. *Nature Neuroscience, 8*(4), 505–511. https://doi.org/10.1038/nn1430.

Tunik, E., Rice, N. J., Hamilton, A., & Grafton, S. T. (2007). Beyond grasping: representation of action in human anterior intraparietal sulcus. *NeuroImage, 36 Suppl 2*, T77–86. https://doi.org/10.1016/j.neuroimage.2007.03.026.

Umilta, M. A, Brochier, T., Spinks, R. L., & Lemon, R. N. (2007). Simultaneous recording of macaque premotor and primary motor cortex neuronal populations reveals different functional contributions to visuomotor grasp. *Journal of Neurophysiology, 98*(1), 488–501. https://doi.org/10.1152/jn.01094.2006.

van Nuenen, B. F. L., Kuhtz-Buschbeck, J., Schulz, C., Bloem, B. R., & Siebner, H. R. (2012). Weight-specific anticipatory coding of grip force in human dorsal premotor cortex. *The Journal of Neuroscience: The Official Journal of the Society for Neuroscience, 32*(15), 5272–5283. https://doi.org/10.1523/JNEUROSCI.5673-11.2012.

Westling, G., & Johansson, R. S. (1984). Factors influencing the force control during precision grip. *Experimental Brain Research. Experimentelle Hirnforschung. Expérimentation Cérébrale, 53*(2), 277–284.

Winges, S. A., & Santello, M. (2004). Common input to motor units of digit flexors during multi-digit grasping. *Journal of Neurophysiology, 92*(6), 3210–3220. https://doi.org/10.1152/jn.00516.2004.

Winges, S. A., Kornatz, K. W., & Santello, M. (2008). Common input to motor units of intrinsic and extrinsic hand muscles during two-digit object hold. *Journal of Neurophysiology, 99*(3), 1119–1126. https://doi.org/10.1152/jn.01059.2007.

Wolpert, D. M., Diedrichsen, J., & Flanagan, J. R. (2011). Principles of sensorimotor learning. *Nature Reviews Neuroscience, 12*(12), 739–751. https://doi.org/10.1038/nrn3112.

Zatsiorsky, V. M., & Latash, M. L. (2004). Prehension synergies. *Exercise and Sport Sciences Reviews, 32*(2), 75–80. https://doi.org/10.1097/00003677-200404000-00007.

Zatsiorsky, V. M., & Latash, M. L. (2008). Multifinger prehension: an overview. *Journal of Motor Behavior, 40*(5), 446–476. https://doi.org/10.3200/JMBR.40.5.446-476.

Zhang, W., Gordon, A. M., Fu, Q., & Santello, M. (2010). Manipulation after object rotation reveals independent sensorimotor memory representations of digit positions and forces. *Journal of Neurophysiology, 103*(6), 2953–2964. https://doi.org/10.1152/jn.00140.2010.

PART IV

Reach-to-Grasp in Developmental Robotics

Issues and Modelling

12

REACHING FOR OBJECTS

A Neural Process Account in a Developmental Perspective

Gregor Schöner, Jan Tekülve, and Stephan Zibner

12.1 Introduction

Reaching and grasping objects is an evolutionarily late achievement. Throughout the animal kingdom, much object-oriented action is achieved with the whole body, in mammals often with the head and snout. Because the vision sensor is anchored in the head, such object-oriented movements can be achieved with simple control strategies, such as visual servoeing (Ruf & Horaud, 1999), that make limited demands on perception, estimation, and movement planning. Reaching with an actuator that is separate from the main visual sensor is prevalent in primates. Humans excel at object-oriented manipulation tasks, much exceeding the skills of other primates. This is a developmental achievement as witnessed by the long and intense period of learning to reach (Berthier & Keen, 2006; Thelen, Corbetta, & Spencer, 1996; von Hofsten, 1984, 1991).

Spatial orientation and navigation may be achieved based on unsegmented visual information (Hermer & Spelke, 1994; Schöne, 1984). In contrast, reaching makes considerable demands on object perception. (1) To reach and grasp an object, that object must be visually segmented against the background and its pose be estimated. (2) Spatial information about the object must be transformed from a visual reference frame into a reference frame in which motor commands to the hands may be formed. Such coordinate transforms are computationally demanding (Pouget & Snyder, 2000; Schneegans, 2015) and learning them is a challenge (Chao, Lee, & Lee, 2010; Sandamirskaya & Conradt, 2013). (3) Motor commands must be generated that drive the hand toward the object and bring it into contact with the object with a small enough terminal velocity that enables grasping. This is particularly challenging as infants are weak relative to the mass of their limbs. Because the force/weight relationship changes during growth, the motor commands must be updated over development. (4) Generating and controlling successful reaches also entails solving the degree of freedom problem, that is, distributing to the many muscles that contribute to arm movement a coordinated set of commands that achieve contact with the hand (Latash, 2008; Sporns & Edelman, 1993). Little is known to date about how this problem is solved during development. (5) Finally, reaching movements must be initiated and then terminated when the hand makes contact with the object. Typically, successful reaches and grasps are followed by further actions like mouthing or banging the object on a table. The sequential organization of such component movements is part of the motor skill involved in reaching.

This broad range of component processes that span perception, cognition, and motor control, may be a reason why many questions about object-oriented reaching and grasping behavior remain unanswered (see, e.g., section on motor control in Lisman, 2015). At one level, a developmental perspective may be particularly attractive. In order for an infant to successfully reach and grasp an object, all processes must be in place and coordinated with each other. Observing how reaching is assembled over development may help in understanding how the pieces come together.

Alas, studying the development of reaching empirically is very hard. Babies do not follow task instructions. They must be seduced to repeatedly reach and parametric manipulations of the reaching movements are difficult to impose. Reaches are highly variable and individual differences important (Thelen et al., 1996). It is hard to measure movement trajectories in detail, as preparing infants with markers takes time and patience, inducing attrition. Once movement trajectories have been obtained, their analysis is hampered by the difficulty to reliably detect onsets and offsets of individual movements and thus, to align trajectories across trials (see, however, Corbetta & Thelen, 1995) for a systematic trial-by-trial solution). In fact, young infants often move continuously, sometimes in approximately periodic fashion.

To date, three data sets about the development of reaching are most elaborate (Berthier & Keen, 2006; Thelen et al., 1996; von Hofsten, 1991). Their properties are reviewed in detail and commented upon in Caligiore, Parisi, and Baldassarre, 2014. Unfortunately, the data are not strongly constraining for theoretical accounts as of now, even though they include longitudinal studies that provide samples of reaching behavior in individuals along developmental time. A principle problem of all accounts of motor learning during development is, of course, that the actual process of motor learning is difficult to observe, because it takes place every hour of the awake and behaving time when infants work on their motor skills with intensity and dedication.

The development of reaching is also hard to study theoretically. First, there is currently no complete neurally grounded theory of reaching! A model from our own group (Martin, Scholz, & Schöner, 2009) integrates process accounts for the generation of timed movement commands, for movement initiation and termination, for solving the degrees of freedom problem, and for muscular control. An account for the scene and object perception, and the processes through which such perceptual information is transformed into motor frames is missing. Also, the solution of the degree of freedom problem is not neurally grounded and questions persist about how well the muscle model solves control problems in reaching. A model based on the framework of optimal control (Shadmehr & Krakauer, 2008) does a better job at addressing these control problems, but is far from neurally grounded, does not address muscle properties, and is also missing a perceptual and movement preparation component. A recent model of learning to reach (Caligiore et al., 2014) takes muscle properties into account while providing an account for control, and addresses the neural grounding of movement parameters, but is not addressing movement initiation and termination nor scene perception and the associated coordinate transformations. The most neurally grounded models of reaching are still those from the Grossberg school (Cisek, Grossberg, & Bullock, 1998), which have been mapped in quantitative detail to neural data for some components (Cisek, 2006). They fall short of addressing scene and object perception as well as the initiation and termination of movements. Closest to integrating all processes come recent proposals based on Dynamic Field Theory (Fard, Hollensen, Heinke, & Trappenberg, 2015; Strauss & Heinke, 2012; Strauss, Woodgate, Sami, & Heinke, 2015). The synthesis (Tekülve, Zibner, & Schöner, 2016; Zibner, Tekülve, & Schöner, 2015a, 2015b) we review here is very similar in spirit and strongly overlaps with these proposals with respect to scene perception and motor planning. We elaborate movement timing, motor control, and the sequential organization of movement in more detailed ways, using different process models.

Second, understanding how reaching develops requires an understanding not just of motor learning, but of learning in the many different component processes from perception to control. Current accounts typically focus on an individual component. The most complete model so far (Caligiore et al., 2014), for instance, focuses on learning to control the arm by predicting the torque profiles needed to reach. Other models look at the learning of kinematic movement plans (Schlesinger, Parisi, & Langer, 2000), of kinematic models (Herbort & Butz, 2009; Narioka & Steil, 2015; Sun & Scassellati, 2005), or of feedback control parameters (Savastano & Nolfi, 2013).

Finally, accounts for learning are often based on a learning regime in which the organization in time of the learning process is not achieved autonomously by the learning system itself, but imposed from the outside (see Sandamirskaya & Storck, 2015, for discussion). In particular, autonomously learning from experience requires much processing "infrastructure." For instance, to learn from the correlation between a motor command and the perceived outcome, as assumed in many of the "motor babbling" accounts, the movement system must be able to keep the motor command in working memory to bridge the considerable delay between the two neural events.

12.1.1 Goals of This Chapter

In this chapter our goal is to review the component processes that must be in place to successfully reach for and grasp objects. The review will be conceptual first, but will also discuss relevant neural principles. We will illustrate the concepts through a concrete neural dynamic model that provides a process account from sensory inputs to generating movement of a simulated biomechanical plant. The model is not quantitatively anchored in experimental data, but we do discuss the link to experimental signatures.

A second step we want to make is expose what happens when some of the components are not in place. By comparing the various ways the model may break down in a form of reverse development, we aim to uncover potential behavioral signatures of development, and also highlight all the problems the human nervous system must solve to reach successfully.

A key contribution of the model is that the reaching behavior itself is entirely autonomous. There is no hidden "script" that activates each component process as required (as is the case for many robotic demonstrations of reaching). In particular, we will show how correction movements emerge from the time-continuous neural dynamics when an initial reach is not complete, and use this observation to account for the emergence of multiple movement units.

Finally, within a limited setting we will show how the organization of motor acts can be learned autonomously from behavior. This demonstration of principle will highlight all the processing infrastructure required to achieve such autonomous learning and provides a perspective for what a full account for learning to reach may need to address.

12.2 A Neural Process Account of Reaching

In a neural process account of reaching, the neural networks of the brain are linked to sensory and motor systems to bring about the motor behaviors that achieve the reaching act. Such an account may be contrasted to abstract "curve fitting" models, which by themselves do not explain how behavior emerges. For instance, to postulate that an infant selects the interpretation of a stimulus that maximizes its likelihood does not explain how its nervous system actually does that even while such a model may provide a fit to certain psychometric curves.

Theoretical models that are process accounts may ultimately be linked to real sensors and real actuators, in robotic demonstrations of reaching. Of course, such robotic demonstrations are never perfectly consistent with what is known about how organisms generate movement. No actuator perfectly mimics real muscles nor do technical sensors mimic the neural function of the retina or other sensory organs.

A neural process account is always based on the choice of a particular level of description that entails a particular level of neurally mechanistic detail. That choice is constrained by what is known about the neural substrate of motor behavior. In spite of enormous progress, the exact neural circuitry underlying motor behavior remains unknown (Lisman, 2015). Neural process accounts are thus usefully founded on neural principles, rather than on a detailed description of the neural substrate and associated neural mechanisms. We know, of course, that the central nervous system is not a digital computer and does not have a CPU, into which it can load data to operate on. Thus, process accounts that depend conceptually on algorithms are not by themselves compatible with neural principles.

The most commonly accepted level of description at which neural principles can be articulated and used to develop process accounts of behavior is the level of neural population activity (Bullock, Cisek, & Grossberg, 1998; Cisek & Kalaska, 2005; Georgopoulos, 1986). The framework of Dynamic Field Theory (DFT) is positioned at this level (Erlhagen & Schöner, 2002; Schöner, Spencer, & the DFT Research Group, 2015). In DFT, properties of the strongly recurrent neural networks that generate behavior are formalized into a set of mathematically expressed concepts. Central is the notion of stability, the capacity of neural activation patterns to resist change. Rather than review yet again the neural principles formalized in DFT, we introduce them as we go through the neural process model of reaching behavior and its development.

What is entailed in making a reaching movement oriented toward an object? Figure 12.1 illustrates the five processes that are minimally required to generate such behavior.

12.2.1 Scene Perception

Object-oriented movement behavior requires, first of all, perception of the environment and attentional selection of the object, toward which the behavior

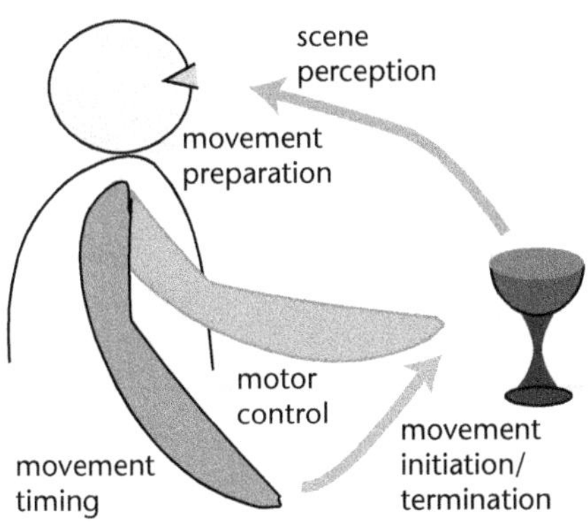

FIGURE 12.1 A schematic illustration of five component processes entailed in generating a reaching movement aimed at an object

is directed. Such scene perception is predominantly visually driven, although reaching toward haptically identified or memorized object locations is possible. Humans are very good at perceiving and memorizing visual scenes, much better than at memorizing arbitrary material. For instance, individuals who looked at 10 objects per naturalistic scene, for 10 seconds each, had 90% recall of object identity or pose even a week later (Hollingworth, 2004). Our visual cognition is particularly well tuned to this problem.

Scene perception clearly is tightly linked to looking and attention. Only portions of a scene that have been attended to are memorized well enough to detect change (when transients are masked; Simons, 2000). The attentional selection of visual objects as well as the selection of gaze locations is commonly accounted for through the notion of visual salience that characterizes the stimulus properties determining the probability of attraction gaze or covert attention (Itti & Koch, 2000). The process of such selection decisions is captured by strongly recurrent neural networks, formalized as neural dynamics (Kopecz & Schöner, 1995; Schöner et al., 2015), as illustrated in Figure 12.2. In the neural field version of neural dynamics, visual space is represented by an activation field in which localized peaks of activation indicate a selected visual location. Such peaks may be induced by localized input, which may reflect the saliency of the visual array, and are stabilized against decay by local excitatory interaction. Selection is enabled by longer-range inhibitory interaction. In Figure 12.2, the field selects the left-most local maximum of input through local excitatory

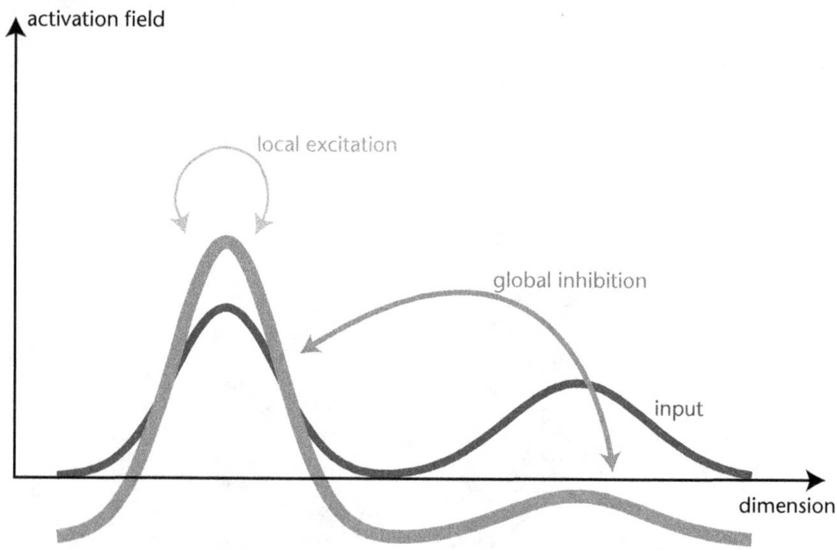

FIGURE 12.2 A neural activation field is defined over some dimension and receives bimodal input, and generates a peak of activation over one of the two stimulated locations

interaction, which lifts activation to above threshold values at that location (more activation than input) and global inhibition which suppresses activation at other locations (below the level of input). When neural interaction within the activation field is sufficiently strong, peaks may be sustained even as localized input is removed. Sustained activation provides an account for visual working memory (Johnson, Spencer, Luck, & Schöner, 2009).

Here is a very brief mathematical tutorial on dynamic neural fields, expressed for a neural activation field, $u(x, t)$, defined over a single dimension, x (in the model, the fields comprise multiple dimensions such as the two-dimensional visual array or two movement parameters). Activation evolves according to this integro-differential equation:

$$\tau\dot{u}(x,t) = -u(x,t) + h + s(x,t) + \int dx' w(x - x')g(u(x',t)). \quad (1)$$

The terms up to the integral are a time-continuous version of an input-driven neural dynamics as it is commonly used to model the time courses of neural activation. Without input, s, activation is in a resting state, $h<0$, that is stable with time constant, τ. Weak localized input, $s(x)$, shifts that stable state to $u(x)=h+s(x)$. The neural dynamics (through the "$-u$" term) creates time courses through which activation tracks changes in input $s(x, t)$ as an exponential low-pass filter.

This input-driven regime becomes unstable once the activation level at any location, x, approaches the threshold at zero defined by the sigmoidal nonlinearity, $g(u)$ (in the modeling below chosen as $g(u) = (1 + \beta u/(1 + \beta|u|))/2)$. This is when neural interaction within the field engages as $g(u)$ becomes positive. Locations, x', close to the activated location, x, provide excitatory input ($w(x - x') > 0$). Locations, x', far from the activation location, x, provide inhibitory input ($w(x - x') < 0$). This pattern of neural interaction within the field stabilizes peaks of activation centered on the location at which input first pushed activation beyond threshold. When activation is pushed above threshold at more than one location, such peaks may instantiate selection decisions in which one activation peak suppresses peaks at alternate locations. Which location is selected depends on the timing and strength of input, on the prior patterns of activation, and on random fluctuations of activation. When neural interaction is sufficiently strong, peaks may be sustained even after the inducing localized input is removed. The location of the peak then encodes a working memory for the past selection decision.

Because scene perception involves gaze and attentional shifts, only a part of the visual array is in the attentional foreground at any moment in time. Scene perception is thus largely based on memory (Hollingworth, 2004), in what may be more appropriately called scene representation.

Early forms of reaching are not always associated with looking at the object (von Hofsten, 1984). As reaching develops, that link becomes closer. Coordinating attention and reaching is, therefore, a developmental achievement rather than a logical necessity. After reaching has been established, the relationship between looking and reaching remains complex, however. Corbetta and associates (Corbetta, Thurman, Wiener, Guan, & Williams, 2014) presented infants with object large enough for them to reach toward different locations on the object. These infants were followed longitudinally from about 2 to 12 months of age. Early in that developmental window, the location at which infants looked did not predict the location toward which infants reached. Over time, infants looked more to the location on the object to which they reached.

In this chapter, we will use a simplified neural process model of reaching to illustrate ideas. Figure 12.3 provides an overview over the architecture. The entire topic of scene perception has been trivialized in this model by assuming that a distribution of activation defined in body-centered coordinates is available to the processes of movement preparation. Visual locations of reachable objects are marked by localized maxima of that distribution of activation.

12.2.2 Movement Preparation

There is ample behavioral and neural evidence that movements are prepared ahead of their initiation (see Erlhagen & Schöner, 2002 for review). Thus, for instance, the time needed to initiate movement after a stimulus has specified the movement goal, the reaction time, reflects the metrics of the movement alternatives. If those alternatives are metrically closer to each other, reaction time is shorter, reflecting more overlap between the neural activation states that correspond to either movement (McDowell, Jeka, Schöner, & Hatfield, 2002). In fact, these neural activation states can be directly observed in motor and premotor cortex in the form of peaks of activation in a distribution of population activation (Bastian, Schöner, & Riehle, 2003; Cisek & Kalaska, 2005; Georgopoulos, 1986). Activation even in motor cortex precedes movement initiation and predicts movement parameters. Finally, at the kinematic level, reaching movements directed at an object start out in the direction of the object in adults, so that from the first milliseconds of its trajectory, the path of the hand and the movement time can be predicted. This is a consequence of the robust kinematic regularity that characterizes adult movement (Soechting & Lacquaniti, 1981).

It is the movement parameters, that characterize movements as a whole, that have specific values from the very start of the motor act. Most prominent among these is the direction of the hand's movement in space, the extend of the movement of the hand through space (amplitude), the overall duration of the movement (movement time) and other parameters such as the anticipated level

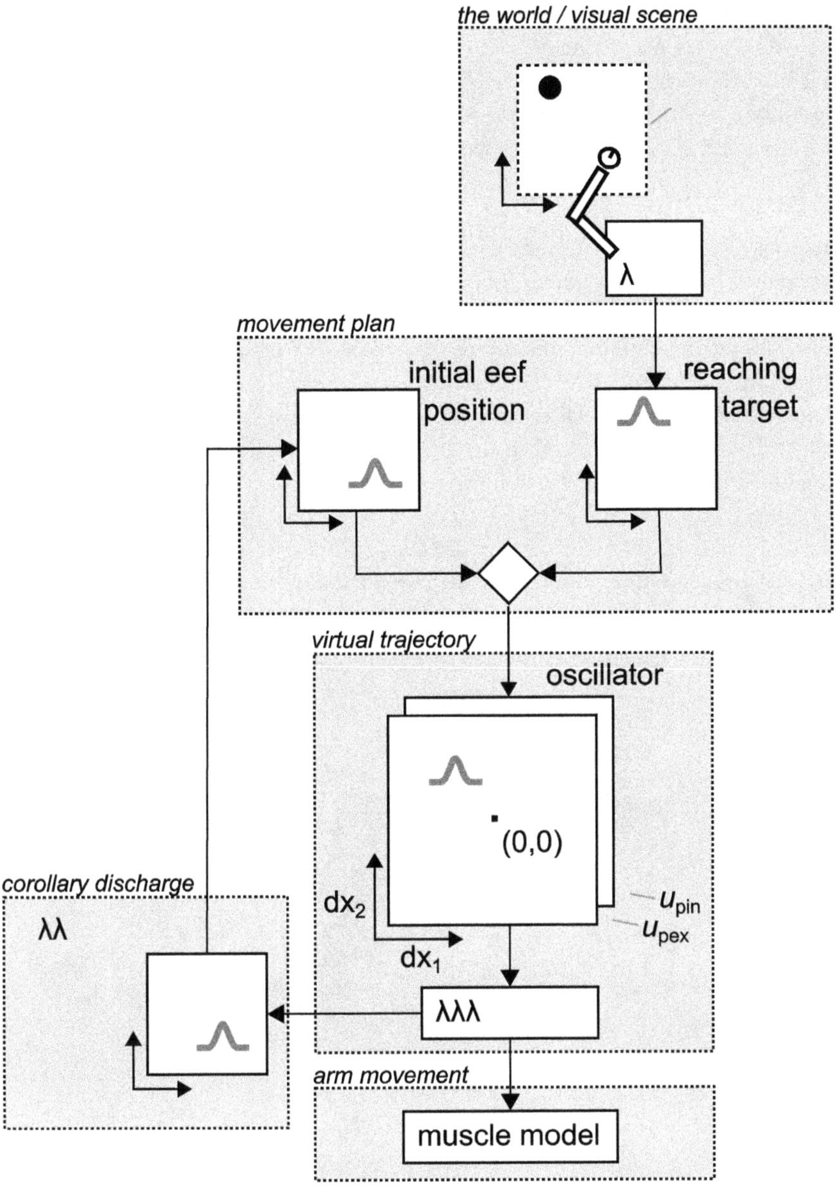

FIGURE 12.3 A survey over the neural dynamic architecture used in this chapter to illustrate the five key processes of movement generation

of resistance to the movement. Movement preparation thus means determining the values of such movement parameters.

Once an object has been selected in the scene representation as the target of a reaching movement, specifying kinematic movement parameters like movement direction and amplitude involves some simple geometrical computations (Bullock & Grossberg, 1988). The direction, for instance, is the angle that the line connecting the initial position of the hand to the object forms relative to some reference axis (see Figure 12.4). Such computations are trivial to implement on a computer, but not in a neural network. Neural networks do not take an "argument" and "operate" on it. They need a particular pattern of connectivity that brings about the computation and that is linked to a neural representation of the argument. In this instance, the computation amounts to a coordinate transform (Figure 12.4): If the spatial representation of the target is transformed into a reference frame that is centered in the initial position of the hand, then the direction and amplitude of the reaching movement can be read off with a fixed neural mapping. For instance, all field locations along a line from the center out vote for the corresponding movement direction. Similarly, all field locations on a circle of a given radius vote for the matching movement amplitude.

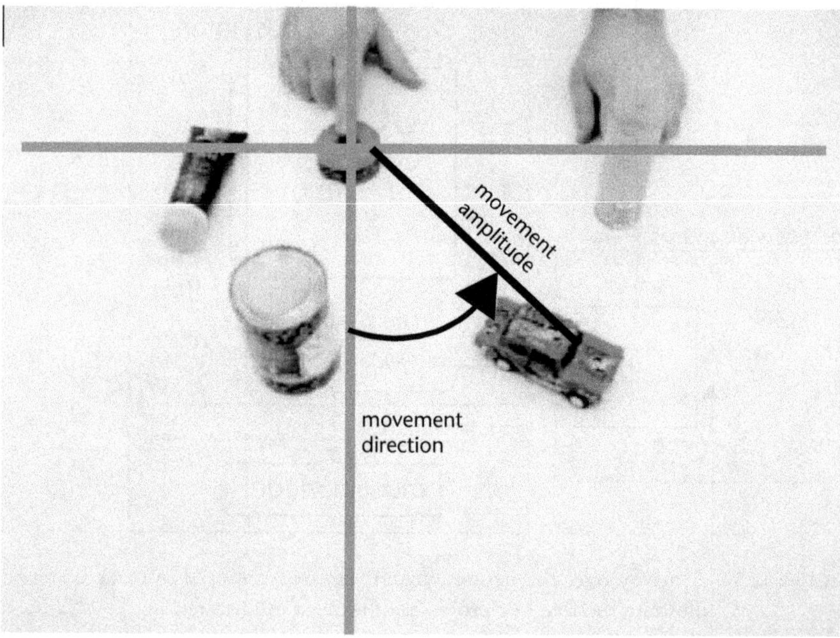

FIGURE 12.4 The kinematic movement parameters that characterize a reach toward an object may be obtained by transforming the neural representation of the location of the object into a frame of reference anchored in the initial position of the hand

We sketch here how the coordinate transform from a visual reference frame to a frame anchored in the initial position of the hand can be formalized mathematically. Let $u_{ini}(x, t)$ be a neural field that represents the initial position of the hand through a peak at the appropriate location (x is a two-dimensional position vector that spans all possible initial positions of the hand). The location of the target object is represented as a peak in the neural field, $u_{tar}(x), t)$, defined over the same two-dimensional space (both spatial representations could be thought of being body-centered). A hand-centered neural representation of the target object can be obtained from this neural map:

$$s_{plan}(x,t) = \iint dx_1' dx_2' \sigma(u_{ini}(x - x'))\sigma(u_{tar}(x',t)) \qquad (2)$$

Every location, $x' = (x_1', x_2')$, in the target field has a connection to every location, x in the movement planning field. At any given moment, only a subset of these connections is active, those selected by the peak in the field that represents the initial position of the hand. The selection mechanism formalized here is a form of "shunting" in which the activation in one set of neurons can turn off and on the connection between two other sets of neurons (the map is also referred to as "sigma-pi" connectivity). This is not the only possible neural implementation of the transform (see Pouget & Snyder, 2000; Schneegans, 2015 for more). The transform is essentially a "steerable" neural map: the mapping from one space, in which the visual object is represented, onto another space, in which the movement direction is represented, is steered by the initial position of the hand. Such transforms are costly: many connections and many neurons are needed and their connectivity pattern needs to be just right. The maps create invariance: no matter where the hand is initially posted, the direction of the movement can be determined. The learning of such transforms is a likely developmental challenge (Sandamirskaya & Storck, 2015).

12.2.3 Movement Timing

Reaching movements, like most voluntary human movements, are timed. That is, the hand's trajectory and velocity profile has a characteristic shape that is reproducible across repetitions and scales with movement amplitude (Morasso, 1981). If a movement sequence (for example, in hand writing) is performed at different spatial scales, the relative timing of the corresponding pieces of movement (the arc forming the letter "e," for instance), remains invariant (the "e" takes up the same percentage of the total movement time) (Viviani & Flash, 1995). Reaching movements are coordinated across limbs so that when timing demands on one limb are varied, the movement timing of the other limb adjusts to retain the same relative timing (Kelso, Southard, & Goodman, 1979). Reaching movements may also be coordinated with perceived events such as when moving objects are intercepted. That coordination is stable in the sense that the reaching movement is updated in response to visual information about the timing of the movement target (Brenner, Driesen, & Smeets, 2014). Coordination is maintained even where it is not strictly necessary. The opening and closing movement that achieves grasping is coordinated with the transport component that moves the hand toward the grasping object (Jeannerod, 1984). In principle, the hand could just open and stay open until the object is reached. Coordination means, however, that a later reach is accompanied by a later opening of the hand for the grasp as well.

The theoretical understanding of timing is based on the notion of an oscillator (Schöner, 2002). More precisely, clocks are dynamical systems that generate reproducible and stable time courses formalized as stable limit cycles. The stability of limit cycles means that the time courses generated resist change. Coupling multiple limit cycles generically leads to their synchronization, an account for coordination.

The concept of a dynamical system generating stable time courses is not restricted to accounts of periodic movement. Temporally discrete motor acts such as a single reaching movement can be understood on the same basis (Schöner, 1990). The idea is that the dynamical system generates a stable time course, which ends, however, in a stable stationary state. (We will address below the processes of initiation and termination required in such a view.) More specifically, a neural oscillator consists of an excitatory population that is reciprocally coupled to an inhibitory population (Amari, 1977; Ermentrout, 1998). An initial stimulus sets in motion an "active transient" in which activation first rises, then falls as inhibition cancels activation. Different dynamic properties of such neural oscillators may generate trajectories of differing amplitude and duration. Our account of reaching movement postulates an ensemble of such neural oscillator. When a peak forms that represents a movement plan, it drives a subpopulation of these neural oscillators that has the appropriate amplitude and movement time to ultimately generate a successful reaching movement (after going through kinematic transformations, muscle activation, and biomechanics, to be discussed below). The projection from these neural oscillators to the downstream neural processes is learned so as to achieve these movement goals (Figure 12.3).

There are good arguments as to why the timing signals are generated in spatial terms, such as the direction and speed of the hand's movement in space. Neural data are supportive of that idea (Moran & Schwartz, 1999; Schwartz, 1994) as are data about movement coordination (Mechsner, Kerzel, Knoblich, & Prinz, 2001) that are consistent with the ideomotor principle according to which movement is generated in the same reference frame in which it is perceived. The ease with which we coordinate the hand's timing with the motion of perceived objects is suggestive of such a reference frame as is the invariance of movement timing with scale (see above).

That movement timing poses a developmental challenge is intuitive, although there seems to be little direct empirical evidence about the development of movement coordination. Counter-intuitively, analysis of spontaneous kicking movements in infants revealed that in-phase and alternating coordination patterns are observable from earliest infancy (Thelen, 1981).

A possible mathematical form of the neural dynamics of movement timing invokes two fields that together form neural oscillators. The excitatory layer, $u_{ex}(x, t)$, and the inhibitory layer, $u_{in}(x, t)$ are coupled as follows:

$$\begin{aligned}\tau_{ex}\dot{u}_{ex}(x,t) = -u_{ex}(x,t) + h + s_{plan}(x,t) + c_{mov}g(u_{mov}^{int}(t)) \\ - \int dx' w_{ex,in}(x-x')\rho(u_{in}(x',t))\end{aligned} \quad (3)$$

$$\tau_{in}\dot{u}_{in}(x,t) = -u_{in}(x,t) + h + s_{plan}(x,t) + c_{mov}g(u_{mov}^{int}(t)) \quad (4)$$

where $\tau_{ex} < \tau_{in}$ and $w_{ex,in}(x-x')$ is an interaction kernel with local excitatory and lateral inhibitory interaction. Anywhere across these homogeneous fields, a local region may generate an active transient, once it receives input from the motor plan, $s_{plan}(x, t)$. Activation in both excitatory and inhibitory layers then rises. The slower rise of inhibition begins to suppress activation in the excitatory layer, eventually suppressing activation below the threshold of the semi-linear transfer function

$$\rho(u(x,t)) = \begin{cases} u(x,t) & \text{for } u(x,t) > 0 \\ 0 & \text{else} \end{cases} \quad (5)$$

This dynamics thus performs a one-shot active transient in response to input. Input by a neural node, u_{mov}^{int}, is required to induce the active transient. This node is part of a neural dynamics of behavioral organization, that generates the sequence of initiation and termination events that organizes reaching movements (see Section 12.2.5).

The field of neural oscillator generates a (virtual) velocity vector

$$v(t) = \iint dx_1 dx_2 \omega(x)\rho(u_{ex}(x,t)). \quad (6)$$

as its output. The weight function, $\omega(\bullet)$, is learned using gradient descent as a linear function of $x = (x_1, x_2)$. These weights define tuning curves of the neural oscillator field, in which each field site is sensitive to both the orientation and the distance (and thus peak velocity) of the movement.

12.2.4 Motor Control

To physically generate the movement, actuators must be given motor commands in such a way that deviations from average trajectory are limited by stability properties of the plant or the control. In human reaching movements, the actuators are, of course, muscles attached to the skeletal musculature. The fact that limbs are actuated at the joint level alone generates challenges for motor control: As a joint is accelerated, interaction torques act on other joints that may move these off a commanded path. For instance, if you rapidly move your elbow while relaxing your wrist, your hand flaps around. This movement at the wrist is a consequence of interaction torques. The wrist is in an accelerating reference frame, inducing inertial interaction torques, that also rotates, inducing coriolis and centrifugal interaction torques.

Conventional robot arms deal with such interaction torques with "brute force": servo controllers keep each joints velocity and position close to the commanded values, applying higher forces to control away these disturbances. Human motor control is very different. Muscles are relatively weak. As elastic elements, they are generally weaker than the inverse pendulum that body segments form with the vertical gravitational form (Loram & Lakie, 2002). Muscles do have feedback control loops, including those through the spinal cord. But the delay in these reflex loops, on the order of about 50 ms, are relatively long compared to movement times as short as 300 ms.

In what form do motor commands ultimately bring about muscle activation? Because the activation of muscles depends on their length and inner state, motor commands do not directly specify muscle forces. More appropriately one may conceive of muscle commands as parameters modulating the peripheral reflex loops that establish the length dependency of muscle activation (Feldman, 1986, 2011). Specifically, the descending motor command could be viewed as setting a threshold for the activation of the muscle. When the muscle length is above that threshold, muscle activation is induced that leads to the generation of muscle force that contracts the muscle. That relationship between muscle length and muscle force implies that muscles are effectively non-linear springs, whose resting length can be varied by the descending motor command.

Because motor commands are thus spatial in nature, they may be derived from the spatial timing signal through a kinematic transform: from the hand's movement direction and speed to muscle lengths and their rates of change. This transformation is complex and contains problems not fully understood at this time. First, multiple muscles act on each joint. The muscles that act in opposing directions of movement, agonists and antagonists, may have different levels of co-contraction that do not affect the joint angle, but joint stiffness. Joint stiffness is not specified by the kinematic transformation and the control of the related joint compliance is one of many open issues. In our model, we assume a particular level of stiffness that remains constant through movement, and thus neglect the known variation of joint stiffness during movement. Second, the redundancy of muscles acting in the same direction on a joint is also a topic of much research. We neglect that issue by lumping all muscles acting on a joint into a single muscle model. The muscle lengths are thus combined into a single joint angle. This is a "virtual" joint angle that describes the motor command given to this ensemble of muscles. The real joint angle then evolves in a way that may not always be close to the virtual joint angle depending on interaction torques and other factors. Third, the kinematic transform faces kinematic redundancy: In most movement systems and certainly for reaching, there are more joints available to achieve the movement of the relevant end-effector (here the hand in space) than necessary. There is extensive research on how these many degrees of freedom are harnessed to realize a particular trajectory of the hand in space (Latash, Scholz, & Schöner, 2007; Martin et al., 2009; Todorov & Jordan, 2002). We condense all that into a kinematic transform that at any moment in time selects one particular virtual joint angle velocity vector across all joints.

The mathematical formalization of the sketched ideas is as follows: The virtual velocity vector, $\boldsymbol{v}$, is transformed into a virtual joint velocity vector, $\dot{\boldsymbol{\lambda}}$, using the pseudo-inverse (upper index $^{+}$) of the manipulator Jacobian, $\boldsymbol{J}$, that depends on the current virtual joint configuration $\boldsymbol{\lambda}(t)$:

$$\dot{\boldsymbol{\lambda}} = \boldsymbol{J}^{+}(\boldsymbol{\lambda}(t))\boldsymbol{v}(t) \tag{7}$$

The virtual joint trajectory, $\boldsymbol{\lambda}(t)$, is obtained by integrating over time.

The virtual joint trajectory defines a set of motor commands to all muscles that transitions from an initial to a final equilibrium length of each muscle. In a simplified, second order linear model of the muscle and its reflex loops, the virtual joint trajectory sets the time-varying equilibrium point for each joint, combining the contributions of all agonist and antagonist muscles converging on each joint:

$$\ddot{\theta} = -K(\bar{\theta} - \lambda) - B\dot{\theta} \tag{8}$$

where θ is the joint angle vector, and $\boldsymbol{K}$ and $\boldsymbol{B}$ are the effective stiffness and viscosity matrices of the muscle groups.

The simple model does not address biomechanics. The kinematics of a seven degrees-of-freedom arm was solved numerically to compute the hand's trajectory in space (see Section 12.3).

12.2.5 Movement Initiation and Termination

All voluntary movement consists of sequence of events. A reach is initiated and ends, but may also entail the initiation of the opening movement of the hand and then its closing for a grasp. Other movements may follow once an object has been grasped. A reaching movement may be interrupted or aborted at any time (Bullock & Grossberg, 1988).

The sequential organization of movement is a special case of the problem we like to call *behavioral organization*, a problem much overlooked in neural models of behavior and cognition (Graziano, 2006; Pezzulo & Cisek, 2016; Sandamirskaya, 2015). In motor control, models will often focus on a single phase of the motor act, such as the phase of movement generation itself. Cognitive models may talk about the serial order of events, but fail to address how these events are organized in time and interact with the time course of behavior generation between events (Botvinick & Plaut, 2004).

An account for behavioral organization is necessary to endow neural models with a form of autonomy: No outside "script" or algorithm must be invoked to understand how neural systems move from one task to another. To date, many connectionist models are limited in their degree of process autonomy. This is true for learning processes as well, in which a learning protocol is often imposed from outside of the connectionist model (we will come back to this topic in Section 12.4).

One particular challenge to a neural process account for behavioral organization is the need to understand the relationship between events that occur at discrete moments in time, the transitions from one action to another, and the movements themselves that take varying amounts of time. For instance, when a reaching movement takes longer because an obstacle is avoided (Grimme, Lipinski, & Schöner, 2012), then the switch to the next phase of the object-oriented action must occur later.

Clearly, behavioral organization is a developmental challenge. Infants do not get the orders of movement tasks right a lot of the time (von Hofsten, 1984). And getting them right is a prerequisite to success.

In saccadic eye movement, the neural basis of aspects of behavioral organization is understood somewhat. A population of neurons in the superior colliculus is known to be active, while the animal fixates gaze on a visual structure. A saccade is initiated when the fixation population becomes deactivated. Theoretical accounts for this form of transition (Kopecz, 1995; Wilimzig, Schneider, & Schöner, 2006) were the origin of the ideas we now sketch (Richter, Sandamirskaya, & Schöner, 2012). The key idea is that motor behaviors are activated or deactivated by dedicated neural nodes that reflect the activation in small populations of neurons that have common input. These nodes may shunt neural connections or act by boosting or suppressing activation in a population. Only neural populations in the self-stabilized state of supra-threshold activation are "active" in the sense that they can impact on the downstream processes. The transition from inactive to active takes the form of the detection instability discussed earlier. The transition from active to inactive takes place through a separate, reverse detection instability. In both cases, even graded changes of input lead to events at a discrete moment in time, when the instability is induced.

The neural nodes modulate the conditions at which these instabilities may occur. They are themselves governed by the same type of neural dynamics and thus their own activation may undergo the dynamic transition from "off" to "on" and vice versa. Three types of neural nodes are distinguished based on their functional role in organizing sequences of neural states. "Intention" nodes represent neural populations that enable a behavior by boosting its processes to push them through the detection instability if the appropriate sensory inputs are available. "Condition of satisfaction" (CoS) nodes become activated when input from an associated intention node is matched by input from sensory or internal sources predicted to signal the successful completion of the intended motor behavior (Sandamirskaya & Schöner, 2010). "Condition of dissatisfaction" (CoD) nodes becomes activated when input from an associated intention node is matched by input from sensory or internal sources predicted to signal the failure of the intended motor behavior.

Mathematically, the sequential organization of movement is achieved by a neural dynamic architecture illustrated in Figure 12.5 (and expanded in Section 12.4 to include grasping and looking behaviors). While the intention node, $u^{\text{int}}_{\text{reach}}$, is active, reaching is being pursued. A single timed movement act is initiated when the intention node, $u^{\text{int}}_{\text{mov}}$, becomes active. The neural dynamics of this node

$$\begin{aligned} \tau \dot{u}^{\text{int}}_{\text{mov}}(t) = &-u^{\text{int}}_{\text{mov}}(t) + h + c_{u,u} g(u^{\text{int}}_{\text{mov}}(t)) + c_{\text{mov,reach}} g(u^{\text{int}}_{\text{reach}}(t)) \\ &- c_{\text{int,CoS}} g(u^{\text{CoS}}_{\text{mov}}(t)) - c_{\text{int,CoD}} g(u^{\text{CoD}}_{\text{mov}}(t)) \end{aligned} \quad . \tag{9}$$

is analogous to that of activation fields (Equation 1), with the interaction kernel being replaced by self-excitation of strength, $c_{u,u}$. The *activated* state of the node with an activation level above the threshold of the sigmoid function, $g(\bullet)$ is analogous to the self-stabilized peaks of activation in neural fields. The node becomes activated when input from the reaching intention node, $u^{\text{int}}_{\text{reach}}$, pushes the sub-threshold state, $u^{\text{int}}_{\text{mov}} \approx h +$ inputs, to the detection instability. Inhibitory coupling from Condition of Satisfaction (CoS) and Condition of Dissatisfaction (CoD) nodes may induce loss of stability of the activated state in the reverse detection instability.

Once the intention to move is activated, it enables the neural timing field to generate a virtual velocity trajectory: In Equation 4, an activated $u^{\text{int}}_{\text{mov}}$ is necessary for localized input from the movement plan, $s_{\text{plan}}(x, t)$, to drive the neural oscillator field above threshold to generate an active transient (somewhat analogously to the "go" signal of Bullock et al. (1998), although it does not shape the velocity profile).

The end of a single time movement is signaled by the activation of the condition of satisfaction, $u^{\text{CoS}}_{\text{mov}}$.

$$\tau \dot{u}^{\text{CoS}}_{\text{mov}}(t) = -u^{\text{int}}_{\text{CoS}}(t) + h + c_{u,u} g(u^{\text{CoS}}_{\text{mov}}(t)) - c_{\text{CoS,ex}} g(u^{\text{pd}}_{\text{ex}}(t)) - c_{\text{CoS,in}} g(u^{\text{pd}}_{\text{in}}(t)) \tag{10}$$

that is driven by the oscillator field through peak-detectors (upper index pd) that spatially integrate over the output of an activation field. A peak in the inhibitory layer excites, a peak in the excitatory layer inhibits. Thus, the CoS is activated once a peak in the inhibitory, but not in the excitatory layer, signals that a single movement event has taken place.

The intention to move may also be deactivated if the neural representations of the initial position of the hand or the target are lost (Figure 12.5), signaled by the CoD. The position of the hand is continuously updated by integrating in time the virtual velocity. This corollary discharge signal updates the neural representation of the initial position of the hand when the movement has been terminated. The reaching intention remains activated until a match is detected between the updated initial hand position and the target position, so that the hand has reached the target.

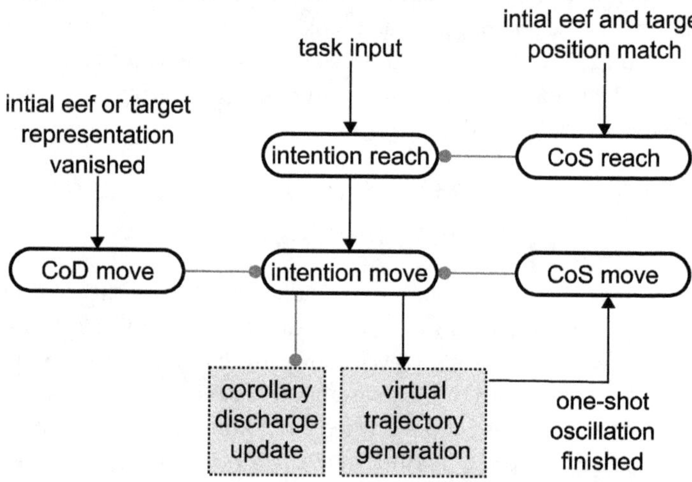

FIGURE 12.5 A part of the neural dynamic network that organizes reaching movements

Note: Arrows indicate excitatory input, lines with a circle at its end indicate inhibitory input.

12.3 Results

To illustrate properties of the neural process model of movement generation and its developmental implications we show simulation results of the complete model. For all experiments, we use artificial visual inputs in the form of localized peaks of activation in a simulated visual array. Compared to using real camera input, this gave us full control of stimulus strength and position for reproducibility. We used the simulation solution Webots (www.cyberbotics.com) to execute the movements with the seven degrees-of-freedom arm. This made it possible to simulate neural dynamics that reflected imperfect control (the "impaired" system, see Section 12.3.2) and produced jerky motor commands without risking damage to robotic hardware. The correctly tuned architecture was tested on hardware as well (using a RGB camera and the Kuka lightweight arm), but this will not be discussed here (see Zibner et al., 2015a for more details).

12.3.1 Making a Movement

Figure 12.6 illustrates how the model makes individual movements from an initial hand position to three different visual targets. The parameters of the model were tuned to put the different components of the model into the dynamic regime described. The movement paths (left) are approximately straight with slight outward curvature. They are generated by the virtual hand velocity profile (right panel, left-most curves). Because movement time was set to be equal across the three movements, peak velocity scaled with movement amplitude. The resulting real velocity profiles of the hand's movement in space (right panel, right-most curves) are considerably delayed over the virtual velocity profiles due to the low-pass filter properties of the muscle model. Near peak real velocity, the virtual velocity has already returned to zero. This is consistent with experimental estimates of the timing relationship between virtual and real trajectories (Ghafouri & Feldman, 2001). Note that in more realistic muscle models there are additional sources of delay such as transmission delays in the reflex loop and the slow buildup of muscle force due to calcium dynamics (Gribble, Ostry, Sanguineti, & Laboissière, 1998). The qualitative picture is, therefore, adequate.

Such offsets between the timing signal generated by the neural oscillator system and the realized movement trajectory raise a range of interesting challenges. In particular, the organization of movement sequences must take into account such delays. Moreover, coordination among different effector systems or with external timing events need to be robust against such timing offsets.

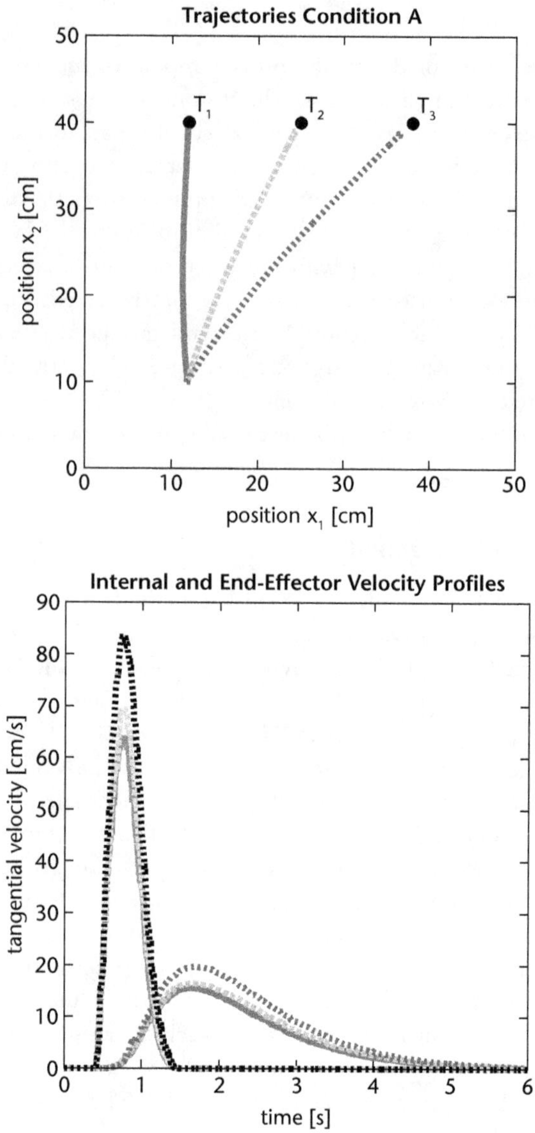

FIGURE 12.6 Exemplary trajectories (left) and profiles of tangential velocity (right) for virtual movements (early large peaks) and end-effector movements (late smaller peaks) for different movement targets

12.3.2 Developmental Signatures

We argued that all components of the movement generation system must be in place and functional in order to bring about reaching movements directed at objects. The simplified integrated model enables us now to examine the ways object-oriented movement fails when different components of the model are not in place. We examine three different ways the model may be "degraded" to explore potential developmental signatures of early stages of reaching. (A) Movement plan: The lateral neural interaction within the neural field that represents the initial position of the hand was weakened. As a result, the working memory of that initial position required for the duration of the movement becomes unstable, leading to a potential loss of the movement plan. The virtual movement trajectory may then be terminated prematurely. (B) Movement timing: Neural connectivity from the field of neural oscillators that generates the timing of the virtual trajectory may be not be appropriately tuned so that virtual trajectories do not reach the target in a single cycle. This was simulated by setting some of the weights from the oscillator field to random values. (C) Motor control: The kinematic mapping from the virtual hand trajectory to the joint-level motor commands may not be appropriately tuned. This was modeled by adding a random offset to the inverse kinematic mapping.

We performed experiments in each of these three forms of "impairment" as well as when all three forms were combined (A + B + C). In each condition, we simulated movements from three starting positions to three target positions, repeating each simulation once with different random values of different forms of degradation. This made for a total of 2.9 = 18 repetitions. We analyzed the resulting trajectories and velocity profiles for the number of movement units (MUs) and for the straightness of the movement path and computed the variance of either measure across the 18 trials. Movement units are segments of the real hand velocity profile between two minima of velocity. Straightness was measured as done in experimental analyses (Thelen et al., 1996). Table 12.1 provides an overview of the results. Individual sample trajectories from one starting position to three target positions are shown in Figure 12.7, which may be compared to the sample trajectories of the intact system in Figure 12.6.

TABLE 12.1 Mean and variance of number of movement units and straightness index for different impairment conditions

Conditions	*A*	*B*	*C*	*A + B + C*	*N*
Mean # of mus	3.00	4.00	1.89	7.56	1.00
Variance	0.00	0.00	0.61	7.53	0.00
Mean straightness	0.998	0.983	0.845	0.941	0.989
Variance	<0:001	<0:001	0.017	0.002	<0:001

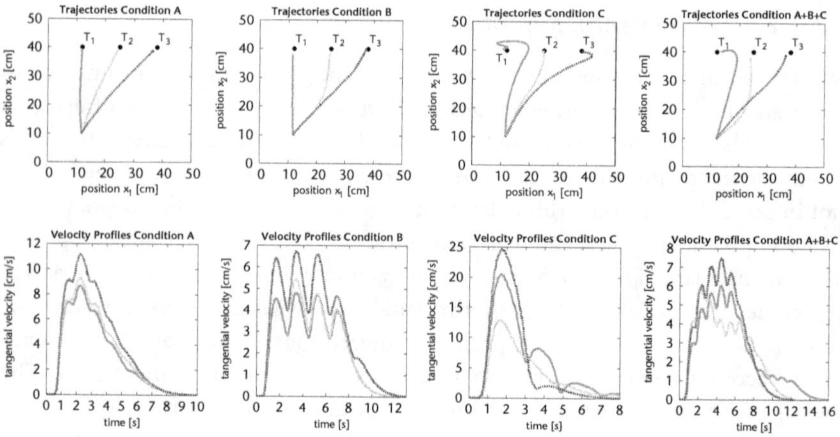

FIGURE 12.7 Trajectories (top) and velocity profiles (bottom) for starting position S_1 to the three targets (solid for T_1, bold-dotted for T_2, dotted for T_3) and, from left to right, the conditions A, B, C, and the combination of all three

When the movement planning system is impaired (A), the velocity profiles are less smooth, featuring more than one movement unit. This is due to the intermittent interrupt of movement generation when a planning peak is lost. The system autonomously resets the movement plan and makes a new movement attempt. When movement timing is impaired (B), the virtual trajectory generated typically does not reach the target. This number of movement units is large and the movement path is less smooth as a result. When the motor control system is impaired (C), movement paths are no longer approximately straight. Multiple movement units are observed as the system does not reach the target in response to a single virtual velocity profile. The combination of all impairments generates variance in the movement path and velocity profile that may begin to approximate reaches of young infants. These observations are borne out by the statistical analysis of the two indices analyzed here (Table 12.1). The developmental path suggests that impairment in motor control induces the strongest variance, followed by impairments in movement timing, followed by impairments in movement planning. It is actually difficult to generate the kind of macroscopic variance of movement paths and trajectories observed in infants from a neural processing model while still reaching the target. Noise added to the model at the processing level (e.g., in the muscle model) will typically show up as high-frequency fluctuations of the movement trajectories around a mean that is close to the intact model. We believe that the structural sources of variance high-lighted here point the way toward a correct account for movement variance in development.

12.4 The Developmental Learning Process

We have sketched how different stages of development may be associated with graded differences in the underlying neural processing structure of movement generation. How may changes in these processes emerge from experience? This is the question of autonomous learning: How may learning processes unfold in a spontaneously behaving system rather than in a special "learning mode" in which actions are elicited and rewarded from the outside?

We all know that this is a difficult question. It is difficult in experiment because we cannot readily observe infants and children as they gather experience, practice skills, and learn. Typically, even in longitudinal studies, we may only have a chance to occasionally take a measurement that reflects the current level of competence. But this is also difficult in theoretical modeling. Autonomous learning is not well understood. The majority of neural models are "learned up" in special learning scenarios (Sandamirskaya & Storck, 2015).

We illustrate the conceptual issues in learning from experience by restricting ourselves to one particular subproblem, the problem of learning to sequentially organize different elementary behaviors. We expand the reaching model of the previous section to include looking behavior, opening and closing the hand, as well as moving the hand back to a resting position. We know that the sequential order of these elementary movement behaviors is not a given as young infants do not necessarily get the order right. Von Hofsten (1984) reported that infants of up to 4 weeks of age generated pre-reaches, movements of the hand in the general direction of an object, even though they often did not visually fixate the object. The frequency of such pre-reaches then decreased while the total time of looking at objects increased. At around 10 weeks of age, pre-reaches re-emerged, now often accompanied by visual fixation of the object. Later, a similar transition led to serial organization of reaching and hand opening and closing.

Structural support for the sequential organization of different elementary behaviors is a prerequisite for learning that organization. Figure 12.8 illustrates the neural dynamics for two elementary behaviors, A and B. In the simple model (Tekülve et al., 2016) we use five such elementary movement behaviors: visual fixation, bringing the hand to a resting position, reaching, opening the hand, and closing the hand. Each elementary behavior consists of two neural nodes that directly interact with sensorimotor processes: (1) an intention node, that projects onto the sensorimotor layer and initiates the motor behavior, and (2) a condition of satisfaction (CoS) node that receives input from the sensorimotor layer that signals the completion of the motor behavior. The intention node is self-excitatory and coupled to an inhibitory node. Together, these two nodes may be in different dynamic regimes, including bi-stable "on" and "off" states, a sustained "on" state, and a neural oscillation between "on" and "off." This is the basis for autonomous activation of an elementary behavior, that plays

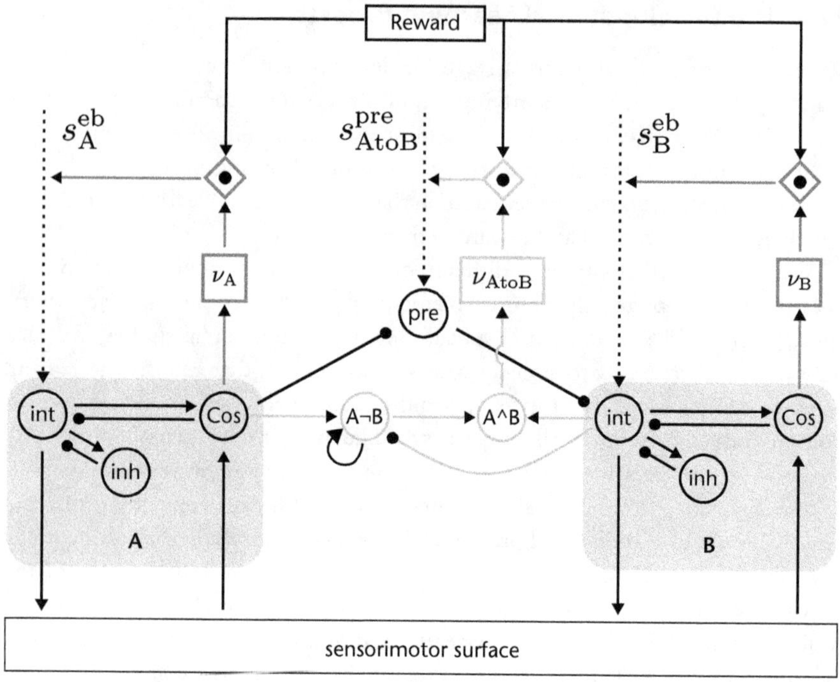

FIGURE 12.8 The processing structure for learning the behavioral organization of movement is illustrated for a pair of two elementary movement behaviors, A and B

Notes: The same structure exists for all pairs of elementary behaviors. Solid lines are fixed weights assumed to exist from the outset. Dashed and dotted lines illustrate weights that are learned autonomously from an initial value of zero. The dashed lines support learning the boosts of each elementary behavior. The dotted lines support learning the sequential order of the two elementary behaviors.

a role in learning: In a form of "behavioral babbling," the intention node may activate itself in the presence of noise, and remain activated for a certain amount of time as illustrated in Figure 12.9.

We also postulate structural support for behavioral organization by assuming that each behavior's condition of satisfaction is coupled to each other behavior's intention node. This connection is doubly inhibitory, consistent with the organization of behavioral selection in the basal ganglia (Chevalier & Deniau, 1990), through an intermediary "pre-condition" node (Figure 12.8). If the pre-condition node between behavior A and behavior B is activated, it inhibits its target, behavior B. If the condition of satisfaction of the source behavior, A, then becomes activated, it inhibits the pre-condition node and releases the target from inhibition. This double inhibitory coupling from behavior A to behavior B thus organizes a sequential pattern of activation in which behavior B can only become active once behavior A has been previously activated. The

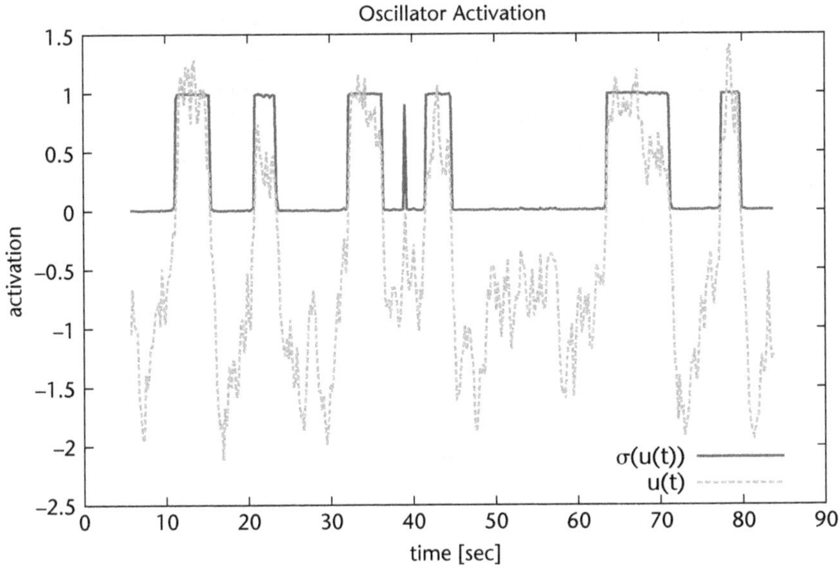

FIGURE 12.9 Activation trace of a single intention node (dotted line) and its output (solid line) during behavioral "babbling." The neural dynamics generates a highly unstable oscillatory pattern

connectivity is effective only if the pre-condition neuron is active in the absence of input from its source.

The learning process varies the resting level of activation of any elementary behavior, s_A^{eb}, and of the pre-condition nodes, s_{AtoB}^{pre} and thus modulates how easy it is to activate which elementary behavior in which order. We were led to introduce yet more specific processing structure for the learning process, inspired by Houk's model of the basal ganglia (Beiser & Houk, 1998). Two additional interneurons represent different patterns of ordering. One node receives excitatory input from the condition of satisfaction of behavior A and inhibitory input from the intention node of behavior B. It becomes active, if behavior A was completed, but behavior B was not activated. Through self-excitation this node sustains its activation which then serves as a working memory. This node provides excitatory input to a second node which becomes activated once behavior A is activated. That second node thus reflects a particular sequential history of activation, first A, then B.

Autonomous learning faces two conceptual problems. First, activation patterns may come and go. Unless they are actively maintained as working memory (Sandamirskaya & Storck, 2015), these patterns may not necessarily be around when a learning event occurs. This problem is solved, we propose, by keeping memory traces of activation patterns (Erlhagen & Schöner, 2002). These act as low-pass filters, that keep activation patterns available to learning processes for a certain amount of time, that may vary depending on what other, perhaps competing activation patterns arise. Second, autonomously learning from behavior is mediated by neuromodulators. Only at certain moments in time, that are linked to reward, does an "update" of the learning process occur. There is a rich literature on the relevance of timing of reward (Gallistel & Gibbon, 2000) and this problem is central to the mathematical concept of reinforcement learning. In most models of learning, that learning takes place at discrete moments in time is taken for granted and is modeled by using discrete iterative time in the learning rules. Because behavior and neural processes unfold continuously in time, however, this is a problem. One solution lies in the concept of eligibility traces (Sutton & Barto, 1990), in which a population of neurons is transiently activated by each detected "event," generating a window within which the learning step can take place (Kazerounian, Luciw, Richter, & Sandamirskaya, 2013). Learning is then modulated by the presence of reward and a matching activation pattern.

We provide a brief review of the mathematics of three elements of autonomous learning. (1) The memory trace, $v(t)$, reflects the recent history of activation, $u(t)$. The memory trace builds up (on a faster time scale, τ_{build}) whenever the activation, $u(t)$, is above threshold, and decays (on the slower time τ_{mem}) when the activation variable falls below the threshold:

$$\dot{v}(t) = \frac{1}{\tau_{\text{build}}}\{-v(t) + g(u(t))\}g(u(t)) + \frac{1}{\tau_{\text{mem}}}\{-v(t)(1-g(u(t)))\} \quad (11)$$

(2) An eligibility trace defines the time window in which learning takes place by building a transient reward signal, $r(t)$. In the model, an neural field, u_{rew}, receives input from both the visual perception of the hand and the target. Whenever the two come sufficiently close, a peak is formed in a detection instability. A neural dynamics analogous to the timing model (Equation 4).

$$\tau_{\text{ex}}\dot{u}^{\text{elig}}_{\text{ex}}(t) = -u^{\text{elig}}_{\text{ex}}(t) + h + g(u_{\text{rew}}) - c_{\text{ex,inh}}g(u^{\text{elig}}_{\text{inh}}(t)) \quad (12)$$

$$\tau_{\text{inh}}\dot{u}^{\text{elig}}_{\text{inh}}(t) = -u^{\text{elig}}_{\text{inh}}(t) + h + g(u_{\text{rew}}) \quad (13)$$

(where $\tau_{\text{ex}} < \tau_{\text{inh}}$) translates that discrete movement in time into a transient, time-continuous reward signal $r(t) = g(u^{\text{elig}}_{\text{ex}}(t))$.

(3) The learning rule updates the resting levels of elementary behaviors, s^{eb}, and of pre-condition nodes, $s^{\text{pre}}_{\text{AtoB}}$, whenever their memory traces and the reward signal are simultaneously activated:

$$\dot{s}^{\text{eb}}_{\text{A}}(t) = r(t)\lambda^{\text{eb}}\{v_{\text{A}} - (1 - H(v_{\text{A}}))\} \quad (14)$$

$$\dot{s}^{pre}_{AtoB}(t) = r(t)\lambda^{pre}\{v_{AtoB} - (1 - H(v_{AtoB}))\} \quad (15)$$

where $H(\bullet)$ is the Heaviside step function, and $\lambda_{eb} < \lambda^{pre}$ are the learning rates. Thus, in the presence of a reward signal, only resting levels with matching memory traces are strengthened ($v - (1 - H(v)) = v$), all others decay ($v - (1 - H(v)) = -1$). In the absence of a reward signal, all resting levels are left unchanged.

We performed simulations in which either of two possible targets were presented at all times (Figure 12.10). Movements were autonomously generated by behavioral babbling. If the reaching behavior was activated, but now peak in the target field was present because the fixation behavior had not previously been activated, then a peak was induced in the target field at a random location. The hand returned to the resting position when the associated intention node was activated. Paths generated at three phases of the learning process are illustrated. Figure 12.11 shows how the system first learns to boost the precondition node that reflects that visual fixation precedes reaching. This is because trials in which that order occurred were most likely to achieve reward as the visual target impacted on the movement planning. Once this is learned, the system learns to boost the elementary behaviors' fixation and reach. The boost to reaching precedes the boost to visual fixation, because reaching is activated closer in time to the reward and thus has a stronger memory trace when updating occurs.

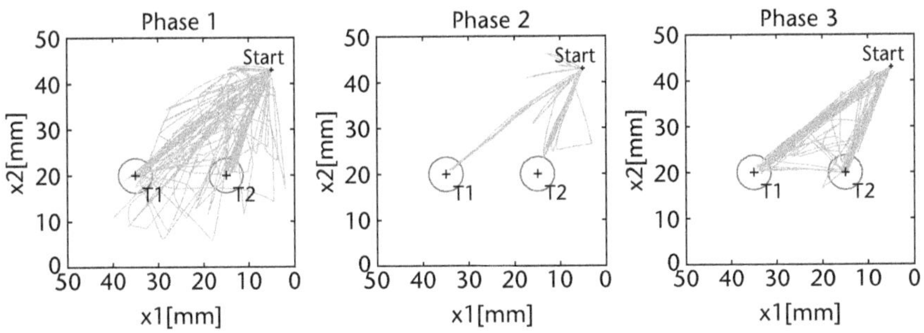

FIGURE 12.10 Movement paths generated by the model in three phases of the learning process

Notes: Early in learning (left), the reach behavior becomes activated with or without a prior fixation behavior. As a result, movements may go to either the visual targets or other random targets. Once the sequential organization "fixate-then-reach" has been learned (middle), reaching movements occur less frequently, because they are inhibited unless a fixation preceded reaching. Reaches tend to be successful, however. After the two behaviors fixate and reach have been boosted (right), reaches are more frequent and are directed at the visual targets.

The movement paths in Figure 12.10 reflect that pre-reaches first go down in frequency. This is because the newly learned ordering now makes activation of reaches less likely if they are not preceded by visual fixation. Once visual fixation picks up, so does reaching. That pattern matches the observations of von Hofsten.

The opening and closing of the hand is not rewarded in these simulations and, therefore, their activation is not promoted during the learning process (Figure 12.11). This was meant to illustrate that the learning process can autonomously "pick out" the relevant behaviors. In reality, grasping the object may also be a rewarding event, of course.

12.5 Discussion

In this chapter, we have reviewed the five basic processes that are minimally required to direct movements at objects. To illustrate the style in which these processes may be neurally implemented, we have formulated and then simulated a simple neural dynamic model of movement generation. This model served as a platform for discussing the many developmental challenges that infants face as they learn to reach for objects.

Clearly, this model is as yet very rudimentary. Many components are mere place holders for neurally more realistic accounts. In many cases, more behavioral data may be needed to test or revise the assumptions we have made. We step through the components to discuss these various limitations and point to open questions and future directions of research.

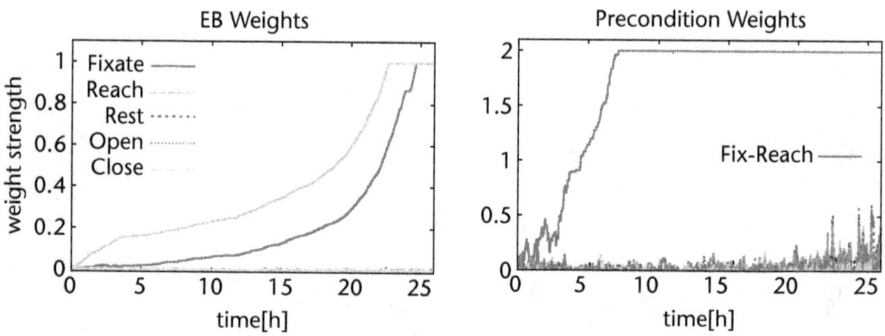

FIGURE 12.11 Evolution over learning time of the neural weights

Notes: Left: The boosts to elementary behaviors "fixate" and "reach" increase over time. Right: Only the boost to the pre-condition neural node for "fixate-then-reach" increases over learning time, all other pre-condition nodes remaining near zero.

12.5.1 Scene Perception

We have only sketched the whole problem of scene perception and not modelled the underlying processes here (but see Zibner, Faubel, Iossifidis, & Schöner, 2011).

Looking behavior is extensively used to assess visual attention and perception in infants, both experimentally (Colombo & Mitchell, 2009) and theoretically (Perone & Spencer, 2013b; Schöner & Thelen, 2006). Typically, these studies make use of dwell time, that is, the duration of infants' looks at particular parts of the visual array. In both habituation and preferential looking paradigms, the visual array is changed all the time. So if infants are building scene representations, those are constantly being challenged as the scene is changing. Scene perception in early infancy itself is not well studied and understood.

Scene representations must naturally be invariant against gaze shifts. A representation in retinal coordinates does not enable redirecting gaze or attention to an object from different viewing angles. Representing the visual array in a coordinate frame that is centered in the body enables such gaze shifts. The transformation from visual input in retinal coordinates to such a body-centered reference frame is neuronally costly (Pouget & Snyder, 2000; Schneegans, 2015). It requires projections from every retinal location to every possible location in a body reference frame. The current gaze angle selects the subset of these projections that reflects the current mapping at a given gaze direction. The demands on the neural machinery required to make coordinate transforms may be reflected in developmental processes that support the learning of such coordinate transforms (Sandamirskaya & Storck, 2015). At this time, little is known about how infants learn coordinate transforms.

On the other hand, infants clearly need to know about the objects they reach to. The intriguing results of Corbetta et al. (2014) suggest, that while infants may be paying attention to an object they reach to on a global scale, the exact location on the object toward which they reach is not predicted by where they look. Is it possible, that learning to reach may impact on learning to look? Because the hand is attached to the body and the visual scene may be represented in a gaze-invariant, but body-related reference frame, it is thinkable that sensory and efference copy processes engaged in reaching contribute to the learning of spatial scene representations and of the associated coordinate transforms.

12.5.2 Movement Preparation

We have assumed that reaching movements are prepared before they are initiated and that this preparation is based on information about both the movement target and the initial position of the hand. This is consistent with a broad range of empirical evidence, from neurophysiology to kinesiology and cognitive psychology: neural populations encoding the hand's movement direction in space, approximately straight movement paths of the hand in space with smooth and invariant velocity profiles, and the dependence of reaction times on movement parameters.

On the other hand, some data point toward a special role of the end-point of a movement compared to the initial posture. Stimulating small neural populations in motor cortex, Graziano (2006) has often seen movements of the hand to particular locations on the body from wherever the hand happens to be initially, for instance. Infants learning to reach do not look at their hands prior to movement initiation and do not seem to pay attention to where their hands are (Corbetta et al., 2014). In fact, their hands seem to be all over the place as they attempt to reach repeatedly (Thelen et al., 1996). So is it possible that there would be an alternative approach to movement, one in which movement consists of moving directly to a new postural state irrespective of initial state and the spatial path required? We do not think that this is plausible. Timing and control problems are really quite different when the movement originates at other locations. A different angle may be this: Perhaps the current position of the hand is at all times neurally represented (as it is in our model). In Graziano's experiment, the activated neural populations first drive an update of a neural representation of the initial position of the hand based on this ongoing estimate. The same population of neurons may thus drive different movements depending on the initial position of the hand. This would be consistent with neural data from Scott and Kalaska (1997) who found that neural tuning curves in motor cortex depend on the arm's kinematic configuration.

Infants may have a difficulty organizing this sequence of processing steps. That difficulty may account for some part of the variance of their early reaches. Learning to reach may thus entail learning to temporally organize this sequence of processing steps (see Section 12.5.5).

12.5.3 Movement Timing

Looking at infants' hand movements, it is obvious that one of their major difficulties is to get the timing of their reaches right. Some infants reach the target at high speeds, others make meek movements. Over development, these patterns of timing converge (Thelen et al., 1993, 1996).

We have talked about and modeled timing in an abstract way: A field of neural oscillators generates a virtual velocity profile that was assumed to be isomorphic in some way to the velocity of the hand's trajectory in space. Clearly, there is more to timing in movement generation. The time course of motor commands needs to be adapted and fine-tuned not only to assure coordination with other actions and perceived events, but also to support control, overcoming variations of inertia, interaction torques, or external forces acting on the moving limb. This is linked to the problem of optimal control: determining the right control signals that move an actuator to a desired state while satisfying constraints.

There is a huge literature on how adults learn to adapt motor commands so as to compensate for external force fields of a different kinds, including force fields that vary with velocity (Shadmehr & Mussa-Ivaldi, 2012). Interestingly, that experimental work has typically imposed force-fields on the end-effector, often the hand, rather than on individual joints (in part, for technical reasons, perhaps). An idea could be that an adaptation of the field of neural oscillators and of its projections onto the motor control level may be capable of modulating timing signals to accommodate the demands of these tasks. This is an open research question that may have impact for development because the changes in strength, mass, and geometry that occur as infants grow requires adaptation.

12.5.4 Motor Control

Learning to time motor commands to reach targets is closely related to the issue of motor control, of course. In fact, researchers embracing the language of the theory of optimal control and of control theory do not typically break up the system into "timing" and "control" (although one could think of the optimal control signal as reflecting the timing level and the feedback controller as reflecting the control level).

One reason we propose to break up the system into "timing" and "control" is because of the particular way in which control of movement and control of posture are unified in human movement. The most general and consequential insight from the equilibrium point account of muscle neurophysiology is that due to the peripheral spinal circuitry (in particular, the stretch reflex) the descending control signal is essentially spatial in nature (Feldman, 2011). Muscles are activated as a function of how long they are (as sensed by muscle spindles, for instance) relative to a threshold. It is this threshold that is set by the descending motor command (and its rate of change), which can therefore be thought of as a length (and a velocity).

This solves a fundamental problem of muscle neurophysiology: After bursts of activity during a movement, muscles return to silence (or to an equilibrium level of activity that may reflect stiffness demands and gravitational forces) even though the lengths of different muscles are no different than they were before the movement was started. So, the threshold of the stretch reflex must be reset during the movement. That requirement is sometimes left out of accounts that focus on the computation of the right torque profile that may overcome the various biomechanical interaction torques that plague movement. Computed torques as signals to the muscle are not only physiologically unrealistic, but also fail to address the problem of resetting the equilibrium length of muscles (which does not mean that motor commands may not be adjusted to deal with forces, e.g., Tee, Burdet, Chew, & Milner, 2004).

One hypothesis that we embrace, but have not elaborated here, is that the peripheral neural circuitry also simplifies the motor control problem (Gribble et al., 1998; Raphael, Tsianos, & Loeb, 2010). This may alleviate some of the problems of dealing with interaction torques or the torques experienced at individual joints due to external forces acting on the hand. A critical issue is how to resolve co-contraction and the redundancies among muscles systems.

The wager of equilibrium point thinking is that much of the remaining problem of control can be achieved through kinematic transformations that shape virtual joint trajectories (or muscle length trajectories). Kinematic transformations from an abstract "hand-in-space" virtual trajectory, $\boldsymbol{r}(t)$, to joint- or muscle-level virtual trajectories, $\boldsymbol{\lambda}(t)$, may go a long way toward addressing control problems if we go beyond the simplistic ideas of inverse Jacobians that we employed here (Eq. 7). This is particularly attractive in a dynamical system (or neural dynamics) approach, in which the virtual joint trajectory is represented as the state of a dynamical system which may be used to express the configuration and velocity dependent modulation of motor commands, e.g.,

$$\dot{\lambda}(t) = \hat{\boldsymbol{J}}\,(\lambda(t), \dot{\lambda}(t)) \cdot \dot{\boldsymbol{r}}(t) \tag{16}$$

where the function, $\hat{\boldsymbol{J}}$, may serve as an internal model that must be learned. (Note that this implicit dynamical system needs to be resolved first.) Such a generalized kinematic transform could be thought of as a steerable neural mapping, analogous to the coordinate transforms we discussed earlier (Eq. 2). This mapping is from the spatial representation of the hand's virtual velocity profile, $\dot{\boldsymbol{r}}(t)$ to the virtual joint level trajectory, $\boldsymbol{\lambda}(t)$ and $\dot{\boldsymbol{\lambda}}(t)$, steered by these same variables. Understanding how such kinematic transforms may be learned autonomously is one of the outstanding problems of developmental motor control (see Guenther, Bullock, Greve, & Grossberg (1994) for early attempts to achieve such).

This is linked to the "uncontrolled manifold" (UCM) effect, a signature of how the neural processes of movement generation deal with the inherent redundancy of the kinematics and dynamics of reaching (Scholz & Schöner, 1999; Tseng, Scholz, Schöner, & Hotchkiss, 2003). The UCM effect is the observation that there is more variance in those directions of joint space along which the hand position remains invariant than in directions along which the hand position varies. The UCM effect has been interpreted as revealing of a control strategy that focuses on the control of the hand in space (Latash et al., 2007; Martin et al., 2009; Todorov & Jordan, 2002;). Such UCM structure of variance may thus serve to probe the structure of the kinematic transformation. In particular, self-motion within the UCM is an index to incomplete compensation for interaction torques (Martin et al., 2009). Such analyses are difficult to perform for children, and very difficult for infant reaching movement because many of the technical prerequisites of estimating the structure of variance are not met in these cases (e.g., temporal alignment of trajectories across different movement). It would be very attractive if these difficulties could be overcome.

12.5.5 Movement Initiation and Termination and Learning to Organize Movements

The autonomous behavioral organization of reaching movement is a key problem, that is, perhaps, least well understood of all component processes. Behavioral organization is critical to integration, in which the component processes are initiated at the right time and hand on their output to downstream components when required. We have argued that sequentially organizing components of object-oriented action is a developmental achievement, referring to the data of von Hofsten (1984) for exemplary demonstration.

The critical issue is how the discrete moments in time, at which transitions among different components or different elementary movements occur, emerge from the underlying continuous time in which neural processes unfold. This is where the concepts of dynamical field theory provide a key element, the detection instability. At that instability, graded changes of input induced a categorical change that occurs at a discrete moment in time.

The emergence of such discrete events is also a key element of autonomous learning. We have argued that in autonomous learning, updates of neural connectivity only happen at particular points in time, at which a rewarding achievement is detected. Patterns of activation that lead up to such achievements must be kept around to properly assign credit at these learning events. We suggested that memory and eligibility traces were the simplest mechanism that may serve such function. Clearly, much more needs to be understood about autonomous learning.

We have also illustrated some of the processing structure that enables reaching movements, introducing the notions of intentional nodes, conditions of satisfaction, and of dissatisfaction. Learning requires even more prior structure, essentially the substrate to extract cues about the sequence of events, to link motor commands to their consequences, and to explore the behavioral options. Autonomous learning behavior from experience is, in our view, a major frontier of theoretical thinking in development (Perone & Spencer, 2013a; Sandamirskaya & Storck, 2015). We are only at the very beginning.

References

Amari, S. (1977). Dynamics of pattern formation in lateral-inhibition type neural fields. *Biological Cybernetics, 27*, 77–87.

Bastian, A., Schöner, G., & Riehle, A. (2003). Preshaping and continuous evolution of motor cortical representations during movement preparation. *European Journal of Neuroscience, 18*, 2047–2058.

Beiser, D. G., & Houk, J. C. (1998). Model of cortical-basal ganglionic processing: encoding the serial order of sensory events. *Journal of Neurophysiology, 79*(6), 3168–3188.

Berthier, N. E., & Keen, R. (2006). Development of reaching in infancy. *Experimental Brain Research, 169*(4), 507–518.

Botvinick, M., & Plaut, D. C. (2004). Doing without schema hierarchies: a recurrent connectionist approach to normal and impaired routine sequential action. *Psychological Review, 111*(2), 395–429.

Brenner, E., Driesen, B., & Smeets, J. B. J. (2014). Precise timing when hitting falling balls. *Frontiers in Human Neuroscience, 8*, 342.

Bullock, D., Cisek, P., & Grossberg, S. (1998). Cortical networks for control of voluntary arm movements under variable force conditions. *Cerebral Cortex, 8*, 48–62.

Bullock, D., & Grossberg, S. (1988). Neural dynamics of planned arm movements: emergent invariants and speed-accuracy properties during trajectory formation. *Psychological Review, 95*, 49–90.

Caligiore, D., Parisi, D., & Baldassarre, G. (2014). Integrating reinforcement learning, equilibrium points, and minimum variance to understand the development of reaching: a computational model. *Psychological Review, 121*(3), 389–421.

Chao, F., Lee, M., & Lee, J. (2010). A developmental algorithm for ocularmotor coordination. *Robotics and Autonomous Systems, 58*(3), 239–248.

Chevalier, G., & Deniau, J. M. (1990). Disinhibition as a basic process in the expression of striatal functions. *Trends in Neurosciences, 13*(7), 277–280.

Cisek, P. (2006). Integrated neural processes for defining potential actions and deciding between them: a computational model. *Journal of Neuroscience, 26*(38), 9761–9770.

Cisek, P., Grossberg, S., & Bullock, D. (1998). A cortico-spinal model of reaching and proprioception under multiple task constraints. *Journal of Cognitive Neuroscience, 10*(4), 425–444.

Cisek, P., & Kalaska, J. F. (2005). Neural correlates of reaching decisions in dorsal premotor cortex: specification of multiple direction choices and final selection of action. *Neuron, 3*(45), 801–814.

Colombo, J., & Mitchell, D. (2009). Infant visual habituation. *Neurobiology of Learning and Memory, 92*(2), 225–234.

Corbetta, D., & Thelen, E. (1995). A method for identifying the initiation of reaching movements in natural prehension. *Journal of Motor Behavior, 27*, 285–293.

Corbetta, D., Thurman, S. L., Wiener, R. F., Guan, Y., & Williams, J. L. (2014). Mapping the feel of the arm with the sight of the object: on the embodied origins of infant reaching. *Frontiers in Psychology, 5*, 576.

Erlhagen, W., & Schöner, G. (2002). Dynamic field theory of movement preparation. *Psychological Review, 109*(3), 545–572.

Ermentrout, B. (1998). Neural networks as spatio-temporal pattern-forming systems. *Reports on Progress in Physics, 61*, 353–430.

Fard, F. S., Hollensen, P., Heinke, D., & Trappenberg, T. P. (2015). Modeling human target reaching with an adaptive observer implemented with dynamic neural fields. *Neural networks: the official journal of the International Neural Network Society, 72*, 13–30.

Feldman, A. G. (1986). Once more on the equilibrium point hypothesis (λ-model) for motor control. *Journal of Motor Behavior, 18*, 15–54.

Feldman, A. G. (2011, May). Space and time in the context of equilibrium-point theory. *Wiley Interdisciplinary Reviews: Cognitive Science, 2*(3), 287–304.

Gallistel, C. R., & Gibbon, J. (2000). Time, rate, and conditioning. *Psychological Review, 107*(2), 289–344.

Georgopoulos, A. P. (1986). On reaching. *Annual Reviews of Neuroscience, 9*, 147–170.

Ghafouri, M., & Feldman, A. G. (2001). The timing of control signals underlying fast point-to-point arm movements. *Experimental Brain Research, 137*, 411–423.

Graziano, M. S. A. (2006). The organization of behavioral repertoire in motor cortex. *Annual Reviews of Neuroscience, 29*, 105–134.

Gribble, P. L., Ostry, D. J., Sanguineti, V., & Laboissière, R. (1998). Are complex control signals required for human arm movement. *Journal of Neurophysiology, 79*, 1409–1424.

Grimme, B., Lipinski, J., & Schöner, G. (2012). Naturalistic arm movements during obstacle avoidance in 3D and the identification of movement primitives. *Experimental Brain Research, 222*(3), 185–200.

Guenther, F., Bullock, D., Greve, D., & Grossberg, S. (1994). Neural representations for sensory-motor control. III. Learning a body-centered representation of 3-D target position. *Journal of Cognitive Neuroscience, 6*, 341–358.

Herbort, O., & Butz, M. (2009). The SURE REACH model for motor learning and control of a redundant arm: from modeling human behavior to applications in robotics. *From Motor Learning to Interaction Learning in Robots*, 85–106.

Hermer, L., & Spelke, E. S. (1994). A geometric process for spatial orientation in young children. *Nature, 370*, 57–59.

Hollingworth, A. (2004). Constructing visual representations of natural scenes: the roles of short- and long-term visual memory. *Journal of Experimental Psychology: Human Perception and Performance, 30*, 519–537.

Itti, L., & Koch, C. (2000). A saliency-based search mechanism for overt and covert shifts of visual attention. *Vision Research, 40*(10–12), 1489–1506.

Jeannerod, M. (1984). The timing of natural prehension movements. *Journal of Motor Behavior, 16*, 235–254.

Johnson, J. S., Spencer, J. P., Luck, S. J., & Schöner, G. (2009). A dynamic neural field model of visual working memory and change detection. *Psychological Science, 20*, 568–577.

Kazerounian, S., Luciw, M., Richter, M., & Sandamirskaya, Y. (2013). Autonomous reinforcement of behavioral sequences in neural dynamics. In *Proceedings of the international joint conference on neural networks (IJCNN 2013).*

Kelso, J. A. S., Southard, D. L., & Goodman, D. (1979). On the nature of human interlimb coordination. *Science, 203*, 1029–1031.

Kopecz, K. (1995). Saccadic reaction times in gap/overlap paradigms: a model based on integration of intentional and visual information on neural, dynamic fields. *Vision Research, 35*, 2911–2925.

Kopecz, K., & Schöner, G. (1995). Saccadic motor planning by integrating visual information and pre-information on neural dynamic fields. *Biological Cybernetics, 73*, 49–60.

Latash, M. L. (2008). *Synergy*. New York: Oxford University Press.

Latash, M. L., Scholz, J. P., & Schöner, G. (2007). Toward a new theory of motor synergies. *Motor Control, 11*, 276–308.

Lisman, J. (2015). The challenge of understanding the brain: where we stand in 2015. *Neuron, 86*(4), 864–882.

Loram, I. D., & Lakie, M. (2002). Direct measurement of human ankle stiffness during quiet standing: the intrinsic mechanical stiffness is insufficient for stability. *Journal of Physiology, 545*, 1041–1053.

Martin, V., Scholz, J. P., & Schöner, G. (2009). Redundancy, self-motion and motor control. *Neural Computation, 21*(5), 1371–1414.

McDowell, K., Jeka, J. J., Schöner, G., & Hatfield, B. D. (2002). Behavioral and electrocortical evidence of an interaction between probability and task metrics in movement preparation. *Experimental Brain Research, 144*(3), 303–313.

Mechsner, F., Kerzel, D., Knoblich, G., & Prinz, W. (2001). Perceptual basis of bimanual coordination. *Nature, 414*, 69–73.

Moran, D. W., & Schwartz, A. B. (1999). Motor cortical representation of speed and direction during reaching movement. *Journal of Neurophysiology, 82*, 2676–2692.

Morasso, P. (1981). Spatial control of arm movements. *Experimental Brain Research, 42*, 223–227.

Narioka, K., & Steil, J. J. (2015). U-shaped motor development emerges from goal babbling with intrinsic motor noise. *Proceedings of the 5th Joint IEEE International Conference on Development and Learning and on Epigenetic Robotics ICDL-EpiRob 2015*, 55–62.

Perone, S., & Spencer, J. P. (2013a). Autonomous visual exploration creates developmental change in familiarity and novelty seeking behaviors. *Frontiers in Psychology, 4*(648), 1–21.

Perone, S., & Spencer, J. P. (2013b). Autonomy in action: linking the act of looking to memory formation in infancy via dynamic neural fields. *Cognitive Science, 37*(1), 1–60.

Pezzulo, G., & Cisek, P. (2016). Navigating the affordance landscape: feedback control as a process model of behavior and cognition. *Trends in Cognitive Sciences, 20*(6), 414–424.

Pouget, A., & Snyder, L. H. (2000). Computational approaches to sensorimotor transformations. *Nature, 3*, 1192–1198.

Raphael, G., Tsianos, G. A., & Loeb, G. E. (2010). Spinal-like regulator facilitates control of a two-degree-of-freedom wrist. *The Journal of Neuroscience: the official journal of the Society for Neuroscience, 30*(28), 9431–9444.

Richter, M., Sandamirskaya, Y., & Schöner, G. (2012). A robotic action selection and behavioral organization architecture inspired by human cognition. In *IEEE/Rsj international conference on intelligent robots and systems* (pp. 2457–2464).

Ruf, A., & Horaud, R. (1999). Visual servoing of robot manipulators Part I: projective kinematics. *The International Journal of Robotics Research, 18*(11), 1101–1118.

Sandamirskaya, Y. (2015). Autonomous sequence generation in dynamic field theory. In G. Schöner, J. P. Spencer, & DFT Research Group (Eds.), *Dynamic thinking: a primer on dynamic field theory* (pp. 353–368). New York: Oxford University Press.

Sandamirskaya, Y., & Conradt, J. (2013). Learning transformations with dynamic neural fields. In *ICANN 2013*.

Sandamirskaya, Y., & Schöner, G. (2010). An embodied account of serial order: how instabilities drive sequence generation. *Neural Networks, 23*(10), 1164–1179.

Sandamirskaya, Y., & Storck, T. (2015). Learning to look and looking to remember: a neural-dynamic embodied model for generation of saccadic gaze shifts and memory formation. In P. Koprinkova-Hristova, V. Mladenov, & N. K. Kasabov (Eds.), *Artificial neural networks* (Vol. 4, pp. 175–200). Dordrecht: Springer International Publishing.

Savastano, P., & Nolfi, S. (2013). A robotic model of reaching and grasping development. *IEEE Transactions on Autonomous Mental Development, 5*(4), 326–336.

Schlesinger, M., Parisi, D., & Langer, J. (2000). Learning to reach by constraining the movement search space. *Developmental Science, 3*, 67–80.

Schneegans, S. (2015). Sensori-motor and cognitive transformation. In G. Schöner, J. P. Spencer, & DFT Research Group (Eds.), *Dynamic thinking: a primer on dynamic field theory* (pp. 169–196). New York: Oxford University Press.

Scholz, J. P., & Schöner, G. (1999). The uncontrolled manifold concept: identifying control variables for a functional task. *Experimental Brain Research, 126*, 289–306.

Schöne, H. (1984). *Spatial orientation: the spatial control of behavior in animals and man*. Princeton, NJ: Princeton University Press.

Schöner, G. (1990). A dynamic theory of coordination of discrete movement. *Biological Cybernetics, 63*, 257–270.

Schöner, G. (2002). Timing, clocks, and dynamical systems. *Brain and Cognition, 48*, 31–51.

Schöner, G., Spencer, J. P., & the DFT Research Group. (2015). *Dynamic thinking: a primer on dynamic field theory*. Oxford: Oxford University Press.

Schöner, G., & Thelen, E. (2006). Using dynamic field theory to rethink infant habituation. *Psychological Review, 113*(2), 273–299.

Schwartz, A. B. (1994). Direct cortical representation of drawing. *Science, 265*, 540–542.

Scott, S. H., & Kalaska, J. F. (1997). Reaching movements with similar hand paths but different arm orientations. I. Activity of individual cells in motor cortex. *Journal of Neurophysiology, 77*, 826–852.

Shadmehr, R., & Krakauer, J. W. (2008). A computational neuroanatomy for motor control. *Experimental Brain Research, 185*, 359–381.

Shadmehr, R., & Mussa-Ivaldi, F. (2012). *Biological learning and control: how the brain builds representations, predicts events, and makes decisions*. Cambridge, MA: MIT Press.

Simons, D. J. (2000, Jan). Current approaches to change blindness. *Visual Cognition, 7*(1–3), 1–15.

Soechting, J. F., & Lacquaniti, F. (1981). Invariant characteristics of a point movement in man. *Journal of Neuroscience, 1*(7), 710–720.

Sporns, O., & Edelman, G. M. (1993). Solving Bernstein's problem: a proposal for the development of coordinated movement by selection. *Child Development, 64*(4), 960–981.

Strauss, S., & Heinke, D. (2012). A robotics-based approach to modeling of choice reaching experiments on visual attention. *Frontiers in Psychology, 3*, 1–11.

Strauss, S., Woodgate, P. J., Sami, S. A., & Heinke, D. (2015). Choice reaching with a LEGO arm robot (CoRLEGO): the motor system guides visual attention to movement-relevant information. *Neural Networks, 72*, 3–12.

Sun, G., & Scassellati, B. (2005). A fast and efficient model for learning to reach. *International Journal of Humanoid Robotics, 2*(4), 391–414.

Sutton, R. S., & Barto, A. G. (1990). Time-derivative models of Pavlovian reinforcement. In M. Gabriel & J. Moore (Eds.), *Learning and computational neuroscience: foundation of adaptive networks* (pp. 497–537). Cambridge, MA: MIT Press.

Tee, K. P., Burdet, E., Chew, C. M., & Milner, T. E. (2004). A model of force and impedance in human arm movements. *Biological Cybernetics, 90*, 368–375.

Tekülve, J., Zibner, S. K. U., & Schöner, G. (2016). A neural process model of learning to sequentially organize and activate pre-reaches. *Proceedings of the 6th Joint IEEE International Conferences on Development and Learning and on Epigenetic Robotics (ICDL-Epirob) 2016*.

Thelen, E. (1981). Rhythmical behavior in infancy: an ethological perspective. *Developmental Psychology, 17*(3), 237–257.

Thelen, E., Corbetta, D., Kamm, K., Spencer, J., Schneider, K., & Zernicke, R. (1993). The transition to reaching: matching intention and intrinsic dynamics. *Child Development, 64*, 1058–1098.

Thelen, E., Corbetta, D., & Spencer, J. P. (1996). The development of reaching during the first year: the role of movement speed. *Journal of Experimental Psychology: Human Perception and Performance, 22*, 1059–1076.

Todorov, E., & Jordan, M. I. (2002). Optimal feedback control as a theory of motor coordination. *Nature Neuroscience, 5*(11), 1226–1236.

Tseng, Y.-W., Scholz, J. P., Schöner, G., & Hotchkiss, L. (2003). Effect of accuracy constraint on joint coordination during pointing movements. *Experimental Brain Research, 149*(3), 276–288.

Viviani, P., & Flash, T. (1995). Minimum-jerk, two-thirds power law, and isochrony: converging approaches to movement planning. *Journal of Experimental Psychology. Human Perception and Performance, 21*(1), 32–53.

von Hofsten, C. (1984). Developmental changes in the organization of prereaching movements. *Developmental Psychology, 20*(3), 378–388.

von Hofsten, C. (1991). Structuring of early reaching movements: a longitudinal study. *Journal of Motor Behavior, 23*(4), 280–292.

Wilimzig, C., Schneider, S., & Schöner, G. (2006). The time course of saccadic decision making: dynamic field theory. *Neural Networks, 19*, 1059–1074.

Zibner, S. K. U., Faubel, C., Iossifidis, I., & Schöner, G. (2011). Dynamic neural fields as building blocks for a cortex-inspired architecture of robotic scene representation. *IEEE Transactions on Autonomous Mental Development, 3*(1), 74–91.

Zibner, S. K. U., Tekülve, J., & Schöner, G. (2015a). The neural dynamics of goal-directed arm movements: a developmental perspective. In *Proceedings of the 5th Joint International Conference on Learning and Development and on Epigenetic Robotics ICDL-EpiRob 2015*.

Zibner, S. K. U., Tekülve, J., & Schöner, G. (2015b). The sequential organization of movement is critical to the development of reaching: a neural dynamics account. In *Proceedings of the 5th Joint International Conference on Learning and Development and on Epigenetic Robotics ICDL-EpiRob 2015*.

13

THE DEVELOPMENT OF REACHING AND GRASPING

Towards an Integrated Framework Based on a Critical Review of Computational and Robotic Models

Daniele Caligiore and Gianluca Baldassarre

Introduction

The development of reaching and grasping skills in infants is a complex process that involves several interdependent brain structures and learning processes and takes 2–3 years to accomplish the final efficient, effective, and flexible performance of the two behaviours. This development process is of critical importance for the child as reaching and grasping represent key building blocks for the acquisition of more complex motor behaviours and for the development of higher-level cognition.

Reaching and grasping initially appear as simple, relatively jerky reflex-like behaviours (McCarty & Ashmead, 1999; von Hofsten & Rönnqvist, 1993). Shortly after birth, for example, neonates generate spontaneous arm and hand movements, possibly visually elicited (Ennouri & Bloch, 1996; von Hofsten, 1984). These "prereaching" movements are usually directed toward nearby objects and most of the times fail to contact the target. Over the next two months, "prereaches" decline in frequency. With time, reaching and grasping actions gradually become skilled and smooth volitional behaviours adapted to the infant's ongoing goals and varied as a function of the task conditions (e.g. the location and size of the target object; Erhardt, 1994; Thelen et al., 1996; von Hofsten, 2004; Williams et al., 2015). Around the age of three months, an increasingly robust and organized set of reaching behaviours begins to emerge, which includes several important new properties, such as the grasp hand preshape and corrective "submovements" (von Hofsten, 1979; von Hofsten & Rönnqvist, 1993), but the behaviours achieve a mature form and variety only after several further months of learning.

In this chapter we discuss some selected computational and robotic models on reaching and grasping drawn from our work and the literature. Our objective is to contribute, with our perspective, to the formulation of an integrated view on the computational mechanisms that might underlie the development of reaching and grasping and possibly other more sophisticated motor and cognitive abilities.

The chapter in particular proposes that six key computational ingredients are important for the development of reaching and grasping. These ingredients, summarised in Figure 13.1, are: (1) *ecological active vision*: this involves the development of the capacity to actively gather visual information supporting the manipulation of the environment in ecological conditions (Ballard, 1991; Ognibene & Baldassarre, 2015); (2) *motor babbling and associative learning processes*: these involve the production of simple exploratory movements and basic learning processes that are particularly important in the early phases of development of reaching and grasping (Caligiore et al., 2008; von Hofsten, 1982); (3) *trial-and-error learning*: this more sophisticated learning process is critical for the acquisition

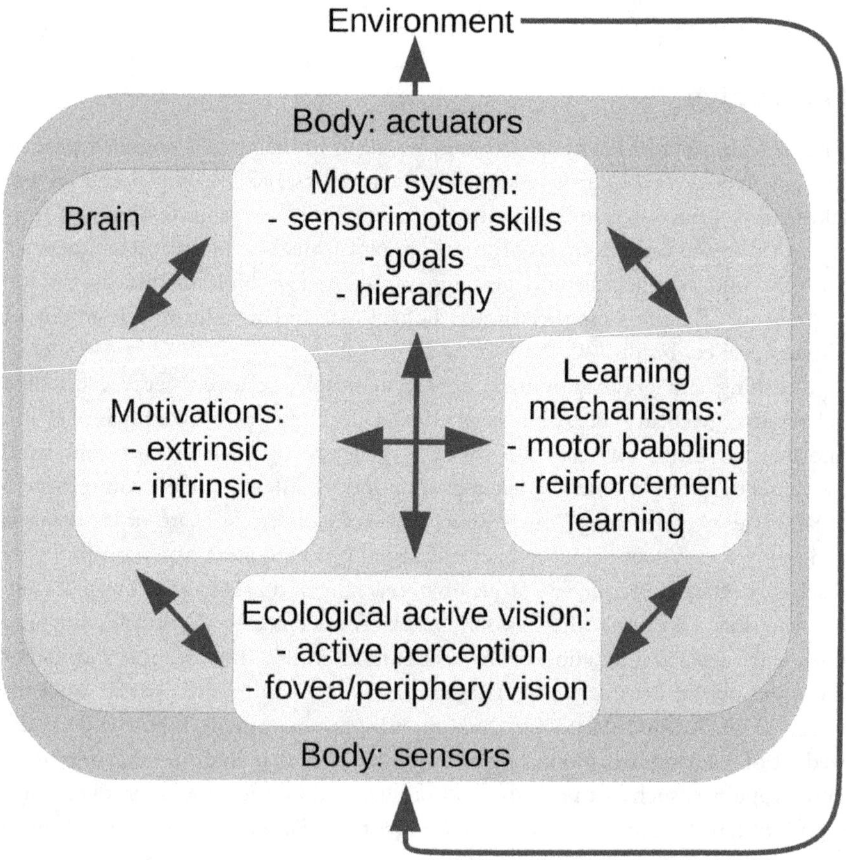

FIGURE 13.1 Components and processes of infant cognition considered in the chapter to analyse the development of reaching and grasping behaviours

of increasingly sophisticated, flexible, and accurate reaching and grasping skills (Berthier et al., 2005; Caligiore et al., 2014a); (4) *modular and hierarchical control architectures*: these are important to acquire and harmonise the storing and performance of reaching and grasping variants, and to combine them to form more complex motor skills (Baldassarre & Mirolli, 2013b; Caligiore et al., 2014b); (5) *intrinsic motivations*: these are the core motivational engines guiding infant autonomous cumulative learning of an increasingly large collection of skills (Baldassarre & Mirolli, 2013a; Santucci et al., 2014); (6) *embodiment*: this involves the idea that motor development crucially relies on the mind-body-environment circular causations (Cangelosi & Schlesinger, 2015; Pfeifer & Gómez, 2009).

In the remainder of the chapter, each section expands one of the six ingredients. Overall the sections present our view for which the development of reaching and grasping, and possibly of more complex motor and cognitive processes building on them, relies on the synergistic and dynamical interplay of all such ingredients.

Ecological Active Vision: Actively Gathering Information to Support Reaching and Grasping

Notwithstanding increasing evidence indicates that the initial development of reaching and grasping is loosely coupled to vision and overt attention/gaze control (henceforth simply "attention"), we believe that they represent a "game-changer" element for understanding how such behaviours develop into the adult form. Indeed, vision and attention represent a pivot of adult manipulation of objects as on one side they offer unexpected important opportunities to support its acquisition and functioning, and on the other they produce important cognitive challenges solved during development in interesting subtle ways. This section analyses some of these opportunities and challenges.

Contrary to what previously thought (e.g. von Hofsten, 1979), evidence is starting to show that initially reaching and grasping development is only loosely coupled with vision. For example, Corbetta et al. (2014) tracked object-directed looking behaviour of infants over an 11-week period covering the emergence of reaching and showed that infants do not learn to align their reach to where they look, but rather learn to align their look to where they reach. On this basis, the authors propose that reaching onset is the product of a deeply embodied process, in which infants first learn how to direct their movement in space using proprioceptive and haptic feedback from self-produced movement contingencies with the environment, and only afterwards do they learn to map visual attention onto these bodily centred experiences. Also, often infants, unlike adults, grasp objects while looking at locations different from the object contact points (McMahon et al., 2017).

Manipulation of the environment by human adults is strongly based on vision and attention. Primates are "visual animals", with a large part of the brain cortex dedicated to visual processing (Barton, 1998). During evolution, the emergence of sophisticated visual and attention capabilities created new adaptive opportunities for solving cognitively complex problems (Passingham & Wise, 2012). During ontogenesis, reaching and grasping movements increasingly depend on attention and vision to achieve the accuracy level of adulthood. In adults, while the initial part of the reaching movement is largely ballistic, the object contact reaching part and its later manipulation are strongly guided by vision (Shadmehr & Wise, 2005). Thus, although attention and manipulation might initially develop in a partially decoupled fashion, they progressively come to form an increasingly integrated system where attention and vision play a key role in guiding the skilled manual interaction with objects (Hayhoe & Ballard, 2005) and, at the same time, the effects of manipulation on the world cause feedback (rewards) important for the further development of attention control (Hikosaka et al., 2000).

Although the visual stream is highly informative, it is also computationally extremely demanding to be processed. An evolutionary "solution" to this problem (cf. Ballard, 1991), which has a primary importance in late primates, has been the emergence of a visual system characterised by a small high-acuity fovea and a low-resolution periphery (Hughes, 1977; Passingham & Wise, 2012). This arrangement abstracts away a large part of visual information and drastically lowers the computational burden. This solution, however, also introduces the need to actively control the target of the high-resolution fovea (gaze) through eye movements ("active vision", Ballard, 1991), and also to *learn* to do so based on environmental feedback. At the same time, however, gaze direction also provides important information, through eye proprioception, on the spatial location of relevant elements located in the environment: this information is very important to guide action, for example to establish clear targets for reaching and contact points for grasping (Shadmehr & Wise, 2005). Moreover, the need to learn to control gaze based on the feedback rewards caused by the manipulation of the world through limbs, and the reliance of adult-like fine manipulation on visual/attentional guidance, creates an important attention-manipulation dependency that strongly affects the development and functioning dynamics of reaching, grasping, and attention control. In Ognibene and Baldassarre (2015), we refer to these issues as *ecological active vision*, where the word "ecological" intends to stress the prominent role that vision and attention have for the guidance of concrete manipulation actions versus a sheer information gathering process. Such issues, faced in various computational models, can be clustered into four main sets (see Ognibene & Baldassarre, 2015) that we now consider in more detail.

A first set of issues is related to the fact that the control of gaze needs to be learned. Whitehead and Ballard (1991) (see also Balkenius, 2000) proposed the possibility that attention learning processes rely on trial-and-error learning mechanisms analogous to those guiding the development of limb behaviour. These processes can be modelled with reinforcement learning (RL) algorithms, for example with actor-critic models (Sutton & Barto, 1998). These models have been proposed to capture the trial-and-error learning processes typical of a brain area critical for motor learning, the basal ganglia (Hikosaka et al., 2000; Houk et al., 1995), and have been used to implement learning gaze controllers that select saccade targets through neural maps analogous to the superior colliculus, a pivotal brain structure underlying attention control (Sperati & Baldassarre, 2014, 2017). These models and other attention learning models (Whitehead & Ballard, 1991) show that the dependence of saccade control on rewards leads them to form internal representations that are strongly action oriented and depend on the agent's tasks (Ognibene et al., 2010) and goals (Sperati & Baldassarre, 2017).

A second set of issues is related to the fact that the feedback, in particular rewards, that guide learning of gaze control is generated by the manipulation of the environment, not from just looking around (Ognibene & Baldassarre, 2015; rewards coming from just looking around are rare, e.g. they happen in social contexts or in psychological experiments such as the gaze contingency test illustrated below). This might cause the reward feedback to fail to arrive in correspondence to useful "looking actions" (saccades). A solution to this problem could be a strong coupling between where the agent looks and where the agent manipulates the environment. In adults, this coupling is very strong, with gaze focused on the main targets of manipulation actions, or preceding them when they are performed in sequence (Hayhoe & Ballard, 2005; Johansson et al., 2001; Sailer et al., 2005). Instead, as we have seen above in infants the attention focus is only loosely coupled to manipulation actions. The coupling might thus emerge later during development, driven by the trial-and-error learning processes governing gaze-control: since "coupling saccades" get rewarded with a higher chance than uncoupled ones, they might be progressively strengthened. The coupling might not only be a solution to a problem, but also offer an important opportunity as it might allow the encoding of actions with a reference frame centred on the gaze direction, and hence on target objects (Ballard et al., 1997). In this respect, evidence shows that many representations in brain are "eye-centred": not only those involving saccades (Hikosaka et al., 2000) but also those governing manipulation actions, for example reaching (Shadmehr & Wise, 2005). The adaptive advantage of this is that the use of gaze-centered (and hence target/object-centered) reference frames allows the learning and production of actions that generalise over the spatial location, and possibly orientation, of the targets of manipulation.

A third set of issues involves a "chicken-and-egg problem": as seen above, the learning of attention top-down control needs to be guided by effective manipulation actions; at the same time, the learning of manipulation actions depends on an effective attention finding useful information to guide it. From the perspective of manipulation, a solution might come from the fact that basic reaching/grasping abilities could initially develop without a strong support from vision, as seen above. This might explain why at their onset reaching and grasping are only loosely coupled to the attention focus (see above). Once some initial approximate manipulation abilities have emerged, they can produce an initial feedback supporting the learning of suitable visual routines (see above) in turn supporting the later development of more accurate manipulation behaviours. With regard to attention control, a solution might come from the existence of a bottom-up, innate component guiding saccades on the basis of task-independent salient features of the scene such as high contrasts or changes (Itti et al., 1998). This bottom-up component might guide saccades on potentially interesting parts of the environment, for example on objects, in the absence of a top-down looking competence. As shown in computational models (Ognibene and Baldassarre, 2015; Ognibene et al., 2010: see Figure 13.2), this can generate interesting developmental trajectories involving bottom-up attention, top-down goal-related attention learning, and manipulation learning.

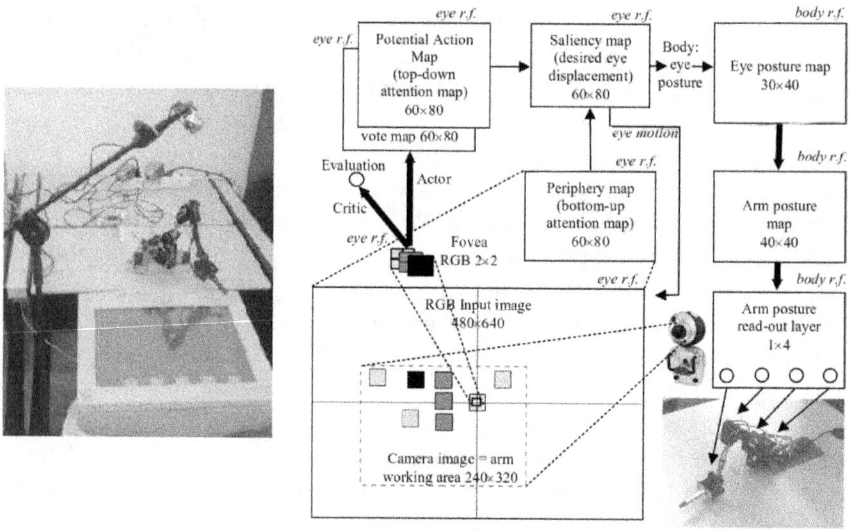

FIGURE 13.2 The BITPIC robot architecture ("Bottom-up attention, Top-down attention, PAM, and Coupling Integrated system")

Source: Reprinted with permission from Ognibene and Baldassarre (2015).

Notes: Left: Experimental set-up involving a robot arm and a camera looking down to a computer screen, forming the working plane of the arm, where different images are produced at each trial; each image involves a black target that needs to be reached by the arm to gain a reward, some dark grey cue objects always located at the right of the target, and some light grey distractors randomly scattered in the image. Right: The system architecture: the real camera image is used to produce the image collected by a "virtual" pan-tilt actuated visual sensor formed by a high-resolution RGB fovea, fed to a top-down RL attention component, and a grey low-resolution periphery, fed to a bottom-up hardwired attention component; the two attention components "vote" for different possible targets of saccades, in turn possibly triggering an arm reach directed to the gaze location.

These processes offer possible explanations of various empirical findings. First, they might explain the U-shaped performance typically exhibited by infants during motor development, for which movements temporarily deteriorate after the initial successes and before the later maturation (Caligiore et al., 2014a; Cangelosi & Schlesinger, 2015). This decline may be related to the transition from a proprioceptive/haptic-based guidance of movements to a more sophisticated one, based on a close-loop multi-sensory integration of vision, proprioception, and touch, which is more powerful but also more difficult to control (Schlesinger et al., 2000). Second, Wang et al. (2012) show that infants are very fast in learning to look at relevant locations in space by using a gaze contingency experimental paradigm where they are visually rewarded, through interesting pictures, for looking at a "button-like" object on a computer screen. A computational model (Marraffa et al., 2012) explains this fast learning by showing how bottom-up attention produces a "default" visual exploration behaviour based on the image salient elements: top-down attention might either rapidly learn and strengthen (by reinforcement learning mechanisms) such behaviour if it is useful to obtain rewards, or it might learn to override it with a different behaviour if it is not useful to accomplish the task at hand. Computational models (Ognibene & Baldassarre, 2015; Ognibene et al., 2008) show that when manipulation is present and produces reward feedback, such dynamics involving bottom-up and top-down attention components is further shaped in subtle ways so as to produce information useful to support the actions of the arm.

A last set of issue relates to the fact that the high visual acuity limited to a small fovea and surrounding regions creates a strong "perceptual aliasing" (Whitehead & Ballard, 1991): this implies that the partial view offered by single saccades might lead the agent to perceive different parts of the scene as the same. This problem can be faced through memory processes that integrate information from succeeding saccades in time. A naïve implementation of this idea would store all available information coming from a succession of structured exploration saccades, but this would be undesirable as it would return a time-consuming inaccurate proxy of the overwhelming information that could be returned by a whole high-resolution image. Some better principles to follow are instead these ones: (a) saccades should target only locations that are useful to support the agent in the accomplishment of the current task (Whitehead & Ballard, 1991); (b) for each saccade, the agent should store in memory only information useful to face the task at hand. An extreme example of this is the action-oriented memory component ("PAM – Potential Action Map") proposed by Ognibene and Baldassarre (2015) (see Figure 13.2) that stores information in terms of possible relevant areas where to perform saccades. Before each saccade, the system produces a probability distribution over the 2D potential saccade space indicating the probability of finding information useful to solve the current task at the corresponding locations. The probability distribution formulated in correspondence to different saccade decisions is integrated into the memory (through a decaying neural mechanism) and the resulting probability distribution is used to decide the specific saccade to perform at each moment.

Overall, these results highlight the dramatic importance, in terms of opportunities and challenges, that an actively controlled fovea-periphery visual sensor represents for the development of reaching and grasping. We hence believe that a full understanding of such development beyond the initial phases, which might mainly rely on proprioceptive and haptic information, must involve the understanding of how it unfolds together with the development of visual and attentional capabilities.

The Learning Processes Underlying Motor Development: Motor Babbling, Associative Learning, and the Development of Body Knowledge

Motor babbling might play an important role in early infants' motor development. Motor babbling is based on a simple learning process for which the infant generates its own training data by producing random movements (von Hofsten, 1982). As we shall see, the learning processes relying on motor babbling are usually based on associative mechanisms although more sophisticated learning processes, such as trial-and-error learning, usually rely on some forms of random perturbation of movements to discover new behaviours.

In the last three decades, several bio-inspired developmental robotics models have been proposed to model motor babbling (e.g., Caligiore et al., 2007; 2008; Demiris & Dearden, 2005; Kuperstein, 1988; Lee et al., 2007; Rolf et al., 2010; Saegusa et al. 2009). The common idea exploited by these models is that exploratory movements allow the formation of associations between the representation of such movements and the representation of their effects, so that a later activation of the effect representations, when these become desirable, allow the recall of the movements that lead to them. The seminal paper of Kuperstein (1988) demonstrates how the execution of random movements can allow a computational model controlling an eye/arm system to form associations between a perceived object-in-hand position and the corresponding arm postures. The model can later use this knowledge to perform reaching tasks. Caligiore et al. (2007) showed that if some biological constraints are fulfilled when building neural-network models based on motor babbling (leaky-neurons, Amari, 1977; equilibrium-point muscle mechanisms, Feldman, 1986; population codes, Pouget, et al., 2003; Hebb learning, Sejnowski, 1977) interesting phenomena emerge such as lateral connection weights having a Mexican hat-like distribution, and movements with bell-shaped speed profiles similar to those of human movements.

Natale et al. (2005; see also Natale et al., 2007) used the Babybot humanoid robotic platform to learn reaching and grasping through motor babbling. The

Babybot first learns to fixate the camera gaze on its hand, which is driven to different locations through motor babbling. Once a repertoire of arm postures are associated with gaze positions, the Babybot learns to grasp seen objects through the following process: (1) an object is placed in the Babybot's hand, and a pre-programmed palmar grasp is executed; (2) the object is brought to the centre of the visual field; (3) the object is returned to the workspace, and the Babybot uses a pre-programmed object-recognition module to search for it; (4) once localized, the previously acquired reaching behaviour is used to guide the hand toward the object and grasp it. The model contributes to show how the mechanisms supporting learning to reach (e.g., motor babbling and visual search) might also support the initial development of grasping.

Most models using motor babbling are used to acquire relatively simple reaching skills. The reason of this is likely that motor babbling and associative learning are suitable for the acquisition of behaviours on the basis of the formation of associations linking sensory and behavioural events which are *synchronous* and can be represented as *points* in space, for example between the current hand/object position in operational space (x, y, z) and the current arm joint angles (e.g., α, β, γ). Instead, they are not suitable for the acquisition of behaviours which require the association of events that take place asynchronously or last for long (e.g. motor *trajectories*). Caligiore et al. (2008) show how the limitations of motor babbling and associative learning can be pushed a bit forward by integrating them with suitable complementary mechanisms. This allows the acquisition of two relatively sophisticated behaviours. The first is reaching with obstacles that needs the production of non-linear trajectories requiring a fine regulation of movement in time. This is possible if non-linear trajectories are generated on the basis of central pattern generators (Ijspeert, 2008; Ijspeert et al., 2002) and motor babbling can lead to form associations between visual percepts and the *parameters* of the central pattern generators. The second is reaching and grasping, which requires the production of a *sequence* of building-block movements in time, in particular: "bringing the hand close to the object" and "close the hand around the object" (Figure 13.3). Motor babbling and simple associative Hebbian rules can produce this behaviour if supported with grasping reflexes similar to those of newborn infants. The model has a higher performance with larger objects vs small ones, in line with the earlier development of power grasps vs. precision ones (McCarty & Ashmead, 1999; Oztop et al., 2004). Although the model performance is bounded, it shows how motor babbling and associative mechanisms suffice to bootstrap the initial emergence of grasping behaviours.

Another set of models that heavily rely on motor babbling to support the development of reaching and grasping have been proposed within the "Lift-Constraint, Act, and Saturate" framework (LCAS; Lee et al., 2007). This framework assumes that infants' staged development involves learning of each competence until its "saturation": when this happens, a "constraint" is lifted and

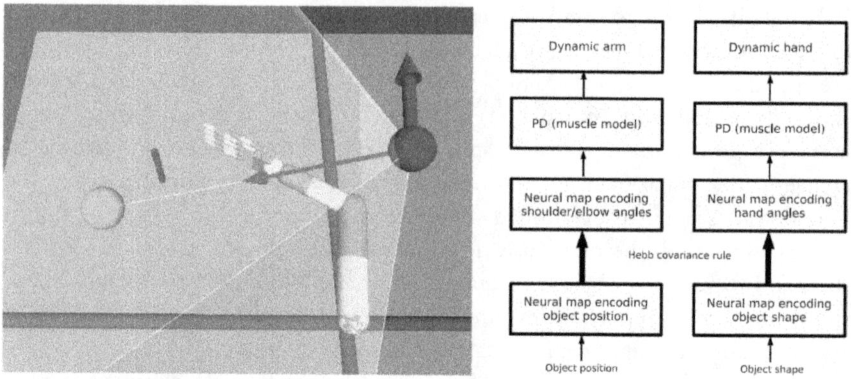

FIGURE 13.3 Model used to study motor babbling and associative learning processes

Source: Reprinted with permission from Caligiore et al. (2008).

Notes: Left: Simulated setup: robot simulated arm, object (sphere), obstacle (small cylinder), and camera (sphere with arrows: the thin arrow indicates the gaze). Right: Model architecture, with a component controlling reaching and a component controlling grasping.

"action" and learning can focus on the acquisition of other competences. The models developed according to this perspective heavily exploit motor babbling to form associations between "fields" (simple kernels, e.g. binary circular kernels) covering the multi-modal/multi-motor spaces of interest. Through these processes humanoid robots mimicking infants' development can acquire various competences, e.g. to look at own hand, reach looked objects, and perform basic object manipulations (Law et al., 2014).

Motor babbling might also contribute to the formation of muscle synergies (Latash et al., 2007). This is a problem that robotics models tend to overlook as they control high-level motor variables such as the angles or torques of the arm joints. In the future, this issue might receive more attention driven by the availability of more biologically plausible robotic plants, real or simulated, encompassing muscles (e.g. see Tommasino et al. 2012b and OpenSim: http://opensim.stanford.edu/).

Overall, for their simplicity motor babbling and associative learning processes, possibly complemented by other supporting mechanisms, might play an important role in the initial phase of motor skill development involving the acquisition of basic sensors-body coordination skills. However, for being able to link only synchronous representations, in the course of development motor babbling and associative mechanisms are replaced by more sophisticated exploration strategies and learning processes.

Learning Processes Underlying Motor Development: The Progressive Refinement of Reaching and Grasping with Trial-and-Error Learning Processes

The limitations of motor babbling and associative learning mechanisms are overcome by reinforcement learning (RL) models (Sutton & Barto, 1998) mimicking trial-and-error animal learning processes. These models exploit the time-prospective nature of RL algorithms to acquire optimized time-extended actions and action sequences (Bonaiuto & Arbib, 2010; Botvinick et al., 2009; Caligiore et al., 2010; Caligiore et al., 2014a; Herbort et al., 2007). Models based on these more sophisticated learning processes are challenged to account for a number of known empirical facts, highlighted by infant longitudinal studies, concerning the evolution of the kinematics and dynamics of reaching and grasping during development. These data are particularly detailed for reaching (Berthier, 2011; Berthier & Keen, 2006). Initially the speed profile of reaching shows multiple peaks, possibly denoting corrective submovements (von Hofsten, 1979; von Hofsten & Rönnqvist, 1993). Berthier and Keen show that submovements progressively disappear until at the end of the second year the overall speed profile of reaching is similar to the typical adult profile where one progressive acceleration is followed by one progressive deceleration that smoothly brings the hand to contact the target. The average and maximum speed decrease indicating that movement slows down with development. The jerk progressively decreases indicating that the smoothness of the movement increases with the age. Finally, the authors found in infants an early fixation, followed by later use, of the elbow: this may represent a solution to the degrees-of-freedom (DOF) problem of motor learning (i.e. Bernstein's problem; Berthier & Keen, 2006).

To the best of our knowledge, only three RL models have been closely confronted with data from empirical experiments on reaching development. The first one (Berthier, 1996) is based on Q-learning (Watkins & Dayan, 1992), an RL algorithm where action selection is based on the estimation of future rewards expected for each state-action pair. The model learns to control the (x, y) planar coordinates of the hand. It incorporates the hypothesis that infant reaching is based on a sequence of submovements (von Hofsten, 1979; von Hofsten & Rönnqvist, 1993) aimed at bringing the hand to the target in the presence of errors affecting movement execution. The model reproduces experimental results on submovements and shows how with age, simulated with a decreasing stochasticity of the RL model, the number of submovements gradually decreases. The second more accurate model (Berthier et al., 2005), uses a RL actor-critic neural network (Sutton & Barto, 1998) to control a two-DOF dynamical arm. On this basis, the model not only reproduces the decrease of submovements during development but also their typical speed profiles and infants' capacity to reach targets in any position, from any initial postures, within the operational space.

The third RL-based model builds on and goes beyond the previous two models by accounting for a considerably larger set of developmental data (Caligiore et al., 2014a). The model, called *iREACH* ("infant Reinforcement learning, Equilibrium point, Accuracy, Control-dependent noise Hypothesis"), is based on three key hypotheses: (a) trial-and-error learning processes drive the progressive development of reaching; (b) the control of the movements based on equilibrium points allows the model to quickly find the initial approximate solution to the problem of gaining a hand contact with the reaching target; (c) the need for an accurate contact with the target drives the progressive refinement of the reaching movement in the presence of control-related muscular noise. The tests of the model, based on a two-DOF simulated dynamical arm, show that together these hypotheses allow the reproduction of a large number of empirical findings on reaching development: the developmental trajectory of the kinematic and dynamical variables mentioned above, the evolution of submovements, the progressive development of a bell-shaped speed profile, and the evolution of the elbow joint use. The model also produces testable predictions on several of these phenomena. Remarkably, the target data sets were accounted for by the same model. The analysis of the model functioning revealed that all data could be ultimately explained by the same two-stage developmental trajectory generated by the interplay of the three aforementioned hypotheses: first the model quickly learns to perform coarse movements that ensure a contact of the hand with the target, and then it slowly refines the detailed dynamical aspects of the movements to increase the object contact accuracy. Figure 13.4 summarizes the integration of the three key hypotheses of iREACH, the developmental trajectory that they generate, and the several empirical findings accounted for by the model.

With respect to grasping, Oztop et al. (2004) proposed the first computational model of infant grasp learning constrained by the literature on infant motor development (ILGM – Infant Learning to Grasp model). In line with empirical evidence, the model assumes that grasping is based on an open-loop movement not guided by on-line visual feedback. The model incorporates possible elements of infant motor behaviour, namely the reliance on elementary motor schemas, the exploratory nature of motor interaction, the inherent variability of motor performance, and the progressive decrease of movement uncertainty following learning. The model also shows how an existing behaviour (reaching) can support the development of a more complex behaviour (grasping) through interactive goal-directed trial-and-error learning (cf. Ansuini et al., 2006). This learning process leads the model to generate grasping movements that match the target object affordances (i.e., its specific possibilities of being grasped, Fagg & Arbib, 1998).

The Integrated Learning of Grasps and Affordances model (ILGA; Bonaiuto & Arbib, 2015) is a recent evolution of the ILGM model. The ILGA model studies, for the first time in a neurobiologically plausible way, the integration of

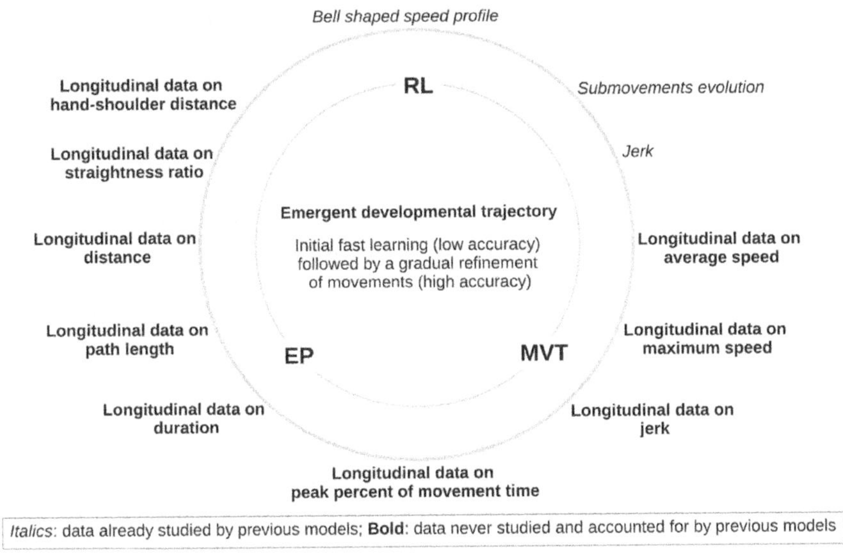

FIGURE 13.4 Key hypotheses incorporated by the iREACH model of reaching development, and target phenomena it accounts for

Source: Reproduced with permission from Caligiore et al. (2014a).

Notes: The integration of reinforcement learning (RL), the equilibrium point hypothesis (EP), and the minimum variance theory (MVT) (linked by the inner circle) leads the model to generate a developmental trajectory that allows the reproduction and prediction of several empirical data (outer circle).

two neural mechanisms: infant grasp learning and affordance learning. The model simultaneously learns grasp affordances from visual object features, and motor parameters for grasp planning using RL. As in the ILGM, the model captures a stage of grasp development prior to the onset of sophisticated visual processing of hand-object relations, and assumes that certain premotor cortex neurons activate neural populations in primary motor cortex that synergistically control different combinations of fingers (Santello, 2015). On this basis, ILGA is able to generalise learned affordances and grasp movements to novel objects.

Combining Reaching and Grasping Movements in a Flexible Way through Modular and Hierarchical Control Architectures

Thus far we have focused on the learning and production of single reaching and grasping movements. Biological evidence on the organisation of the human brain (e.g. Meunier et al., 2010) suggests that its (soft-)modular and hierarchical structure might be critical to support the combination of two or more movements to develop flexible more complex reaching and grasping abilities. Computationally, the key idea behind modular/hierarchical architectures is that chunking behaviour into addressable pieces has two main advantages: (a) single "actions" can be easily learned and recalled, in particular based on their goal (i.e. the set of states of the world that can be produced by their execution), and it is possible to transfer knowledge between actions requiring similar sensorimotor mappings while avoiding memory interference between them; (b) the chunks can be flexibly recombined to form more complex behaviours. Following these ideas, some RL models have scaled up systems' learning capacities by composing motor behaviours into more complex behaviours. In particular, this has been done within the *RL option framework* (Barto & Mahadevan, 2003; Vigorito & Barto, 2010) which pivots on the concept of "options", data structures encoding close-loop behaviours and their goals and conditions for execution. The models proposed within this framework, however, have not been used to directly face developmental issues.

Another approach is based on the use the mixture-of-expert model (Jacobs et al., 1991) together with RL algorithms (Baldassarre, 2002). We used this approach to build a modular hierarchical architecture controlling a simulated 2DOF planar arm that learns to solve different reaching tasks (Tommasino et al., 2012a; Caligiore et al., 2014b; see also Tommasino et al., 2016, for the latest development of the computational aspects of the model). Each expert of the architecture is an actor-critic RL model whose behavioural output is mixed through the mechanisms of the mixture-of-expert model. The architecture proposes a novel interpretation of Piagetian assimilation and accommodation processes (Piaget, 1953) based on the functional organization of the architecture experts allocated to different tasks through learning. Assimilation occurs when in order to solve a novel task the architecture trains an expert by copying another expert previously trained for solving a task requiring a similar sensorimotor mapping. Instead, accommodation occurs when the model trains new experts from scratch to face tasks requiring very different sensorimotor mappings. The architecture hence shows how assimilation and accommodation might be the behavioural counterpart of mechanisms trading off the need to transfer knowledge from one task to another, for the sake of speeding up learning, and the need to avoid interference of knowledge acquired to solve different tasks. The model also provides a novel theoretical framework to investigate possible impairments of assimilation/accommodation processes, possibly related to developmental disorders such as autism.

Another approach uses modules formed by "dynamic movement primitives" (DMPs; Ijspeert et al., 2002; Schaal et al., 2005). These are parameterised dynamical models suitable to produce and learn rhythmic or discrete movement trajectories with RL algorithms. A model developed within such framework (Meola et al., 2016; see also Ciancio et al., 2011, 2013) is based on multiple DMPs with different complexity and a policy search algorithm (these are RL algorithms that are efficient enough to work on real robots, Kober & Peters, 2011). The model is tested with simulated and real robotic hands and is used to study how, through RL, the initial exploratory rhythmic movements produced by infants, studied since Piaget (1953) as "circular reactions", develop into discrete movements (e.g. "touching" and "tapping" an object) or into sophisticated rhythmic movements (e.g. "rotating back and forth" an object) depending on the task at hand. Moreover, the model is also used to study how RL can learn to select DMPs with different complexity based on the task at hand.

DMPs are also being used within another promising approach where whole (non-modular) networks are used to learn multiple tasks that can be represented with continuous features, for example the (x, y) location of an object on a table (e.g. Castro da Silva et al., 2014). These networks, that take as input the task parameters and give as output the parameters of a DMP, are able to transfer previously acquired skill knowledge to the solution of new tasks. Although these models have not been closely tested with developmental data, they propose interesting mechanisms forming a tool-box that in the future might be used to study how infants and children learn multiple tasks in effective ways.

Another related approach proposes models relying on a soft/functional modularity and hierarchy where whole recurrent neural networks develop functional modules to solve different tasks based on parameter inputs. A first example of this approach (Nishimoto & Tani, 2009) is a modular hierarchical model used to control a humanoid robot able to learn multiple goal-directed tasks. The model, formed by a recurrent neural network, is trained with a supervised learning procedure where the experimenter directly moves the robot's arm along the desired movement trajectories leading to the goals. The input units of the model receive the current proprioceptive and visual signals whereas the output units produce the next-step proprioceptive and visual signals then used to produce motor commands. Additional context units, divided into fast- and slow-changing units, furnish information on current state and the goal to pursue. Depending on the slow context units activation, the model learns to produce different sensorimotor mappings ("behaviour primitives") to accomplish different goals. A second example (Mannella & Baldassarre, 2015; see Figure 13.5) is also based on parametrised soft modularity, supervised learning, and recurrent dynamical networks (in this case "echostate networks"). The architecture structure reflects the hierarchical/modular organisation of brain motor cortex and basal-ganglia. The model is able to learn, through a supervised learning algorithm, to perform different classes of movements, for example to draw circles and squares or to produce different hand gestures, based on parameters sent as input to its high-level components. Moreover, it is able to learn to produce variations of those movements, for example to draw circles with different size or location in space, based on parameters sent to its low-level districts. Overall, these results show how parameterised recurrent neural networks are able to learn to solve multiple tasks and to generalise between them through functional modularity and hierarchy by relying on brain-like mechanisms: this might be possibly exploited to capture the mechanisms underlying the same

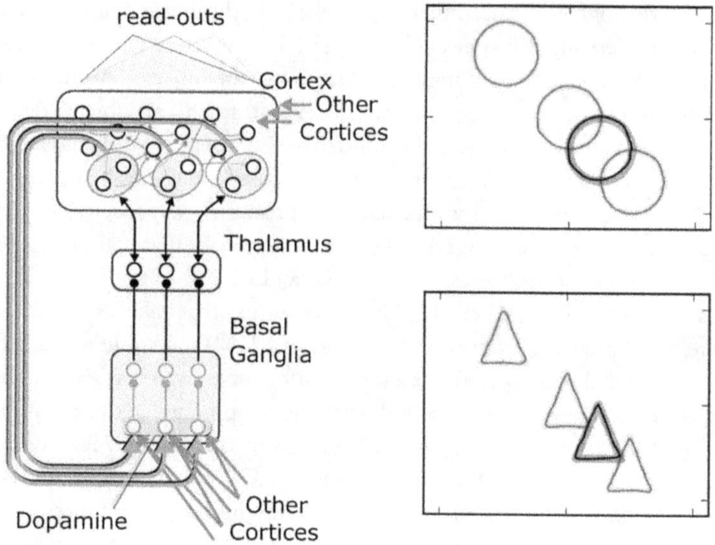

FIGURE 13.5 A bio-constrained model based on echo-state neural networks

Source: Reproduced with permission from Mannella and Baldassarre (2015).

Notes: Left: Architecture of the model. Right: Different inputs to the high-level basal ganglia component of the model can lead the model to learn and produce different movements (e.g. to draw a circle or a triangle), whereas different inputs to the lower-level cortical component can lead the model to produce variations of the same movement, e.g. to draw a circle or a triangle in different positions (the thin triangles/circles are movements corresponding to learned trajectories, whereas the bold ones are trajectories produced by generalisation).

processes in children. To this purpose, however, an important limitation to overcome is to substitute the supervised learning algorithms currently used by these architectures with more plausible RL algorithms.

Extrinsic and Intrinsic Motivations Guiding the Learning Processes of Reaching and Grasping

Motivations are very important for development as they drive exploratory behaviours and furnish the rewards needed to consolidated successful skills (von Hofsten, 2004). Within motivations, we think intrinsic motivations (IMs) play a prominent role, especially during the initial years of development. Indeed, for their nature IMs can lead infants and children, and in part also adults, to engage in activities that ensure a high rate of acquisition of knowledge and skills, thus serving what is by definition a main adaptive function of development. However, we recognise that IMs are not the only drive of development but they work in synergy with other mechanisms, in particular with sensorimotor primitives and embodiment, task space representations, maturational processes of the sensors and actuators, and social guidance (Oudeyer et al., 2013), that although important are not considered in this chapter for the sake of focusing.

IMs have been defined within the psychological literature as motivations driving the performance of actions "for their own sake" (Berlyne, 1966). The concept has been later specified by computational literature (Baldassarre, 2011; Baldassarre & Mirolli, 2013a; Oudeyer & Kaplan, 2009) as involving motivations that drive the performance of actions leading to the acquisition of knowledge that might be useful in a later stage, for example to drive the exploration of novel or surprising stimuli in order to learn object representations and models of the word, or the performance of actions directed to specific goals for improving the actions themselves. IMs differ from extrinsic motivations (EMs) directed to accomplish valuable resources, such as food or pain-avoidance in animals (Deci & Ryan, 1975), or the solution of the user's tasks in robots. Such knowledge and skills could later enhance extrinsically motivated behaviours and learning processes directly satisfying survival and reproduction. Although investigated by psychologists for a long time (Berlyne, 1960; White, 1959), and despite some early pioneering computational explorations (e.g. Schmidhuber, 1991), IMs have fully entered the field of computational modelling and robotics only recently (see Barto et al., 2004, Oudeyer et al., 2007, and Baldassarre & Mirolli, 2013a, for overviews).

Interdisciplinary research involving psychology, neuroscience, and computational modelling has highlighted three main classes of IM mechanisms (Baldassarre and Mirolli, 2013a; Barto et al., 2013; Caligiore et al., 2015; Oudeyer & Kaplan, 2009). The first two classes are linked to the properties of stimuli (Berlyne, 1960) and have for long been considered together under the name of "knowledge-based IMs" (Oudeyer & Kaplan, 2009) to distinguish them from competence-based IMs forming the third class. Recently, the differences between two classes of knowledge-based IMs have been stressed and clarified (Barto et al., 2013). The first class are *prediction-based IMs*. These are related to predictions of possible future states of the world based on its current state and planned actions (Barto et al., 2013; see Mirolli et al., 2013; Oudeyer et al., 2007; Schmidhuber, 1991, for some examples of models). The second class are *novelty-based IMs*. These are elicited by objects that have not been previously experienced and hence are not in the agent's memory. Computational models capturing novelty-based IMs are typically based on the detection of anomalous/unfamiliar stimuli (Nehmzow et al., 2013).

The third class of IMs involves *competence-based IMs*, linked to improving the agent's ability ("competence") to modify the world as desired through own actions (White, 1959). Computational models have implemented the mechanisms behind these IMs in terms of measures of the agent's success in accomplishing desired goals/tasks, and have used them to decide on which skills to focus the agent's activities and learning resources (Barto et al., 2004; Santucci et al., 2013, 2014; Schembri et al., 2007a). In this respect, recent experiments with robots (Santucci et al., 2013) suggest that competence-based IMs are more effective than knowledge-based ones to drive the selection of the task/goal on which to focus skill learning. Knowledge-based IMs might be instead more useful to guide the autonomous formation of tasks/goals (Mirolli & Baldassarre, 2013).

The implementation of intrinsically motivated robots able to discover new tasks/goals and learn the related skills autonomously and in an incremental manner gives interesting suggestions about the mechanisms that could drive infant motor development (Baldassarre & Mirolli, 2013a). For example, Oudeyer et al. (2007) showed how prediction-based IM can lead, in an Aibo dog-like robot, to undergo a developmental trajectory where the robot autonomously focuses its learning processes on increasingly difficult tasks. Schembri et al. (2007a; 2007b; see also Baldassarre & Mirolli, 2013c) show how the TD-error signal computed by RL algorithms, and related to the prediction-error of future rewards, can be used as a competence-based IM mechanism to decide on which task and skill to focus depending on the environmental conditions. Baranes and Oudeyer (2013) show how competence-based IMs can drive a redundant robotic arm to progressively focus a "goal-babbling" algorithm, supporting the solution of multiple reaching tasks, on progressively more complex goals covering the whole reachable space.

IMs might contribute to gradually shift reaching and grasping acquisition mechanisms from initial learning processes related to motor- and goal-babbling, covering large portions of learning spaces related to body postures and trajectories, to trial-and-error learning processes leading to the acquisition of skills related to goals involving the external environment. In this respect, Santucci et al. (2016) proposes an architecture called GRAIL ("goal-discovering robotic architecture for intrinsically-motivated learning", see Figure 13.6), tested in a simulated humanoid robot and environment, that by exploring the external environment discovers that some spherical objects can be "turned-on" if touched. The model architecture uses: (a) novelty-based IMs applied to environmental changes (caused by the robot actions) to autonomously generate goals; (b) competence-based IMs to motivate the selection of the discovered goals on which to focus skill learning. Overall, the model represents an hypothesis of how IMs, rather than directly, might support infants' acquisition of reaching and other types of skills by generating *goals* on the basis of action-outcome contingencies (cf. Williams & Corbetta, 2016). A central tenet behind the architecture is indeed that the self-generation of goals based on IMs is a key element for development (see also Sperati & Baldassarre, 2017, and the European project GOAL-Robots, www.goal-robots.eu).

There are fewer models on IMs driving the acquisition of grasping. An example is Temel et al., (2014) proposing an IM RL model that uses "frustration", computed on the basis of the failure to grasp objects, to determine when giving up the attempts to grasp a specific object. The proposed mechanism is

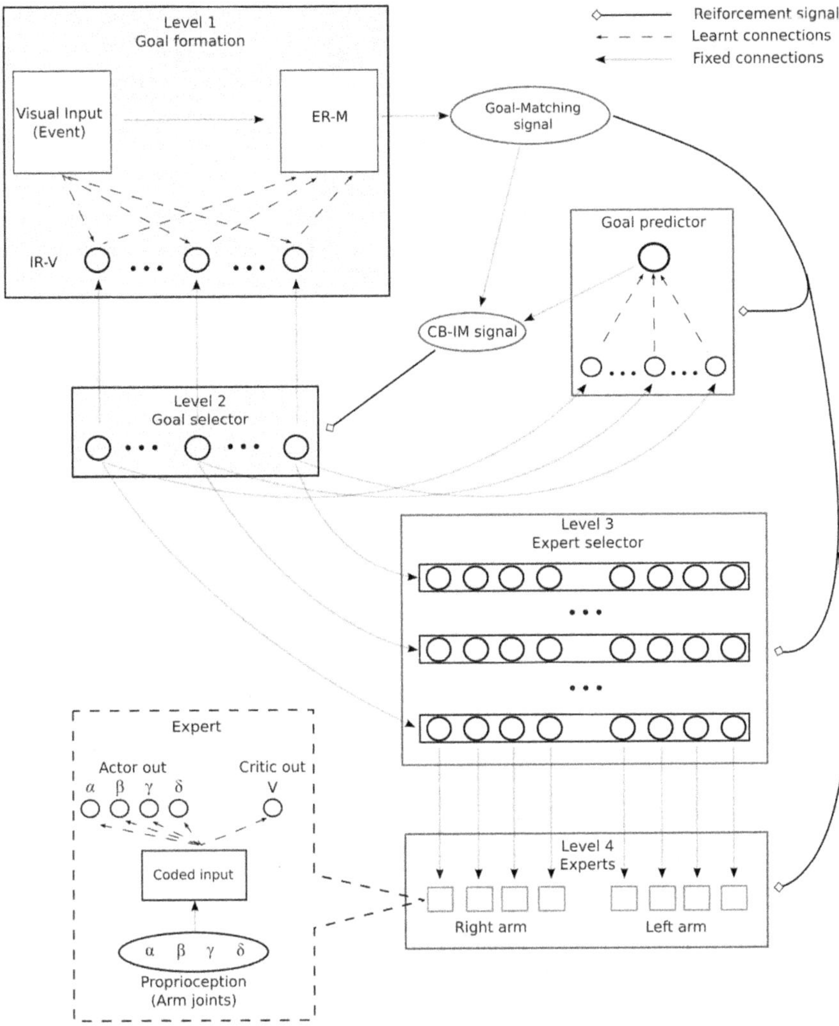

FIGURE 13.6 The GRAIL hierarchical architecture

Source: Reproduced with permission from Santucci et al. (2016).

Notes: The four-level hierarchical architecture of GRAIL: (1) Goal self-generation component: goals are generated based on the progressive encoding and novelty of environment changes; (2) Goal-selector: goals are selected with a probability that increases on the basis of a competence-based IM which is positively related to the rate of improvement of the goal-related skill; (3) RL expert-selector; (4) Actor-critic RL experts: the selected goal-related expert, and the expert selector, learn on the basis of the "matching reward" produced when the system sees a world state similar to the selected goal.

tested with a real robotic arm-gripper using multiple pairs of possible grasping points computed for each object through a simulator. The results show that by using IM mechanisms the system can learn faster than through standard action selection strategies.

IMs have also been used to support the learning of motor sequences and hierarchies of reaching/grasping movements. For example, Mirolli et al. (2013) propose a neural model, tested in a simple simulated 2D arm, that uses IMs to develop a simple sequence of actions involving reaching, magnet-like grasping, and carrying to the mouth. The model shows how IMs, for their transient nature (IM signals fade away after the knowledge to which they relate has been acquired) are suitable to scaffold the discovery and learning of long/difficult sequences of actions that would otherwise be never discovered and learned, e.g. sequences that lead to ingest food and get an extrinsic reward signal. The model links IMs to the brain as the IM and EM learning signals it produces are analogous to the brain phasic dopamine bursts. Hart and Grupen (2011) propose a "control basis framework" where primitive actions are represented by "control potential functions", each being a close-loop policy involving specific sensors and actuators and able to drive a controlled variable to a stable state (here interpretable as a desired goal). Each control function can be in one of four states, depending on the fact that the control variable is unknown, non-recordable, different from the goal, or equal to the goal. The system performs RL to combine the control functions into sequences ("catalogues"): an intrinsic reward is produced when the system discovers a new sequence of control functions that accomplish a goal. These models hence show how IMs can support the learning of new complex actions formed by composing existing reaching and manipulation skills.

The Role of "Embodiment" and the Dynamical Agent-Environment Coupling for the Development of Reaching and Grasping

The importance of *embodiment* as a view that intelligence is acquired and expressed on the basis of a strong interplay between the mind and the body, and between the agent and the environment (often separately referred to as "situatedness"), is being widely accepted in different areas of cognitive science, from artificial life (Floreano et al., 2004; Pfeifer & Gómez, 2009) and evolutionary robotics (Nolfi & Floreano, 2000), to cognitive psychology and cognitive neuroscience (Borghi et al., 2013; Caligiore & Fischer, 2013) and developmental psychology (Corbetta, 2009; Thelen, 2000). In this section, we discuss some of the implications of embodiment for the study of the development of reaching and grasping (see Cangelosi & Schlesinger, 2015, for further information).

An important concept of embodiment is *morphological computation*, referring to the fact that a suitable body hardware can drastically reduce the complexity of the computations needed to perform a given behaviour (Lungarella & Berthouze, 2004; Pfeifer & Gómez, 2009). For example, Caligiore et al. (2014a) show that during reaching development the spring-damper properties of muscles might be important to decrease movement jerk and that redundant degrees of freedom might be initially underused to favour learning and then released at a later stage to refine performance (see also Baranes & Oudeyer, 2011). We have also already seen in the first section how the physical properties characterising the actively controlled fovea/periphery visual sensor can drastically reduce the computational burden (Ballard, 1991), furnish information on the location of targets based on eye proprioception (Ognibene & Baldassarre, 2015), or generate an object-centred reference frame that facilitates action generalisation.

Another important principle of embodiment is *sensorimotor coordination* for which effective behaviours can be produced by exploiting the properties of the sensation-action close loop that an agent engages with the environment: the importance of this principle for the emergence of flexible intelligence has been widely demonstrated with evolutionary robotics models (Nolfi & Floreano, 2000). For example, Ognibene and Baldassarre (2015) showed how a system endowed with active vision can learn to dynamically interact with the environment to seek information needed to solve reaching tasks. Meola et al. (2016) showed how a RL iCub humanoid robot is able to discover and exploit the subtleties of the interaction with the physical world to improve the performance of manipulation tasks. For example, Figure 13.7 shows a sophisticated rhythmic behaviour discovered by the robot where the alternation of the movements of the index and thumb fingers allows a fast rotation of the object. The behaviour is made possible by the fact that the rotated object is anchored to the environment and hence the robot can use one finger at a time to spin the object. This represents an example of the possible opportunities that infants might discover and exploit while learning to manipulate objects in a real physical set-up. Savastano and Nolfi (2013) presented a neurorobotic model, learning to reach and grasp objects based on a trial-and-error learning algorithm, that shows how an infant-like touch-triggered grasping reflex can bootstrap the learning of grasping and how the maturation of body features (e.g. of a higher visual accuracy) can channel grasping development.

An influential view of development that stresses the importance of embodiment is the *dynamical system approach* founded on the idea that development strongly depends on the fact that body, perception, action, and cognition form a closely integrated system (Spencer et al., 2006; Thelen & Smith, 1996). A computational framework used to not only capture such interplay but also to view cognition (i.e. brain) itself as a dynamical system is the *dynamic field approach* (Amari, 1977; Schöner & Spencer, 2015; Spencer & Schöner, 2003). This approach proposes that key common neural states, and the underlying mechanisms, implement perception, cognition, and action processes: "resting states", which encode inputs with neural-assembly stable activity peaks while such

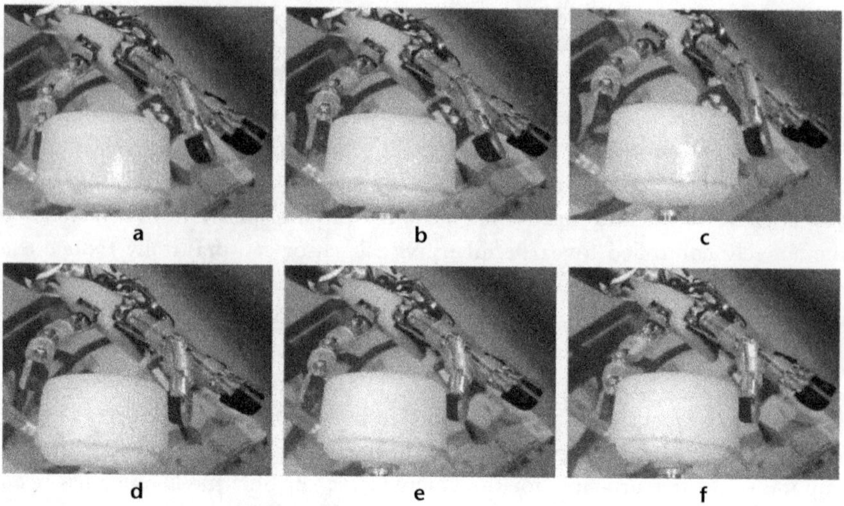

FIGURE 13.7 Top views of the hand movements discovered by an RL robot that learned to rotate an object anchored to the environment as fast as possible

Source: Reprinted with permission from Meola et al. (2016).

Notes: The thumb moves upward by spinning the object while the index moves upward to regain a new contact with the object (a–d). The index moves downward by spinning the object while the thumb moves downward to regain a new contact with the object (e–f).

inputs are in place; "self-sustaining states", which are neural-assembly activations that last in the absence their initial causes; and "dynamical transient states", which are activations involving sequences of resting or self-sustaining states. The mechanisms proposed by the dynamic field approach might allow the study of how initially simple stimulus-response reaching and grasping abilities might progressively become controlled by more sophisticated higher-level cognitive processes such as goal-directed behaviour and planning (Baldassarre et al., 2013; Mannella et al., 2013; Rueckert et al., 2016).

Conclusions

This chapter discussed some computational models, drawn from the authors' work and the literature, contributing to formulate a computationally oriented view on the mechanisms underlying the development of reaching and grasping in infants (Spencer et al., 2006). From this discussion it emerges that active vision, motor babbling and associative learning processes, trial-and-error learning, intrinsic motivations, modular and hierarchical control architectures, and embodiment are critical ingredients to understand the computational mechanisms underlying the development of those behaviours. The framework, although still largely incomplete, is expected to aid the design and interpretation of further experiments and models directed to investigate the development of reaching and grasping, and also of the more complex behaviours and cognitive abilities relying on them. In this respect, recent robotic models have already shown the importance of the ingredients discussed here for the acquisition of language capabilities (e.g. motor babbling and embodiment: Morse & Cangelosi, 2017), goal-directed behaviour (e.g. attention control, intrinsic motivations, trial-and-error learning: Baldassarre et al., 2013; Santucci et al., 2016), mental rotation and decision-making abilities (e.g. modular and hierarchical architectures, and embodiment: Seepanomwan et al., 2013; 2015; Merrick, 2017), and numbering ability (e.g. embodiment: Rucinski et al., 2011). The most important outcome promised by the interdisciplinary effort to formulate the framework is the possible progressive identification of common principles underlying the development of reaching and grasping and of the other multifaceted aspects of human cognition relying on them.

References

Amari, S. (1977). Dynamics of pattern formation in lateral-inhibition type neural fields. *Biological Cybernetics, 27*, 77–87.

Ansuini, C., Santello, M., Massaccesi, S., & Castiello, U. (2006). Effects of end-goal on hand shaping. *Journal of Neurophysiology, 95*(4), 2456–2465.

Baldassarre, G. (2002). A modular neural-network model of the basal ganglia's role in learning and selecting motor behaviours. *Journal of Cognitive Systems Research, 3*(2), 5–13.

Baldassarre, G. (2011). What are intrinsic motivations? A biological perspective. In *Proceedings of ICDL-EpiRob-2011* (pp. E1–8). Piscataway, NJ: IEEE.

Baldassarre, G., Mannella, F., Fiore, V. G., Redgrave, P., Gurney, K., & Mirolli, M. (2013). Intrinsically motivated action-outcome learning and goal-based action recall: a system-level bio-constrained computational model. *Neural Networks, 41*, 168–187.

Baldassarre, G., & Mirolli, M. (Eds). (2013a). *Intrinsically motivated learning in natural and artificial systems*. Berlin: Springer.

Baldassarre, G., & Mirolli, M. (eds) (2013b). *Computational and robotic models of the hierarchical organization of behaviour*. Berlin: Springer.

Baldassarre, G., & Mirolli, M. (2013c). Deciding which skill to learn when: Temporal-Difference Competence-Based Intrinsic Motivation (TD-CB-IM). In G. Baldassarre & M. Mirolli (Eds), *Intrinsically motivated learning in natural and artificial systems* (pp. 257–278). Berlin: Springer.

Balkenius, C. (2000). Attention, habituation and conditioning: toward a computational model. *Cognitive Science Quarterly, 1*, 171–204.

Ballard, D. H. (1991). Animate vision. *Artificial Intelligence, 48*(1), 57–86.

Ballard, D. H., Hayhoe, M. M., Pook, P. K., & Rao, R. P. (1997). Deictic codes for the embodiment of cognition. *The Behavioural and Brain Sciences, 20*, 723–742.

Baranes, A., & Oudeyer, P.-Y. (2011). The interaction of maturational constraints and intrinsic motivations in active motor development. In *IEEE international conference on development and learning (ICDL)*, Vol. 2, pp. 1–8.

Baranes, A., & Oudeyer, P.-Y. (2013). Active learning of inverse models with intrinsically motivated goal exploration in robots. *Robotics and Autonomous Systems, 61*, 49–73.

Barto, A. G., & Mahadevan, S. (2003). Recent advances in hierarchical reinforcement learning. *Discrete Event Dynamic Systems, 13*(4), 341–379.

Barto, A., Mirolli, M., & Baldassarre, G. (2013). Novelty or surprise?. *Frontiers in Psychology, 4*, 907.

Barto, A. G., Singh, S., & Chentanez, N. (2004). Intrinsically motivated learning of hierarchical collections of skills. In *Proceedings of the 3rd International Conference on Development and Learning* (ICDL) (pp. 112–119).

Barton, R. A. (1998). Visual specialization and brain evolution in primates. *Proceedings of the Royal Society of London – Biological sciences, 265*, 1933–1937.

Berlyne, D. E. (1960). *Conflict, arousal, and curiosity*. New York: McGraw-Hill.

Berlyne, D. E. (1966). Curiosity and exploration. *Science, 153*(3731), 25–33.

Berthier, N. E. (1996). Learning to reach: a mathematical model. *Developmental Psychology, 32*(5), 811.

Berthier, N. E. (2011). The syntax of human infant reaching. In *Proceedings of the eighth international conference on complex systems* (pp. 1477–1487). Cambridge, MA: NECSI.

Berthier, N. E., & Keen, R. (2006). Development of reaching in infancy. *Experimental Brain Research, 169*(4), 507.

Berthier, N. E., Rosenstein, M. T., & Barto, A. G. (2005). Approximate optimal control as a model for motor learning. *Psychological Review, 112*(2), 329.

Bonaiuto, J., & Arbib, M. A. (2010). Extending the mirror neuron system model, II: what did I just do? A new role for mirror neurons. *Biological Cybernetics, 102*(4), 341–359.

Bonaiuto, J., & Arbib, M. A. (2015). Learning to grasp and extract affordances: the Integrated Learning of Grasps and Affordances (ILGA) model. *Biological Cybernetics, 109*(6), 639–669.

Borghi, A. M., Scorolli, C., Caligiore, D., Baldassarre, G., & Tummolini, L. (2013). The embodied mind extended: using words as social tools. *Frontiers in Psychology, 4*, 214.

Botvinick, M. M., Niv, Y., & Barto, A. C. (2009). Hierarchically organized behaviour and its neural foundations: a reinforcement learning perspective. *Cognition, 113*(3), 262–280.

Caligiore, D., Ferrauto, T., Parisi, D., Accornero, N., Capozza, M., & Baldassarre, G. (2008). Using motor babbling and Hebb rules for modeling the development of reaching with obstacles and grasping. In R. Dillmann, C. Maloney, G. Sandini, T. Asfour, G. Cheng, G. Metta, & A. Ude (orgs.), *International Conference on Cognitive Systems* (pp. E1–8), University of Karlsruhe, Karlsruhe, Germany, 2–4 April 2008.

Caligiore, D., & Fischer, M. H. (2013). Vision, action and language unified through embodiment. *Psychological Research,* 77(1), 1–6.

Caligiore, D., Guglielmelli, E., Borghi, A. M., Parisi, D., & Baldassarre, G. (2010). A reinforcement learning model of reaching integrating kinematic and dynamic control in a simulated arm robot. In *2010 IEEE 9th international conference on development and learning* (pp. 211–218). IEEE.

Caligiore, D., Mustile, M., Cipriani, D., Redgrave, P., Triesch, J., De Marsico, M., & Baldassarre, G. (2015). Intrinsic motivations drive learning of eye movements: an experiment with human adults. *PloS One, 10*(3), e0118705.

Caligiore, D., Parisi, D., & Baldassarre, G. (2007). Toward an integrated biomimetic model of reaching. In *2007 IEEE 6th international conference on development and learning* (pp. 241–246). IEEE.

Caligiore, D., Parisi, D., & Baldassarre, G. (2014a). Integrating reinforcement learning, equilibrium points, and minimum variance to understand the development of reaching: a computational model. *Psychological Review, 121*(3), 389–421.

Caligiore, D., Tommasino, P., Sperati, V., & Baldassarre, G. (2014b). Modular and hierarchical brain organization to understand assimilation, accommodation and their relation to autism in reaching tasks: a developmental robotics hypothesis. *Adaptive Behaviour, 22*(5), 304–329.

Cangelosi, A., & Schlesinger, M. (2015). *Developmental robotics: from babies to robots.* Cambridge, MA: MIT Press.

Castro da Silva, B., Baldassarre, G., Konidaris, G., & Barto, A. (2014). Learning parameterized motor skills on a humanoid robot. In *IEEE international conference on robotics and automation (ICRA2014)*, 1–6.

Ciancio, A. L., Zollo, L., Baldassarre, G., Caligiore, D., & Guglielmelli, E. (2013). The role of learning and kinematic features in dexterous manipulation: a comparative study with two robotic hands. *International Journal of Advanced Robotic Systems, 10.*

Ciancio, A. L., Zollo, L., Guglielmelli, E., Caligiore, D., & Baldassarre, G. (2011). Hierarchical reinforcement learning and central pattern generators for modeling the development of rhythmic manipulation skills. In *2011 IEEE international conference on development and learning (ICDL)* (Vol. 2, pp. 1–8). IEEE.

Corbetta, D. (2009). Brain, body, and mind: lessons from infant motor development. In J. P. Spencer, M. Thomas, & J. McClelland (Eds), *Toward a unified theory of development: connectionism and dynamic systems theory re-considered* (pp. 51–66). New York: Oxford University Press.

Corbetta, D., Thurman, S. L., Wiener, R. F., Guan, Y., & Williams, J. L. (2014). Mapping the feel of the arm with the sight of the object: on the embodied origins of infant reaching. *Frontiers in Psychology, 5*, 576.

Deci, E. L., & Ryan, R. M. (1975). *Intrinsic motivation.* New York: John Wiley & Sons, Inc.

Demiris, Y., & Dearden, A. (2005). From motor babbling to hierarchical learning by imitation: a robot developmental pathway. In L. Berthouze, F. Kaplan, H. Kozima, H. Yano, J. Konczak, G. Metta, J. Nadel, G. Sandini, G. Stojanov, & C. Balkenius (Eds.), *Proceedings of the Fifth International Workshop on Epigenetic Robotics: Modeling cognitive development in robotic systems* (pp. 31–37). Lund University Cognitive Studies.

Ennouri, K., & Bloch, H. (1996). Visual control of hand approach movements in newborns. *British Journal of Developmental Psychology, 14*(3), 327–338.

Erhardt, R. P. (1994). *Developmental hand dysfunction: theory, assessment, and treatment.* Tucson, AZ: Communication Skill Builders.

Fagg, A. H., & Arbib, M. A. (1998). Modeling parietal-premotor interactions in primate control of grasping. *Neural Networks, 11*(7–8), 1277–1303.

Feldman, A. G. (1986). Once more on the equilibrium-point hypothesis (λ model) for motor control. *Journal of Motor Behaviour, 18*(1), 17–54.

Floreano, D., Mondada, F., Perez-Uribe, A., & Roggen, D. (2004). Evolution of embodied intelligence. In F, Iida, R. Pfeifer, L. Steels, & Y. Kuniyoshi (Eds), *Embodied artificial intelligence* (pp. 293–311). Berlin; Heidelberg: Springer.

Hart, S., & Grupen, R. (2011). Learning generalizable control programs. *IEEE Transactions on Autonomous Mental Development, 3*(1), 216–231.

Hayhoe, M., & Ballard, D. (2005). Eye movements in natural behaviour. *Trends in Cognitive Science, 9*, 188–194.

Herbort, O., Ognibene, D., Butz, M. V., & Baldassarre, G. (2007). Learning to select targets within targets in reaching tasks. In *2007 IEEE 6th international conference on development and learning* (pp. 7–12). IEEE.

Hikosaka, O., Takikawa, Y., & Kawagoe, R. (2000). Role of the basal ganglia in the control of purposive saccadic eye movements. *Physiol Rev, 80*, 953–978.

Houk, J. C., Adams, J. L., & Barto, A. G. (1995). A model of how the basal ganglia generate and use neural signals that predict reinforcement. In J. C. Houk, J. L. Davids, & D. G. Beiser (Eds), *Models of information processing in the basal ganglia* (pp. 249–270). Cambridge, MA: MIT Press.

Hughes, A. (1977). The topography of vision in mammals of contrasting life style: comparative optics and retinal organisation. In F. Crescitelli (Ed.), *The visual system in vertebrates* (pp. 613–756). Berlin: Springer.

Ijspeert, A. J. (2008). Central pattern generators for locomotion control in animals and robots: a review. *Neural Networks, 21*(4), 642–653.

Ijspeert, A., Nakanishi, J., & Schaal, S. (2002). Learning attractor landscapes for learning motor primitives. In M. I. Jordan, Y. LeCun, & S. A. Solla (Eds), *Advances in neural information processing systems 15* (pp. 1523–1530). Boston, MA: MIT Press.

Itti, L., Koch, C., & Niebur, E. (1998). A model of saliency-based visual attention for rapid scene analysis *IEEE Transactions on Pattern Analysis and Machine Intelligence, 20*, 1254–1259.

Jacobs, R., Jordan, M., Nowlan, S., & Hinton, G. (1991). Adaptive mixtures of local experts. *Neural Computation, 3*(1), 79–87.

Johansson, R. S., Westling, G., Bäckström, A., & Flanagan, J. R. (2001). Eye–hand coordination in object manipulation. *Journal of Neuroscience, 21*(17), 6917–6932.

Kober, J., & Peters, J. R. (2011). Policy search for motor primitives in robotics. *Machine Learning, 84*(1–2), 171–203.

Kuperstein, M. (1988). Neural model of adaptive hand-eye coordination for single postures. *Science, 239*(4845), 1308–1311.

Latash, M. L., Scholz, J. P., & Schoner, G. (2007). Toward a new theory of motor synergies. *Motor Control, 11*(3), 276.

Law, J., Shaw, P., Kevin, E., Sheldon, M., & Lee, M. (2014). A psychology based approach for longitudinal development in cognitive robotics. *Frontiers in Neurorobotics, 8*(1).

Lee, M. H., Meng, Q., & Chao, F. (2007). Staged competence learning in developmental robotics. *Adaptive Behaviour, 15*, 241–255.

Lungarella, M., & Berthouze, L. (2004). Robot bouncing: on the synergy between neural and body-environment dynamics. In F. Iida, R. Pfeifer, L. Steels, & Y. Kuniyoshi (Eds), *Embodied artificial intelligence* (pp. 86–97). Berlin; Heidelberg: Springer.

Mannella, F., & Baldassarre, G. (2015). Selection of cortical dynamics for motor behaviour by the basal ganglia. *Biological Cybernetics, 109*, 575–595.

Mannella, F., Gurney, K., & Baldassarre, G. (2013). The nucleus accumbens as a nexus between values and goals in goal-directed behavior: a review and a new hypothesis. *Frontiers in Behavioral Neuroscience,* 7(135), e1–29.

Marraffa, R., Sperati, V., Caligiore, D., Triesch, J., & Baldassarre, G. (2012). A bio-inspired attention model of anticipation in gaze-contingency experiments with infants. In *2012 IEEE international conference on development and learning and epigenetic robotics (ICDL)* (pp. 1–8). IEEE.

McCarty, M. E., & Ashmead, D. H. (1999). Visual control of reaching and grasping in infants. *Developmental Psychology, 35*(3), 620.

McMahon, E., Weiner, R., Connell, J. P., DiMercurio, A., & Corbetta, D. (2017). An analysis of prospective reaching in 9-months-old infants using eye-tracking. *North American Society for Sport Psychology and Physical Activity, 39*(3).

Meola, V., Caligiore, D., Sperati, V., Zollo, L., Ciancio, A., Taffoni, F., Guglielmelli, E., & Baldassarre, G. (2016). Interplay of rhythmic and discrete manipulation movements during development: a policy-search reinforcement-learning robot model. *IEEE Transactions on Cognitive and Developmental Systems, 8*, 152–170.

Merrick, K. (2017). Value systems for developmental cognitive robotics: a survey. *Cognitive Systems Research, 41*, 38–55.

Meunier, D., Lambiotte, R., & Bullmore, E. T. (2010). Modular and hierarchically modular organization of brain networks. *Frontiers in Neuroscience, 4*, 200.

Mirolli, M., & Baldassarre, G. (2013). Functions and mechanisms of intrinsic motivations: the knowledge versus competence distinction. In G. Baldassarre & M. Mirolli (Eds), *Intrinsically motivated learning in natural and artificial systems* (pp. 49–72). Berlin: Springer.

Mirolli, M., Santucci, V., & Baldassarre, G. (2013). Phasic dopamine as a prediction error signal of intrinsic and extrinsic reinforcements: a computational model. *Neural Networks, 39*, 40–51.

Morse, A. F., & Cangelosi, A. (2017). Why are there developmental stages in language learning? A developmental robotics model of language development. *Cognitive Science, 41*(1), 32–51.

Natale, L., Metta, G., & Sandini, G. (2005). A developmental approach to grasping. In *Proceedings of the 2005 AAAI spring symposium*. Palo Alto, CA: AAAI.

Natale, L., Nori, F., Sandini, G., & Metta, G. (2007). Learning precise 3D reaching in a humanoid robot. In *Proceedings of the IEEE 6th international conference on development and learning* (pp. 11–13). London: IEEE.

Nehmzow, U., Gatsoulis, Y., Kerr, E., Condell, J., Siddique, N., & McGuinnity, T. M. (2013). Novelty detection as an intrinsic motivation for cumulative learning robots. In G. Baldassarre & M. Mirolli (Eds), *Intrinsically motivated learning in natural and artificial systems* (pp. 185–207). Berlin; Heidelberg: Springer.

Nishimoto, R., & Tani, J. (2009). Development of hierarchical structures for actions and motor imagery: a constructivist view from synthetic neuro-robotics study. *Psychological Research, 73*, 545–558.

Nolfi, S., & Floreano, D. (2000). *Evolutionary robotics: the biology, intelligence, and technology of self-organizing machines*. Cambridge, MA: MIT Press.

Ognibene, D., & Baldassare, G. (2015). Ecological active vision: four bioinspired principles to integrate bottom–up and adaptive top-down attention tested with a simple camera-arm robot. *IEEE Transactions on Autonomous Mental Development*, 7(1), 3–25.

Ognibene, D., Balkenius, C., & Baldassarre, G. (2008). Integrating epistemic action (active vision) and pragmatic action (reaching): a neural architecture for camera-arm robots. In M. Asada, J. C. T. Hallam, J.-A. Meyer, & J. Tani (Eds), *From animals to animats 10: proceedings of the tenth international conference on the simulation of adaptive behaviour (SAB2008)* (pp. 220–229). Berlin; Heidelberg: Springer.

Ognibene, D., Pezzulo, G., & Baldassarre, G. (2010). How can bottom-up information shape learning of top-down attention-control skills? In B. Kuipers, T. Shultz, A. Stoytchev, & C. Yu (Eds), *Proceedings of 9th IEEE international conference on development and learning (ICDL2010)* (pp. 231–237). Ann Arbor, MA: University of Michigan, 18–21 August 2010.

Oudeyer, P.-Y., Baranes, A., & Kaplan, F. (2013). Intrinsically motivated learning of real-world sensorimotor skills with developmental constraints. In G. Baldassarre & M. Mirolli (Eds), *Intrinsically motivated learning in natural and artificial systems* (pp. 303–365). Berlin: Springer.

Oudeyer, P. Y., & Kaplan, F. (2009). What is intrinsic motivation? A typology of computational approaches. *Frontiers in Neurorobotics, 1*, 6.

Oudeyer, P.-Y., Kaplan, F., & Hafner, V.V. (2007). Intrinsic motivation systems for autonomous mental development. *IEEE Transactions on Evolutionary Computation, 11*(2), 265–286.

Oztop, E., Bradley, N. S., & Arbib, M. A. (2004). Infant grasp learning: a computational model. *Experimental Brain Research, 158*(4), 480–503.

Passingham, R. E., & Wise, S. P. (2012). *The neurobiology of the prefrontal cortex: anatomy, evolution, and the origin of insight.* Oxford: Oxford University Press.

Pfeifer, R., & Gómez, G. (2009). Morphological computation: connecting brain, body, and environment. In B. Sendhoff, E. Körner, O. Sporns, H. Ritter, & K. Doya (Eds.), *Creating brain-like intelligence* (pp. 66–83). Berlin; Heidelberg: Springer.

Piaget, J. (1953). *The origins of intelligence in children.* London: Routledge and Kegan Paul.

Pouget, A., Dayan, P., & Zemel, R. S. (2003). Inference and computation with population codes. *Annual Review of Neuroscience, 26*(1), 381–410.

Rolf, M., Steil, J. J., & Gienger, M. (2010). Goal babbling permits direct learning of inverse kinematics. *IEEE Transactions on Autonomous Mental Development, 2*, 216–229.

Rucinski, M., Cangelosi, A., & Belpaeme, T. (2011). An embodied developmental robotic model of interactions between numbers and space. In *Expanding the space of cognitive science: proceedings of the 23rd annual meeting of the Cognitive Science Society* (pp. 237–242). Austin, TX: Cognitive Science Society.

Rueckert, E., Kappel, D., Tanneberg, D., Pecevski, D., & Peters, J. (2016). Recurrent spiking networks solve planning tasks. *Sci Rep, 6*, 21142.

Saegusa, R., Metta, G., Sandini, G., & Sakka, S. (2009). Active motor babbling for sensorimotor learning. In. *ROBIO 2008. IEEE International Conference on Robotics and Biomimetics, 2008* (pp. 794–799). IEEE.

Sailer, U., Flanagan, J. R., & Johansson, R. S. (2005). Eye–hand coordination during learning of a novel visuomotor task. *Journal of Neuroscience, 25*(39), 8833–8842.

Santello, M. (2015). Getting a grasp of theories of sensorimotor control of the hand: identification of underlying neural mechanisms. *Motor Control, 19*(2), 149–153.

Santucci, V. G., Baldassarre, G., & Mirolli, M. (2013). Which is the best intrinsic motivation signal for learning multiple skills? *Frontiers in Neurorobotics, 7*, 22.

Santucci, V. G., Baldassarre, G., & Mirolli, M. (2014). Autonomous selection of the "what" and the "how" of learning: an intrinsically motivated system tested with a two armed robot. In *4th international conference on development and learning and on epigenetic robotics* (pp. 434–439). IEEE.

Santucci, V. G., Baldassarre, G., & Mirolli, M. (2016), GRAIL: a goal-discovering robotic architecture for intrinsically-motivated learning. *IEEE Transactions on Cognitive and Developmental Systems, 8*(3), 214–231.

Savastano, P., & Nolfi, S. (2013). A robotic model of reaching and grasping development. *IEEE Transactions on Autonomous Mental Development, 5*(4), 326–336.

Schaal, S., Peters, J., Nakanishi, J., & Ijspeert, A. (2005). Learning movement primitives. In *Robotics research: the eleventh international symposium* (pp. 561–572). Berlin; Heidelberg: Springer.

Schembri, M., Mirolli, M., & Baldassarre, G. (2007a). Evolving internal reinforcers for an intrinsically motivated reinforcement-learning robot. *IEEE 6th international conference on development and learning*, 282–287.

Schembri, M., Mirolli, M., & Baldassarre, G. (2007b). Evolving childhood's length and learning parameters in an intrinsically motivated reinforcement learning robot. In *Proceedings of the seventh international conference on epigenetic robotics (EpiRob2007)*, 141–148.

Schlesinger, M., Parisi, D., & Langer, J. (2000). Learning to reach by constraining the movement search space. *Developmental Science, 3*(1), 67–80.

Schmidhuber, J. (1991). Curious model-building control systems. In *IEEE International Joint Conference on Neural Networks* (pp. 1458–1463). IEEE.

Schöner, G., & Spencer, J. (2015). *Dynamic thinking: a primer on dynamic field theory*. Oxford: Oxford University Press.

Seepanomwan, K., Caligiore, D., Baldassarre, G., & Cangelosi, A. (2013). Modelling mental rotation in cognitive robots. *Adaptive Behaviour, 21*(4), 299–312.

Seepanomwan, K., Caligiore, D., Cangelosi, A., & Baldassarre, G. (2015). Generalisation, decision making, and embodiment effects in mental rotation: a neurorobotic architecture tested with a humanoid robot. *Neural Networks, 72*, 31–47.

Sejnowski, T. J. (1977). Storing covariance with nonlinearly interacting neurons. *Journal of Mathematical Biology, 4*(4), 303–321.

Shadmehr, R., & Wise, S. P. (2005). *The computational neurobiology of reaching and pointing: a foundation for motor learning*. Boston, MA: MIT Press.

Spencer, J. P., Clearfield, M., Corbetta, D., Ulrich, B., Buchanan, P., & Schöner, G. (2006). Moving toward a grand theory of development: in memory of Esther Thelen. *Child Development, 77*(6), 1521–1538.

Spencer, J. P., & Schöner, G. (2003). Bridging the representational gap in the dynamic systems approach to development. *Developmental Science, 6*(4), 392–412.

Sperati, V., & Baldassarre, G. (2014). Learning where to look with movement-based intrinsic motivations: a bio-inspired model. In *4th international conference on development and learning and on epigenetic robotics* (pp. 461–468). IEEE.

Sperati, V., & Baldassarre, G. (2017). A bio-inspired model learning visual goals and attention skills through contingencies and intrinsic motivations. *IEEE Transaction on Cognitive and Developmental Systems*, E1–8, DOI:10.1109/TCDS.2017.2772908.

Sutton, R. S., & Barto, A. G. (1998). *Reinforcement learning: an introduction* (Vol. 1, No. 1). Cambridge, MA: MIT Press.

Temel, E., Grzyb, B. J., & Sariel, S. (2014). Learning graspability of unknown objects via intrinsic motivation. *2014 Second Workshop on Artificial Intelligence and Cognition (AIC)* (pp. 98–109).

Thelen, E. (2000). Grounded in the world: developmental origins of the embodied mind. *Infancy, 1*(1), 2–28.

Thelen, E., Corbetta, D., & Spencer, J. P. (1996). Development of reaching during the first year: role of movement speed. *Journal of Experimental Psychology: Human Perception and Performance, 22*(5), 1059.

Thelen, E., & Smith, L. B. (1996). *A dynamic systems approach to the development of cognition and action*. Cambridge, MA: MIT Press.

Tommasino, P., Caligiore, D., Mirolli, M., & Baldassarre, G. (2012a). Reinforcement learning algorithms that assimilate and accommodate skills with multiple tasks. In *2012 IEEE international conference on development and learning and epigenetic robotics (ICDL)* (pp. 1–8). IEEE.

Tommasino, P., Caligiore, D., Mirolli, M., & Baldassarre, G. (2016). A reinforcement learning architecture that transfers knowledge between skills when solving multiple tasks. *IEEE Transactions on Cognitive and Developmental Systems*, E1–8. DOI:10.1109/TCDS.2016.2607018.

Tommasino, P., Caligiore, D., Sperati, V., & Baldassarre, G. (2012b). A McKibben muscle arm learning equilibrium postures. In *2012 4th IEEE RAS & EMBS international conference on biomedical robotics and biomechatronics (BioRob)* (pp. 1229–1234). IEEE.

Vigorito, C. M., & Barto, A. G. (2010). Intrinsically motivated hierarchical skill learning in structured environments. *IEEE Transactions on Autonomous Mental Development, 2*(2), 132–143.

von Hofsten, C. (1979). Development of visually directed reaching: the approach phase. *Journal of Human Movement Studies, 5*, 160–178.

von Hofsten, C. (1982). Eye-hand coordination in the newborn. *Developmental Psychology, 18*, 450–461.

von Hofsten, C. (1984). Developmental changes in the organization of prereaching movements. *Developmental Psychology, 20*(3), 378.

von Hofsten, C. (2004). An action perspective on motor development. *Trends in Cognitive Sciences, 8*(6), 266–272.

von Hofsten, C., & Rönnqvist, L. (1993). The structuring of neonatal arm movements. *Child Development, 64*(4), 1046–1057.

Wang, Q., Bolhuis, J., Rothkopf, C. A., Kolling, T., Knopf, M., & Triesch, J. (2012). Infants in control: rapid anticipation of action outcomes in a gaze-contingent paradigm. *PloS ONE, 7*(2), e30884.

Watkins, C. J., & Dayan, P. (1992). Q-learning. *Machine Learning, 8*(3–4), 279–292.

White, R. W. (1959). Motivation reconsidered: the concept of competence. *Psychological Review, 66*(5), 297.

Whitehead, S. D., & Ballard, D. H. (1991). Learning to perceive and act by trial and error. *Machine Learning,* 7, 45–83.

Williams, J. L., & Corbetta, D. (2016). Assessing the impact of movement consequences on the development of early reaching in infancy. *Frontiers in Psychology,* 7(587).

Williams, J. L., Corbetta, D., & Cobb, L. (2015). How perception, action, functional value, and context can shape the development of infant reaching. *Movement & Sport Sciences, 89*, 5–15.

14

REACHING AND GRASPING

What We Can Learn from Psychology and Robotics

Philippe Gaussier and Alexandre Pitti

Introduction

Grasping an object is an elementary behavior that looks easy for an adult. Yet, grasping an object is still a challenging topic in robotics. Classical approaches consider a sequence of subtasks ranging from object recognition and localization, the planning of the trajectory to reach this object with the correct orientation, and finally the control of the arm movements to grasp securely the object. If this approach has proved to be efficient in simple cases such as reaching a cup on an empty table, a lot of problems remain when the object or the environment is complex. Following Gerardin et al. (2000) and Sirigu et al. (1996), a lot of works have shown the interdependence and even the overlapping between the brain structures involved for real and imagined hand movements. Moreover, some recent works show that the grasping trajectory of an object is impacted by the social environment (Becchio et al. 2008a, 2008b; Lewkowicz et al. 2013; Quesque et al. 2013). Grasping an object in order to give it to somebody else is performed differently than picking it up to place it somewhere else. Even if the deposit place is the same, the global trajectory and especially the preparatory movement to pick the object is different. Hence, our brain is perhaps not planning the grasping as a sequence of elementary and independent subtasks. Such studies on the precise modeling of the human arm control can be found in Bullock and Grossberg 1989 and Bullock et al. 1993, but is out of the scope of the present paper.

Our aim in this review is to present some underlying neuro-computational mechanisms that can serve for such multimodal integration in infants in order to reach and grasp objects. We will show first how reaching a place or an object can be seen as learning sensory-motor attractors. We propose in our framework that recognizing the object we try to grasp is not performed before the grasping but is the result of the whole grasping dynamics (Kelso 1997). The grasping "procedure" can be seen as the building of an attractor basin around the desired object. To succeed in such a task, we will insist on the importance of using the same kind of topological coding from sensory areas to motor areas. We will show also several neural mechanisms found important for multimodal integration, reaching and grasping, as well as the importance of sighted touch for anticipatory movements.

Reaching as an Emergent Property of Visuo-Motor Coordination

According to Bremner, the way infants perceive the space around them (i.e., infants' spatial representation) relies on two different mechanisms that mature separately during the first year (Bremner et al. 2008). One is ego-centric and achieves a spatial correspondence of the default body parts and the other is allo-centric and helps to localize dynamically the position of the limbs. For instance, infants find hard to locate their own hands when their arms are crossed.

It is assumed that the infant brain exploits strongly Hebbian learning to acquire those spatial maps (Del Giudice et al. 2009; Heyes 2010; Keysers et al. 2014). Hebbian learning can provide to neurons the ability to correlate the contingent events, even across multiple modalities. For instance, infants even at birth can connect the modulation of a sound (from low pitch to high pitch) with the modulation of a light intensity (from low brightness to high luminosity; Lewkowicz and Turkewitz 1980), they can relate the contingency between their legs motion and their visual location displayed on a TV monitor (Rochat 1998; Shimada et al. 1995) or between the texture of a pacifier on mouth and its visual shape (Meltzoff and Borton 1979).

Tenants of the embodied system approach suggest that infants learn the regularities in the structure of sensorimotor information to shape the neural activity in the brain. In this line, sensorimotor neurons become sensitive to the geometrical features of objects in space (their position, their orientation) relative to their bodily gestures and postures; what we call sensorimotor primitives. In return, these neurons would guide and constrain the baby behaviors toward its affordances on the environment; e.g. to orient oneself in the environment or to grasp one object in the correct orientation relative to the body.

All-in-all, we propose that the contingency found across the body signals during sensory-motor exploration – coming from the arm muscles spindles, the joint angles from the shoulder-elbow-wrist system, the tactile information from the hand, the sound of a noisy object in the hand, and the eye's vision cells as well as their orientation – may organize the cortical memory into a map of "reachable regions" cells via Hebbian learning. Following this, we can propose that the parietal cortex – which is acknowledged as the center for multi-modal integration and body representation – learns body-place associations by creating parieto-motor "reaching cells" for manipulation tasks (Graziano and Cooke 2006) (see Figure 14.1a and 14.1b respectively).

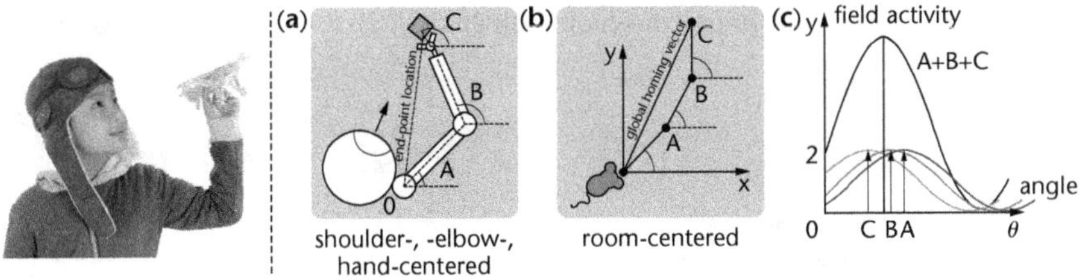

FIGURE 14.1 Child holding a toy in his hand

Notes: Reaching, grasping, and holding objects are non-trivial tasks. In (a), they require to coordinate one's own actions relative to the objects' relative spatial location and orientation in eye-, hand-, or body-centered reference frames. In (b), reaching and grasping with the hand are similar in essence to a navigation task in a third-person perspective toward a place (Gaussier et al. 2007). In (c), grasping can be viewed as performing homing: integrating information by aligning different reference frames through time (Gaussier et al. 2007).

We suppose that the learning of *reach cells* is presumably done by discovering the correct binding that relates each stimulus in its correct reference frame with others in different reference frames; for example this would help for locating the hand in eye coordinates (self-recognition), one object in hand-centered coordinates (relative distance), or one object in shoulder-centered coordinates (body-centered distance).

Those reach cells require to combine contextually or conditionally different sensory signals to represent one location, even if they change dynamically; e.g., if an object is placed at a relative distance of the arm independent of the arm location. For instance, tactile information is dependent on the fingers in the hands, but also on the wrist orientation and arm location, see Figure 14.2a and 14.2b). The visual information is dependent of the eye angle in its orbit (eye-centered) *and* to the head orientation (body-centered), for which the later is only valid for auditory signals.

Considering the neural mechanisms involved in the construction of reach cells, visuo-motor neurons have been found to encode the Preferential visual movement Direction (PD) in extrinsic coordinates (likely shoulder-centered) in 3D space (Blohm and Crawford 2012; Blohm et al. 2008). Kakei and colleagues for instance found some muscle-like neurons based on the composition of cosine functions tuned to visual PD and modulated monotically with the limb posture (see the neuron's activity in Figure 14.2a). When the wrist rotates, the extrinsic neurons discharge proportionally to the preferred visual direction in order to compensate the muscles' displacement (Kakei et al. 2003). Other researchers like Scherberger and colleagues found some hand-centered cells whose activity depends both on the object features like the size and the orientation and on the grip type (pronation or supination) (Fluet et al. 2010; Gail et al. 2009) (see Figure 14.2b). These reach neurons are multimodal and context-dependent. They compute *where* an object is located with respect to the hand

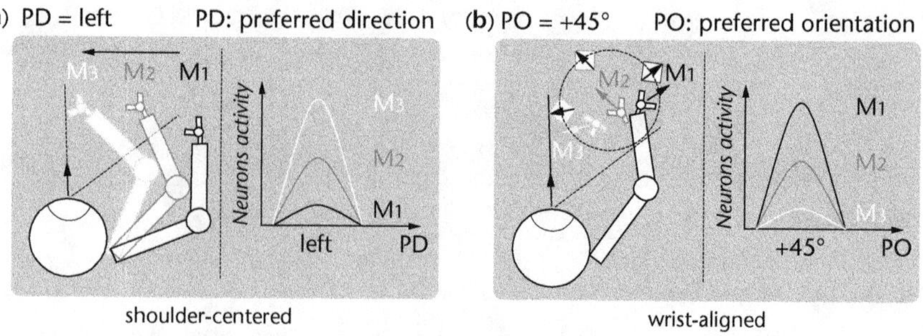

FIGURE 14.2 Principle of gain-modulation in parieto-motor neurons (gain-field neurons) for the arm reaching and orientation

Notes: Gain-field neurons combine visuo-motor features to encode one preferred direction (PD) or one preferred orientation (PO). In (a), exemplar activity of one GF neuron with visual PD and when the arm moves on the left. In (b), exemplar activity of one GF neuron with some visual PO and when the wrist is on pronation.

position and *how* it is oriented with respect to the hand posture. They are called also in the literature gain-field neurons (GF), because their gain-level or amplitude is varying depending on multiple inputs to represent the current object-body relation (Salinas and Sejnowski 2001; Salinas and Thier 2000) as shown in Figure 14.2.

Hand Orientation Based on the Shape of the Object

The gain-field modulation mechanism describes the phenomenon where the motor and the sensor signals (resp. *A* and *B*) mutually influence the amplitude activity of their afferent parietal neurons (resp. *C*) (Andersen and Mountcastle 1983; Andersen et al. 1985). As an example, one possible modulation is the multiplicative effect across the unimodal neurons, *A* and *B*, so that the afferent multimodal neuron *C* possesses the activity $C = A \times B$. Following this, these neurons encode stimulus location simultaneously in more than one reference frame using basis function or “gain fields” (Pouget and Snyder 1997; 2000). For instance, there is a non-linear dependency on eye position for certain visual neurons in posterior parietal neurons (PPNs) whose reference frame is centered on the head, whilst others are found to be influenced more by the coding of somatic information into hand/arm-centered reference frame. The same is found for audio-visual signals integration (Deneve and Pouget 2004; Pitti et al. 2012).

Gain modulation contributes therefore as a major computational mechanism for coordinates transformation and for the compensating of distortions caused by movements (Salinas and Sejnowski 2001). Its role is even broader as PPNs are found also important for reaching targets, goal-directed movements (Chang et al. 2009) and even for intentional acts (Cui and Andersen 2009). Pouget and

Deneve suggest that the parietal neurons behave as a population of basis functions that are continuously adapting their dynamics to the current coordinate frame relative to the task (Deneve et al. 2002; Pouget and Snyder 1997).

Kakei and colleagues (Kakei et al. 2003) proposed a three stages architecture to explain the extrinsic-to-intrinsic transformation necessary for hand reaching using visual directional cells. It is inspired from the model of directional motor cells discovered by Georgopoulos (Georgopoulos et al. 1982, 2007).

At the first stage, extrinsic-like neurons A and B encode some preferred directions (PD) of the movement orientation toward and respectively for the neuron A and B in the form of cosine-like functions:

$$A = \cos(\theta - \theta_A), B = \cos(\theta - \theta_B).$$

At a second stage, the gain levels and receive incoming signals from various modalities (e.g., visual, somatopic, proprioceptive, or auditory) and modulate the amplitude of the extrinsic neurons proportionally to it, so that we have the gain-modulated extrinsic neurons A' and B' equal to:

$$A' = \lambda_A \cos(\theta - \theta_A), B' = \lambda_B \cos(\theta - \theta_B).$$

At the third stage, the gain-modulated extrinsic-like neurons A' and B' are combined linearly to produce the generation of muscle-like neurons C and the encoding of the motor angle into $C = A' + B'$.

In this motor scheme, the input activity of the proprioceptive gain-levels is connected to the preferred direction in the visual space toward objects so that the motor responses of these neurons are linked to the goal-directed actions toward these objects. The muscle response C will be either (1) the shoulder-elbow-hand articular system in the reaching task, (2) the wrist muscle in the close reaching posture of the hand, or (3) the wrist muscle in the close wrist/object relative orientation task (Mahe et al. 2015).

Emergent Control of Different Joints with One Neural Field

When turning our eyes, our head, or our arm in a given direction, some specific subpopulations of neurons are activated (Georgopoulos et al. 1982). These neurons have a tuning curve activity that looks like a bell curve according to their preferred direction. Following the brain anatomical projection from the primary visual area to the parietal cortex (Sirigu et al. 1996) and the motor cortical area, one simple solution is to suppose the topological organization of the information is preserved (or rebuilt). Electrophysiological recording of "motor" neurons activated by somatosensory and visual stimulation (Fadiga et al. 2000) seem to confirm this hypothesis. Using similar coding from the sensory cortical areas to the motor cortical areas can hugely simplify the control problems. For instance, if we suppose the azimuth of a target in the retinal field is projected onto a larger field taking into account the eye orientation and the head orientation thanks to some kind of neuronal shifting mechanisms (Koch and Ullman 1987) the same kind of neural field can be used to code the azimuth of a target in the body space and to control the direction of the movement to reach it. In the dynamical field theory, this population coding is modeled thanks to a field of neurons topologically organized. Recurrent excitatory and inhibitory connections following a Mexican hat shape allow to maintain the goal position as a stable attractor for intermittently detected targets (Amari 1977) and to control the body and arm movements. Once the target has been located in the image, two one-dimensional fields can be used to minimize the angular position between the target and the arm end effector. One interesting readout of the neural field activity is to consider the spatial derivative of the field as the speed set-point (Schöner et al. 1995). Starting from an aligned body configuration (with the eye, head, body, and arm in the same direction), the system will read the same speed set-point for all the different joints. Yet, because of the inertia differences between the different parts of the body, after one step the different joints will end at different azimuths. Hence for the second time step their speed command will become different inducing the feeling that the different degrees of freedom are controlled by different dynamical equations while all their activities are the result of a single one-dimensional field. Generalizing this principle both for the azimuth and the site, it is possible to model a smooth trajectory to reach an arbitrary target with the eyes and the head turning first, followed by the arm and next the whole body reorienting in front of the target (see Figure 14.3). A key point is clearly the importance of the low level control and the use of proprioceptive information (efferent copy of the action) to perform a continuous planning as opposed to a preplanned trajectory.

The proposed strategy is working correctly only if the different degrees of freedoms are well aligned with vision, i.e., if they are either purely in the vertical or horizontal plane. For more complex cases, the visuo-motor transformation must be learned. It can be done using some random motor exploration allowing associating visual position with a specific configuration of the different joints. When the different critical positions have been learned, a neural network can easily interpolate for intermediate positions (Burnod et al. 1995; De Rengervé et al. 2015; Gaussier and Zrehen 1995) allowing the reaching of the visual position. It can be used for gesture imitation or for object grasping.

Building an Attractor Basin for Reaching or Grasping

In real life condition, grasping an object is difficult because learning cannot be perfect. If a known object is presented under an unusual position it can be not trivial to decide about the best way to grasp it. How to orient the hand, how large to open the hand … The problem can become even more challenging for

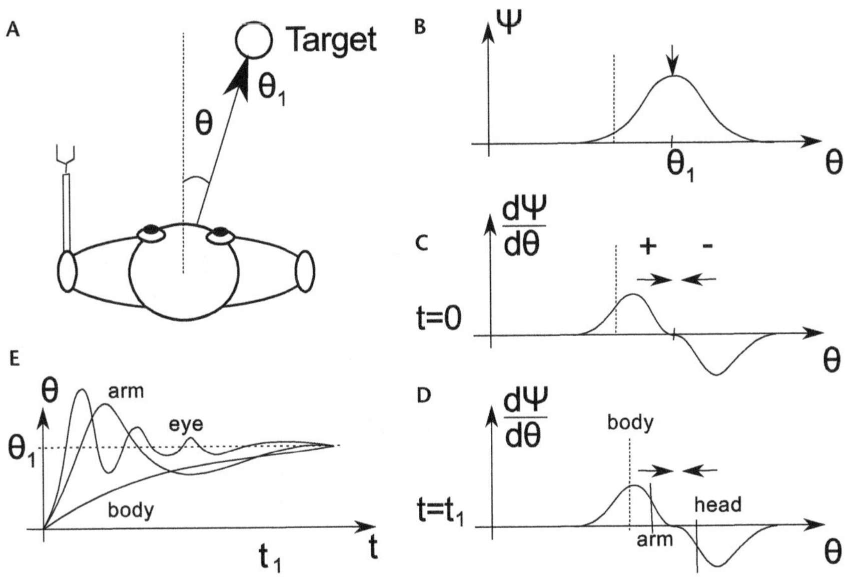

FIGURE 14.3 Different apparent dynamics from the same neural field. The body, arm, and eye appear to have three different dynamics according to their inertia

novel objects. Where is the right place and orientation to grasp it? Generalization capabilities are a key feature to allow grasping. In the architectures we develop, we suppose a pattern (here an object) can never be perfectly recognized (perception is always ambiguous). Hence, recognizing an object cannot be seen as an all or nothing task.

For instance, if we have learned from three places in the vicinity of a goal location what is the action to reach that place, then it is possible to reach that location from any starting place if the environment remains the same (see Figure 14.4). To succeed, the only constraint is to suppose that place or state recognition decreases with the distance to the learned position. In that case, in a two-dimensions environment, learning three places around the goal and taking the action associated to the better recognized place is sufficient to reach the goal location whatever the starting position. The only constraint is that the starting place belongs to the same visual environment (i.e., the starting place is in the neighborhood of the attractor place). The learning creates an attraction basin and the robot is then nothing more than a ball going to the center of the valley defined by the different state/action learned associations. Hence, it is not important to "recognize" one state but to decide the winner state using a competition procedure. At the opposite to classical pattern recognition, we will not suppose that recognizing a shape is performed when the recognition level is higher than a given threshold. Recognizing a shape is related to the

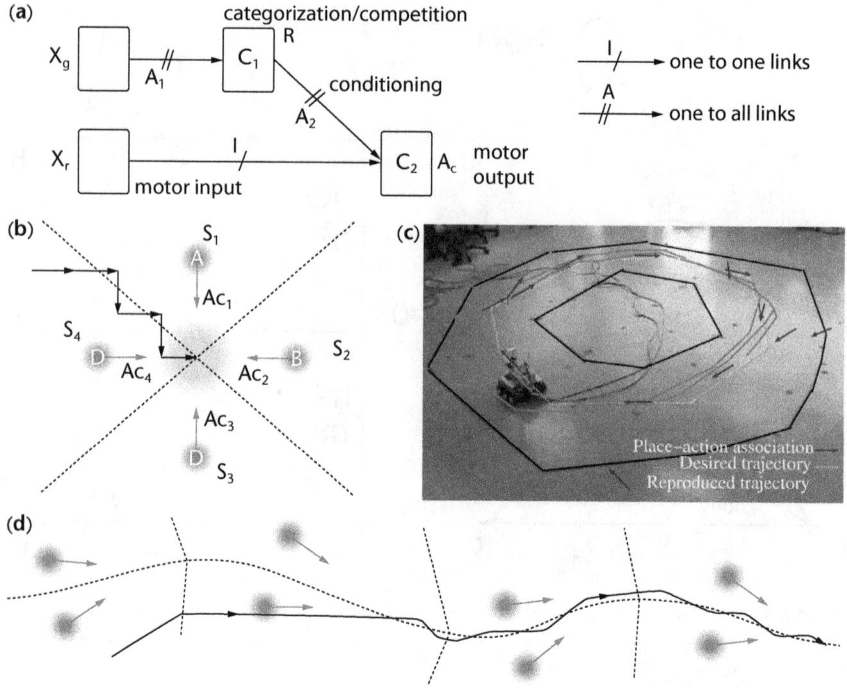

FIGURE 14.4 (a) PerAc model. (b–d) Examples of built dynamics in 2D spaces

Notes: (b) Fixed point attractor. (c) Limit cycle in the case of navigation experiment. (d) Trajectory following. In (b) and (d), the gray dotted lines are the Voronoi boundaries. The plain black line is a trajectory sample

sensorimotor sequence of actions allowing reaching that state. It can be performed even if the different states are badly recognized. The only constraint is that the recognition decreases when moving away from a learned state. The same way reaching an object is equivalent to some extent to recognizing this object and can be performed using the same kind of architecture (see Figure 14.5).

When grasping an object, the mechanical constraints induce some stable proprioceptive-motor states. From the visual information (for instance some Gabor signatures) it is possible to predict a field where the maximum activity corresponds to one of the orientations to grasp the object. As a result, learning the vision of the object will induce a field activity with as many maxima as the number and positions of the possible grasping (see Figure 14.6).

Using this field as the goal for reaching will allow the arm to move in the direction of the nearest "good" grasping location on the object (moving according to the derivative of the field). It is not clear if this field has to be explicit in our brain (what we consider as a field could be the projection on the action space of the high dimension sensory-motor connectivity). Moreover, the learning of two views of the same objects and their associated actions can be

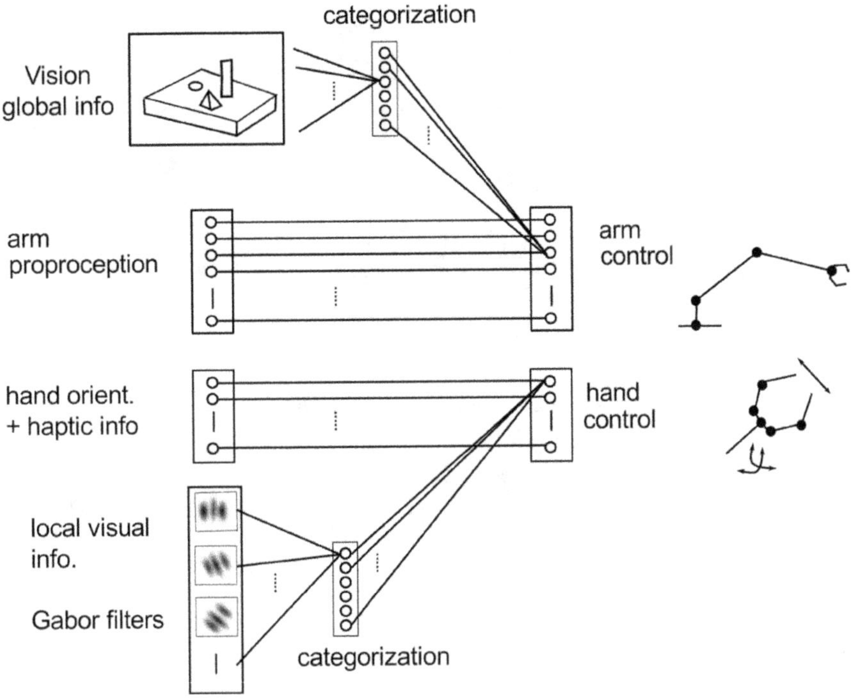

FIGURE 14.5 PerAc architecture for the visual control of object grasping

sufficient to induce a competition allowing the grasping and can be interpreted as the result of a virtual neural field (De Rengervé et al. 2015). Hence the attractor could be completely emergent and not formally endowed in a neural field (the neural field being an abstract way to represent the attractor reached by the arm/hand couple).

It is also clear that recognizing an object involves not only cortical areas but also subcortical areas. Once the object has been grasped, it is possible to use the dynamical expectations of the arm movement as a way to recognize some important feature of an object such as its weight. As a matter of fact, lifting an object will induce different trajectories according to its mass and to the force used. Knowing (and deciding) the force profile, it is possible to predict the timing of the arm proprioception. Then a simple difference between the temporal predictions according to the force used and the recognition of the object is sufficient to decide if the object lifted was the correct object or not (Razakarivony et al. 2010). In our case, we are using a simple model of the cerebellum to predict the temporal dynamics of a given force command in a given visual context (i.e., the recognition of a given object) (Blakemore et al. 2001; Jenmalm et al. 2006). It allows to predict if there is a mismatch between the object visual appearance and the object inertia.

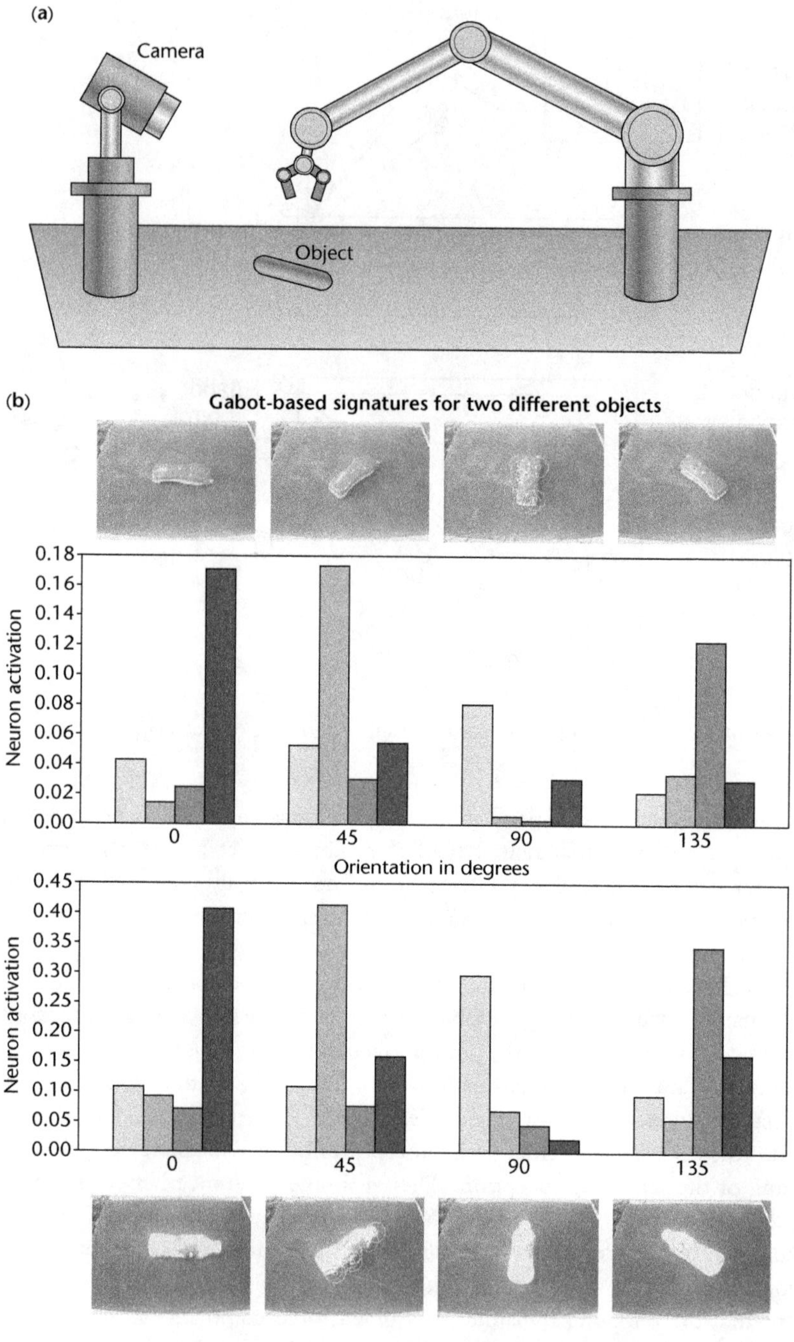

FIGURE 14.6 (a) A simple robotic setup for object grasping. (b) Predicting object orientation for grasping according to the result of Gabor filters on the object image

Multimodal Strategies for Reaching and Grasping

Perceiving objects in space is one of the first tasks babies have to deal with during infancy. It is a rather difficult problem since infants have to represent one object with multiple sensory modalities (vision, sound, touch) encoded in different reference frames (e.g., eye-centered, head-centered, or hand-centered). This curse of dimensionality corresponds to the so-called binding problem across the modalities and requires at least to construct one amodal and unified reference frame relying on coordinates transformation between the senses *or* to form a network of partially interconnected reference frames to represent one stimulus in each modality (Heed et al. 2015).

Touch is often neglected when we consider reaching and grasping tasks. However, these actions are goal-directed toward manipulating objects. Although tactile information is acquired at the very end of these tasks, it has an influence on the whole action sequence from the beginning till its end, by *anticipating* the orientation of the hand with respect to the object to be manipulated and touched. The predictive coding done by the somatosensory neurons, combined with other senses (vision, proprioception), may give rise to the emerging capabilities of the so-called mirror neurons, invisible to the eyes like body image (Caggiano et al. 2009; Maravita et al. 2003; Pitti et al. 2013; Rizzolatti and Craighero 2004), anticipatory touch (Keysers 2004; Pitti et al. 2008), action observation (Keyser et al. 2014), social cognition (Keysers et al. 2010) as well as tool-use (Maravita and Iriki 2004) and language (Rizzolatti and Arbib 1998).

Touching objects permits to discover the physical limits of the body and to construct its spatial representation with the use of vision and proprioception (Pitti et al. 2017) (see Figure 14.7 left graph). This spatial representation is however non-linear, the same tactile stimulus can correspond to different

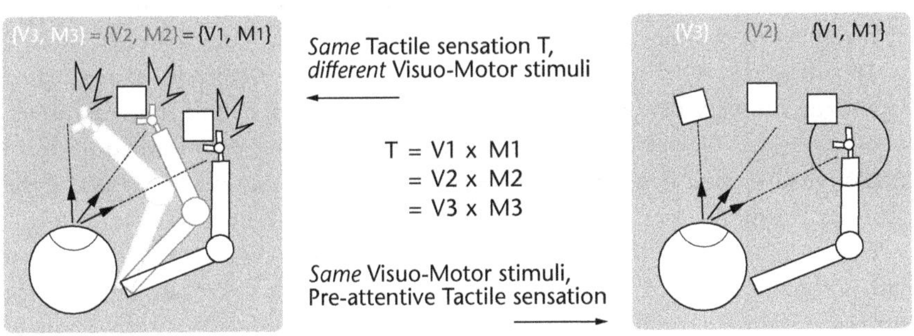

FIGURE 14.7 Tactile information is important for calibrating and anticipating spatial locations for reaching and grasping tasks

Notes: On the left, one task that reach cells have to resolve is to link the same tactile cell with different visuo-motor pairs. On the right, when spatial locations have been learned by reach cells with the mechanisms of multimodal integration explained above, they can serve for anticipating visually one touch; acquisition of body image.

visuo-motor pairs. However, the tactile stimuli can serve to learn the relative locations of these different visuo-motor pairs with respect to the hand motion as in Figure 14.1. Over time, the tactile cells can develop extended visual receptive fields with anticipatory effect (Graziano and Cooke 2006; Graziano et al. 1997; Pitti et al. 2009) (see Figure 14.7 right graph). These effects have been attributed also to the so-called mirror neurons in the premotor area that integrate multisensory information for action purposes (Caggiano et al. 2009; Keysers, 2004; Rizzolatti et al. 1997).

These visuo-tactile receptive fields can be learned from the same gain-modulation mechanism introduced in the previous section. The output motor neuron *C*, which combines visuo-motor information for hand position/orientation, is here replaced by the tactile neuron *T* anchored in one location on the skin and relative to the arm posture.

We present some preliminary results that compile experiments done in one robotic arm, the JACO arm from Kinova Robotics, with vision and one artificial skin for self-representation, reaching and grasping motion; see Figure 14.8. The framework is the one explained above based on gain-modulation for the modeling of different cells that integrate multimodal information for grasping (hand-object alignment), distance (combining visuo-motor information with one artificial skin on the arm), and reaching (object location in shoulder-, elbow-, hand-center reference frame); resp. a–d.

Figure 14.8a presents the model proposed by Kakei and described in previous sections to explain the general architecture for extrinsic-to-intrinsic transformations based on gain-field modulation; e.g., for object reaching (visual direction sensitivity of motor cells and hand-centered visual cells), for object-wrist alignment (relative visual orientation of motor cells). Figure 14.8b displays the neural dynamics of a gain-field neuron sensitive to one object relative orientation to the robot wrist after motor babbling. The graph shows five wrist motor angles with respect to the global visual orientation of the stick (colormap hsv, resp. 0 to 180 degrees). This graph shows how the neuron amplitude is modulated depending on the wrist orientation (intrinsic coordinates) and to the stick orientation (extrinsic coordinates); this neuron permits the transformation from intrinsic to extrinsic coordinates. The shift in preferred orientation (PO) of the neuron corresponds to the graded contribution of the GF neuron.

Figure 14.8c reproduces the experience for body-centered representation with a robotic arm and its artificial skin. With a similar neural architecture for intrinsic-to-extrinsic transformation, the tactile input perceived from a salient object touching the skin surface is used to localize at the same time its visual location in the image for different motor positions. After the learning stage, it is possible to reconstruct back the relative visuo-tactile receptive fields even when an object is approaching the skin without touching it. The length of each arrow indicates the activity level of the nearby neuron anchored at the visual position of the skin for a specific orientation and distance of any approaching objects.

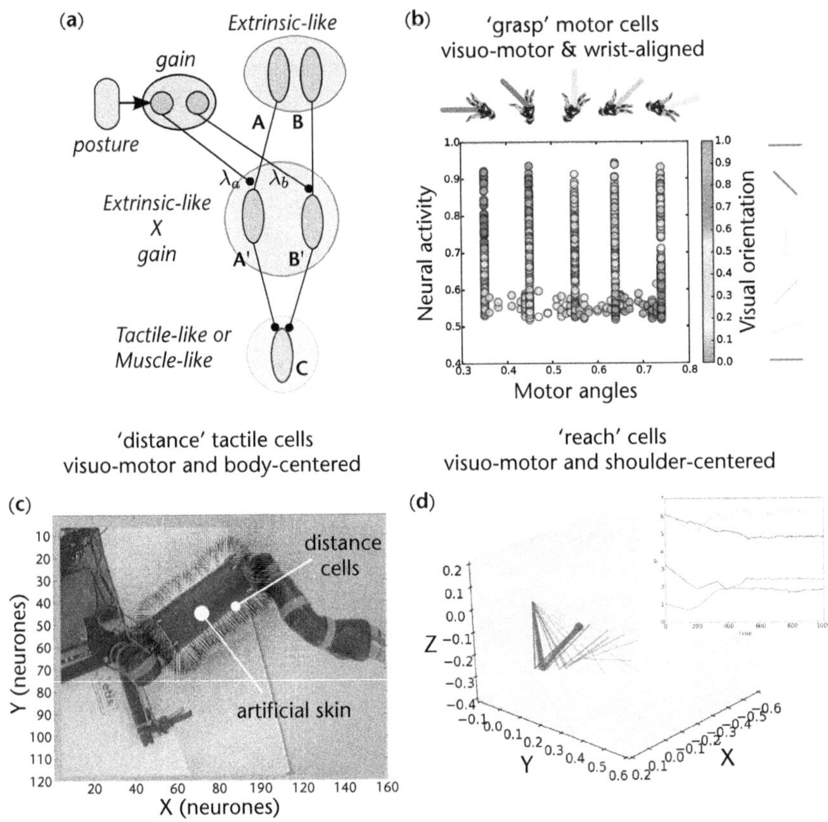

FIGURE 14.8 Gain-field neurons for the modeling of grasping, distal, and reaching cells

Notes: In (a), mechanism of gain-modulation for multimodal integration based on multiplying various signals. In (b), its modeling for the generation of wrist motor "grasp" cells aligned to the visual orientation of one object. The neural activity of the neuron is dependent of the wrist orientation and the visual orientation of the object. In (c), the activity of "touch" cells on the artificial skin are superimposed to integrate distance depending on the visual position on the camera and the arm orientation. In (d), reach cells are modeled in this 3D simulation of the arm by summing up cosine/Gaussian functions as in (a). It is similar to the path integration framework explained in Figure 14.1c).

The last figure – Figure 14.8d – shows a simulation of a reaching task based on the modeling of visual motor cells in 3D. The network learns the relation between posture and visual direction, AxB, to generate a motor command C in that direction. These different experiments based on multiple modalities show that this mechanism of gain-modulation may serve as a unifying principle for spatial representation in order to interact in the world, for robots, brains, and babies.

Discussion

The neurons that learn the proper sensory-motor associations of the body may be useful for perceiving someone else's actions by simulating them. They will react in a similar way to mirror neurons (Gaussier et al. 1998; Pitti et al. 2009). They are either activated by the action or by the recognition of the visual features associated with the object to grasp. Hence, gesture imitation and action imitation could be seen, at first, as the result of the perception ambiguity. Being unable to recognize his/her own body from the body of their parent, could force the baby or the robot to mimic their parent/caretaker gestures as a way to minimize the difference between their own proprioception and the visual position observed hand. Indeed, in this case, the only solution for the system to minimize the discrepancy between vision and proprioception is to move his/her arm in a way inducing an emergent low level imitation behavior (Andry et al. 2001; Gaussier et al. 1998).

For learning to grasp and avoid objects, it is necessary to add emotional mechanisms as a primary conditioning system allowing to associate positive and negative objects with grasping and avoidance behaviors. If we suppose a cascade of sensorimotor architectures (Gaussier and Zrehen 1995) allowing first to associate the adult facial expression with some positive or negative values (Boucenna et al. 2010) then, smiling when a baby (or a robot) is grasping an object or doing an angry face can be used as way to trigger object attraction or repulsion for objects which are primarily neutral for the baby or the robot. Hence, social referencing and complex motor behavior can emerge from the same architecture using some second order conditioning mechanisms (Boucenna et al. 2014; Karaouzene et al. 2013).

For even more complex tasks involving tool use, we can suppose first that the "graspability" of an object can be represented by one or several neural fields representing from where the object can be grasped (the higher, the easier to grasp). The aperture and orientation of the hand can be learned. Next, when an object has been grasped, it can modify the field associated to the hand (or the arm extremity) becoming a tool for the external observer. On the neural field associated to the arm extremity, if the object orientation is on the left, an activity bubble on the left will traduce the capability of the object to act on the left. A "T" object will produce both left and right action capabilities inducing two emergent solutions when grasping is necessary. Hence, if we suppose the arm is composed of several degrees of freedom in a serial order, their associated bubble of activities can be summed. As a result the maximum of activity on the azimuth and site fields is associated to the global arm direction and the intensity of the maximum to the arm elongation. Grasping or using a tool would be the result of the same reconfigurable neural field associated to the arm end effector and learned from the dynamical sensory-motor exploration of the object or tool.

Acknowledgments

We thank Ganna Pugach, Sylvain Mahé, and Kemo Adrian for fruitful work on the experimental setups. This work was partially funded by EQUIPEX-ROBOTEX (CNRS), ANR DIRAC, chaire d'excellence CNRS-UCP, and project Labex MME-DII (ANR11-LBX-0023–01).

References

Amari, S.-i. Dynamics of pattern formation in lateral-inhibition type neural fields. *Biological Cybernetics* **27**, 77–87 (1977).

Andersen, R. and Mountcastle, V. The influence of the angle of gaze upon the excitability of the light-sensitive neurons of the posterior parietal cortex. *Journal of Neuroscience* **3**, 532–548 (1983).

Andersen, R., Essick, G., and Siegel, R. Encoding of spatial location by posterior parietal neurons. *Science* **230**, 450–458 (1985).

Andry, P., Gaussier, P., Moga, S., Banquet, J.-P., and Nadel, J. Learning and communication via imitation: an autonomous robot perspective. *IEEE Transactions on Systems, Man, and Cybernetics – Part A: Systems and Humans* **31**, 431–442 (2001).

Becchio, C., Sartori, L., Bulgheroni, M., and Castiello, U. The case of Dr. Jekyll and Mr. Hyde: a kinematic study on social intention. *Consciousness and Cognition* **17**, 557–564 (2008a).

Becchio, C., Sartori, L., Bulgheroni, M., and Castiello, U. Both your intention and mine are reflected in the kinematics of my reach-to-grasp movement. *Cognition* **106**, 894–912 (2008b).

Blakemore, S.-J., Frith, C. D., and Wolpert, D. M. The cerebellum is involved in predicting the sensory consequences of action. *Neuroreport* **12**, 1879–1884 (2001).

Blohm, G. and Crawford, J. Computations for geometrically accurate visually guided reaching in 3-D space. *Journal of Vision, in special issue "Sensorimotor processing of goal-directed movements"* **7**, 4, 1–22 (2012).

Blohm, G., Khan, A., and Crawford, J. Spatial transformations for eye–hand coordination. *New Encyclopedia of Neuroscience* **9**, 203–211 (2008).

Boucenna, S., Gaussier, P., Andry, P., and Hafemeister, L. Imitation as a communication tool for online facial expression learning and recognition. In *2010 IEEE/RSJ International Conference on Intelligent Robots and Systems (IROS)*, 5323–5328 (IEEE, 2010).

Boucenna, S., Gaussier, P., and Hafemeister, L. Development of first social referencing skills: emotional interaction as a way to regulate robot behavior. *IEEE Transactions on Autonomous Mental Development* **6**, 42–55 (2014).

Bremner, A., Holmes, N., and Spence, C. Infants lost in (peripersonal) space? *Trends in Cognitive Sciences* **12**, 298–305 (2008).

Bullock, D. and Grossberg, S. Vite and flete: neural modules for trajectory formation and postural control. *Advances in Psychology* **62**, 253–297 (1989).

Bullock, D., Grossberg, S., and Guenther, F. H. A self-organizing neural model of motor equivalent reaching and tool use by a multijoint arm. *Journal of Cognitive Neuroscience* **5**, 408–435 (1993).

Burnod, Y., Baraduc, P., Battaglia-Mayer, A. et al. Parieto-frontal coding of reaching: an integrated framework. *Experimental Brain Research* **129**, 325–346 (1999).

Caggiano, V., Fogassi, L., Rizzolatti, G., Thier, P., and Casile, A. Mirror neurons differentially encode the peripersonal and extrapersonal space of monkeys. *Science* **324**, 403–406 (2009).

Chang, S., Papadimitriou, C., and Snyder, L. Using a compound gain field to compute a reach plan. *Neuron* 744–755 (2009).

Cui, H. and Andersen, R. Posterior parietal cortex encodes autonomously selected motor plans. *Neuron* **56**, 552–559 (2009).

Del Giudice, M., Manera, V., and Keysers, C. Programmed to learn? The ontogeny of mirror neurons. *Developmental Science* **12**, 350–363 (2009).

Deneve, S. and Pouget, A. Bayesian multisensory integration and cross-modal spatial links. *Journal of Physiology – Paris* **98**, 249–258 (2004).

Deneve, S., Pouget, A., and Duhamel, J. A computational perspective on the neural basis of multisensory spatial representations. *Nature Reviews. Neurosciences* **98**, 741–747 (2002).

De Rengervé, A., Andry, P., and Gaussier, P. Online learning and control of attraction basins for the development of sensorimotor control strategies. *Biological Cybernetics* **109**, 255–274 (2015).

Fadiga, L., Fogassi, L., Gallese, V., and Rizzolatti, G. Visuomotor neurons: ambiguity of the discharge or "motor" perception? *International Journal of Psychophysiology* **35**, 165–177 (2000).

Fluet, M., Baumann, M., and Scherberger, H. Context-specific grasp movement representation in macaque ventral premotor cortex. *Journal of Neuroscience* **30**, 15175–1518 (2010).

Gail, A., Klaes, C., and Westendorff, S. Implementation of spatial transformation rules for goal-directed reaching via gain modulation in monkey parietal and premotor cortex. *Journal of Neuroscience* **29**, 9490–9499 (2009).

Gaussier, P. and Zrehen, S. PerAc: A neural architecture to control artificial animals. *Robotics and Autonomous Systems* **16**, 291–320 (1995).

Gaussier, P., Banquet, J.-P., Sargolini, F. et al. A model of grid cells involving extra hippocampal path integration, and the hippocampal loop. *Journal of Integrative Neuroscience* **6**, 447–476 (2007).

Gaussier, P., Moga, S., Quoy, M., and Banquet, J.-P. From perception-action loops to imitation processes: a bottom-up approach of learning by imitation. *Applied Artificial Intelligence* **12**, 701–727 (1998).

Georgopoulos, A. P., Kalaska, J. F., Caminiti, R., and Massey, J. T. On the relations between the direction of two-dimensional arm movements and cell discharge in primate motor cortex. *The Journal of Neuroscience* **2**, 1527–1537 (1982).

Georgopoulos, A.P., Merchant, H., Naselaris, T., and Amirikian, B. Mapping of the preferred direction in the motor cortex. *Proc Natl Acad Sci USA.* **104**, 11068–11072 (2007).

Gerardin, E., Sirigu, A., Lehéricy, S., et al. Partially overlapping neural networks for real and imagined hand movements. *Cerebral Cortex* **10**, 1093–1104 (2000).

Graziano, M. and Cooke, D. Parieto-frontal interactions, personal space, and defensive behavior. *Neuropsychologia* **44**, 845–859 (2006).

Graziano, M., Hu, X., and Cooke, D. Visuospatial properties of ventral premotor cortex. *Journal of Neurophysiology* **77**, 2268–2292 (1997).

Heed, T., Buchholz, V., Engel, A., and Röder, B. Tactile remapping: from coordinate transformation to integration in sensorimotor processing. *Trends in Cognitive Science* **19**, 251–258 (2015).

Heyes, C. Where do mirror neurons come from? *Neuroscience and Biobehavioral Reviews* **30**, 575–583 (2010).

Jenmalm, P., Schmitz, C., Forssberg, H., and Ehrsson, H. H. Lighter or heavier than predicted: neural correlates of corrective mechanisms during erroneously programmed lifts. *The Journal of Neuroscience* **26**, 9015–9021 (2006).

Kakei, S., Hoffman, D., and Strick, P. Sensorimotor transformations in cortical motor areas. *Neuroscience Research* **46**, 1–10 (2003).

Karaouzene, A., Gaussier, P., and Vidal, D. A robot to study the development of artwork appreciation through social interactions. In *2013 IEEE Third Joint International Conference on Development and Learning and Epigenetic Robotics (ICDL)*, 1–7 (IEEE, 2013).

Kelso, J. S. *Dynamic patterns: the self-organization of brain and behavior* (Cambridge, MA: MIT Press, 1997).

Keysers, C. Demystifying social cognition: a Hebbian perspective. *Trends in Cognitive Sciences* **8**, 501–507 (2004).

Keysers, C., Kaas, J., and Gazzola, V. Somatosensation in social perception. *Nature Reviews. Neuroscience* 427–428 (2010).

Keysers, C., Perrett, D., and Gazzola, V. Hebbian learning is about contingency, not contiguity, and explains the emergence of predictive mirror neurons. *Behavioral and Brain Sciences* **37**, 205–206 (2014).

Koch, C. and Ullman, S. Shifts in selective visual attention: towards the underlying neural circuitry. In L. M. Vaina (ed.), *Matters of intelligence: conceptual structures in cognitive neuroscience*, 115–141 (Dordrecht: Springer, 1987).

Lewkowicz, D. and Turkewitz, G. Cross-modal equivalence in early infancy: auditory-visual intensity matching. *Developmental Psychology* **16**, 597–607 (1980).

Lewkowicz, D., Delevoye-Turrell, Y., Bailly, D., Andry, P., and Gaussier, P. Reading motor intention through mental imagery. *Adaptive Behavior* 1059712313501347 (2013).

Mahe, S., Braud, R., Gaussier, P., Quoy, M., and Pitti, A. Exploiting the gain-modulation mechanism in parieto-motor neurons: application to visuomotor transformations and embodied simulation. *Neural Networks* **62**, 102–111 (2015).

Maravita, A. and Iriki, A. Tools for the body (schema). *Trends in Cognitive Sciences* **8**, 79–96 (2004).

Maravita, A., Spence, C., and Driver, J. Multisensory integration and the body schema: close to hand and within reach. *Current Biology* **13**, R531–R539 (2003).

Meltzoff, A. and Borton, R. Intermodal matching by human neonates. *Nature* **282**, 403–404 (1979).

Pitti, A., Alirezaei, H., and Kuniyoshi, Y. Mirroring maps and action representation through embodied interactions. *Connectionist Models of Behaviour and Cognition II, World Scientific* 27–37 (2008).

Pitti, A., Alirezaei, H., and Kuniyoshi, Y. Cross-modal and scale-free action representations through enaction. *Neural Networks* **22**, 144–154 (2009).

Pitti, A., Blanchard, A., Cardinaux, M., and Gaussier, P. Gain-field modulation mechanism in multimodal networks for spatial perception. *12th IEEE-RAS International Conference on Humanoid Robots Nov.29–Dec.1, 2012*, 297–302 (Business Innovation Center Osaka, Japan, 2012).

Pitti, A., Kuniyoshi, Y., Quoy, M., and Gaussier, P. Modeling the minimal newborn's intersubjective mind: the visuotopic-somatotopic alignment hypothesis in the superior colliculus. *PLoS ONE* **8**, e69474 (2013).

Pitti, A., Pugach, G., Gaussier, P., and Shimada, S. Spatio-temporal tolerance of visuo-tactile illusions in artificial skin by recurrent neural network with spike-timing-dependent plasticity. *Scientific Reports* **7** (2017).

Pouget, A. and Snyder, L. Spatial transformations in the parietal cortex using basis functions. *Journal of Cognitive Neuroscience* **3**, 1192–1198 (1997).

Pouget, A. and Snyder, L. Computational approaches to sensorimotor transformations. *Nature Neuroscience* **3**, 1192–1198 (2000).

Quesque, F., Lewkowicz, D., Delevoye-Turrell, Y. N., and Coello, Y. Effects of social intention on movement kinematics in cooperative actions. *Frontiers in Neurorobotics* **7**, 14 (2013).

Razakarivony, S., Gaussier, P., and Ouezdou, F. B. From force control and sensory-motor informations to mass discrimination. In *International Conference on Simulation of Adaptive Behavior*, 221–231 (Dordrecht: Springer, 2010).

Rizzolatti, G. and Arbib, A. Language within our grasp. *Trends in Neuroscience* **21**, 188–194 (1998).

Rizzolatti, G. and Craighero, L. The mirror-neuron system. *Annual Review of Neuroscience* **27**, 169–192 (2004).

Rizzolatti, G., Fadiga, L., Fogassi, L., and Gallese, V. The space around us. *Science* **277**, 190–191 (1997).

Rochat, P. Self-perception and action in infancy. *Experimental Brain Research* **123**, 102–109 (1998).

Salinas, E. and Sejnowski, T. J. Gain modulation in the central nervous system: where behavior, neurophysiology and computation meet. *The Neuroscientist* **7**, 430–440 (2001).

Salinas, E. and Thier, P. Gain modulation: a major computational principle of the central nervous system. *Neuron* **27**, 15–21 (2000).

Schöner, G., Dose, M., and Engels, C. Dynamics of behavior: theory and applications for autonomous robot architectures. *Robotics and Autonomous Systems* **16**, 213–245 (1995).

Shimada, S., Hiraki, K., and Oda, I. The parietal role in the sense of self-ownership with temporal discrepancy between visual and proprioceptive feedbacks. *NeuroImage* **24**, 1225–1232 (2005).

Sirigu, A., Duhamel, J.-R., Cohen, L., et al. The mental representation of hand movements after parietal cortex damage. *Science* **273**, 1564 (1996).

INDEX

Page numbers in **bold** denote tables, those in *italics* denote figures.